MW01012293

# ARTHROPOD FOSSILS AND PHYLOGENY

# ARTHROPOD FOSSILS AND PHYLOGENY

Edited by *Gregory D. Edgecombe*

COLUMBIA UNIVERSITY PRESS

New York

Columbia University Press
Publishers Since 1893
New York Chichester, West Sussex



Library of Congress Cataloging-in-Publication Data
Arthropod fossils and phylogeny / edited by Gregory D. Edgecombe.
p. cm.
Includes bibliographical references and index.
ISBN 0–231–09654–2
1. Arthropoda, Fossil. 2. Arthropoda—Phylogeny. I. Edgecombe, Gregory D.
QE815.A78 1998
595'.138—dc21 97–35380

Casebound editions of Columbia University Press books
are printed on permanent and durable acid-free paper.
Printed in the United States of America
c 10 9 8 7 6 5 4 3 2 1

# Contents

# ARTHROPOD FOSSILS AND PHYLOGENY

# Introduction: The Role of Extinct Taxa in Arthropod Phylogeny

*Gregory D. Edgecombe*

What light do fossils shed on the deep history of the Arthropoda and its major components — the chelicerates, crustaceans, and tracheates? Are the interrelationships between major groups of arthropods considered the same whether or not fossils are included? Which, if any, extinct taxa play a decisive role in altering phylogenetic hypotheses based on extant arthropods? How can we integrate disparate lines of evidence from fossil and living taxa in a robust phylogenetic scheme for life's most diverse phylum? These are the underlying questions of the contributions in this volume.

Two topical issues in systematics inspired this book. One is an explosion of research on high-level phylogeny of the Arthropoda. The other is a rekindling of interest in the role fossils play in phylogenetic analysis.

Arthropod phylogeny research has thrived in the last few years, with many important and novel contributions coming from molecular biology. Gene expression studies, for example, have provided new ways of examining classical morphological homologies. Notable examples include appraisals of head segmentation (Schmidt-Ott et al. 1994; Scholtz 1995) and limb development (Panganiban et al. 1995). Old questions — such as whether crustacean and tracheate mandibles are homologous and whether the distinction between "whole limb" and "gnathobasic" jaws has phylogenetic meaning — are being addressed at the genetic level (e.g., Popadić et al. 1996). DNA sequencing studies are now employing sufficiently sophisticated taxonomic sampling for the deep history of arthropods to be appraised by these methods. The widespread acceptance of phylogenetic systematics has also been an important factor in the progression of evolutionary research. Debates that dominated the field twenty years ago, such as whether Arthropoda has a single origin or multiple origins, have largely been resolved on methodological grounds.

Many recent controversies in arthropod evolution have been initiated by new fossil discoveries and new interpretations of early fossils. This book examines the role these discoveries and ideas have played in understanding the deep evolutionary history of arthropods. Authors of papers in the book have been at the forefront of this research.

Patterson (1981) boldly asserted that no instance of fossils overturning a phylogenetic hypothesis formulated upon extant taxa was known. This claim has inspired a number of empirical studies using real data sets for extinct and living taxa to examine what effect the fossils have (e.g., Gauthier et al. 1988; Novacek 1992). Simulations have also been devised to determine those situations in which fossils might be expected to have a decisive role in phylogenetic inference (Huelsenbeck 1991). The arthropods, in light of the quality and duration of their fossil record, provide many appropriate test cases for these sorts of empirical studies. Wills et al. (chapter 2) and Schram and Hof (chapter 6) undertake comparisons between phylogenies (for all arthropods and for crustaceans, respectively) along the lines pioneered by Gauthier et al. (1988). Arthropod fossils dominate diversity as far back as the Early Cambrian, and it is generally considered that the sudden flourishing of the arthropod fossil record is closely connected to the basic splits between the major lineages (but see Fortey et al. 1996 for a contrary view). The antiquity of these divergences, having occurred at least by the Atdabanian Stage of the Early Cambrian — 525 million years ago — suggests that fossils might play a vital role in accurately estimating the deep branch points in the arthropod tree.

## Scope and Organization

Contributors were asked to address the questions that begin this introduction, summarizing their ideas in the form of phylogenetic diagrams. Evidence for phylogenetic hypotheses — characters — must be made explicit if competing ideas are to be appraised. In the end, "Phylogenetics, to put it crudely, is a put-up or shut-up discipline" (Wiley et al. 1991:2).

The "Recent tree" is an obvious starting point for investigating the effect of fossil evidence. Toward this end, Ward Wheeler (chapter 1) summarizes current data from molecular sequences for the major taxa of extant arthropods. He advocates a "total evidence" approach, combining morphological characters with the most exhaustive possible sample of sequence data. This method permits extinct taxa to be coded for the morphological characters and analyzed together with the larger character set for the living taxa.

A broadly sampled cladistic survey of arthropod phylogeny that addresses the role of fossils is undertaken by Matthew Wills and colleagues (chapter 2). Their character/taxon matrix includes a range of Early Paleozoic arthropods, as well as each of the major extant arthropod taxa, and the characters include fossilized and nonfossilized systems. A series of parsimony analyses demonstrates the effects of the fossils on inferred patterns of relationship.

Cambrian lagerstätte, sites of exceptional fossil preservation, have dominated the discussion on paleontology's contribution to metazoan interrelationships. The Early Cambrian Chengjiang fauna of China and the Late Cambrian "Orsten" of Sweden provide vital data on early arthropod history in superb anatomical detail. The evolutionary significance of arthropods in these faunas is presented by Jan Bergström and Hou Xianguang (chapter 4) and Dieter Walossek and Klaus Müller (chapter 5), respectively. Both chapters provide general phylogenetic overviews of the schizoramian arthropods (Crustacea, Chelicerata, and their stem groups) from the perspective of the early fossils as well as ideas on the earliest stages of arthropod evolution.

The position of arthropods within the Metazoa remains a vexing problem (see, e.g., Rouse and Fauchald 1995). The phylogenetic alignment of purported "protarthropods" — the Onychophora and Tardigrada — is critical to an understanding of the evolutionary events at the base of the Arthropoda. In recent years a great quantity of morphological information has come to light on Cambrian lobopodians. Lars Ramsköld and Chen Junyuan summarize current knowledge on the anatomy of these fossils in chapter 3. They undertake a cladistic synthesis of the Cambrian lobopodians and conclude that they are indeed Onychophora. As a consequence, these Cambrian taxa should be considered in future attempts to root the arthropod tree.

Studies by Frederick Schram and Cees Hof (chapter 6) and Paul Selden and Jason Dunlop (chapter 7) focus on the two components of extant arthropod diversity whose stem groups are richly represented in the early record — the Crustacea and Chelicerata, respectively. Schram and Hof review recent theories on the deep history of Crustacea and make new advances by assembling the most comprehensive character set yet available for major groups of crustaceans, living and fossil. Further progress in crustacean phylogeny can be made through the progressive refinement of their character matrix. Selden and Dunlop review the defining characters of major taxa within the Chelicerata and compare cladistic patterns with the known temporal record of each group.

## Controversies and Consensus

Most contributors review the fundamental questions in arthropod phylogeny (Wills et al. provide a succinct summary of competing hypotheses). Some general conclusions may be drawn from the chapters in order to summarize the current state of thought in arthropod paleontology and suggest directions for the future.

### The Mandibulata Hypothesis

The classical "Recent tree" for basic arthropod interrelationships includes a taxon, Mandibulata, that unites crustaceans and atelocerates (myriapods and hexapods) to the exclusion of the chelicerates. Promoted by R. E. Snodgrass and many others, the Mandibulata hypothesis has received new support in some recent morphological and biochemical studies (e.g., Wägele 1993). Wheeler's (chapter 1) literature-derived matrix of morphological characters supports the traditional mandibulate resolutions — monophyletic Crustacea is the sister group of Atelocerata, within which a myriapod grouping of uncertain status is most closely related to the hexapods. Combining the morphological characters with molecular sequence data, Wheeler concludes that Mandibulata remains supported based on total evidence from extant taxa.

It is thus noteworthy that Mandibulata is rejected in those studies (Wills et al.; Bergström and Hou; Walossek and Müller) that test it with extinct taxa. Paleontological data, as summarized by this book's contributors, favor closest relations between Crustacea and Chelicerata — the Schizoramia or "TCC" (trilobite — chelicerate — crustacean) hypothesis.

The results of Wills et al. (chapter 2) are particularly interesting because the authors' taxonomic sample includes a broad range of extant arthropods, indeed comparable in number and scope to that used in Wheeler's synthesis. That Wills et al. reject "Mandibulata" based on the combined fossil/extant sample is significant but not particularly surprising. Walossek and Müller show that some alleged synapomorphies of Mandibulata, such as paired maxillae, are lacking in early-derived crustaceans and must be interpreted as having independently evolved within Crustacea and in Atelocerata. But Wills et al. also retrieve the crustacean-chelicerate alliance when fossil taxa are *excluded* from the analysis.

Future investigations should consider the influence of particular taxa as well as particular characters in determining these differences. In the first instance (effects of taxonomic sampling) the range of Crustacea analyzed may be significant. Proponents of the Mandibulata hypothesis have included entomologists whose comparisons with crustaceans have been limited in scope — an exemplar eumalacostracan or brine shrimp is unlikely to produce an accurate estimate of characters at the base of Crustacea. A second factor to investigate is the nature of the character evidence. Schram and Hof (chapter 6) suggest that missing data — in the form of soft-part characters that cannot be coded for fossils — are significant in producing different resolutions when fossils are or are not included. But a different factor may be contributing to the "TCC" resolution of Wills et al., namely, the predominance of fossilizable characters in their matrix. Evidence for "TCC," even in the absence of the fossils, may be a result of including morphological characters the extinct taxa suggest as having phylogenetic significance.

## The Status of Crustacea

The monophyly of Crustacea is endorsed in every chapter that investigates the issue (Wheeler; Wills et al.; Bergström and Hou; Schram and Hof). Furthermore, the early fossils provide a basis for tracing the acquisition of shared derived characters within the Crustacea. This finding opposes a number of recent studies that have speculated that crustaceans may be a grade group ancestral to the hexapods (e.g., Averof and Akam 1995). Molecular sequence evidence raised in support of the crustacean-hexapod alliance (exclusive of the myriapods) is dismissed by Wheeler as unparsimonious. Again comparisons must include a broader range of Crustacea in order to properly evaluate the meaning of detailed similarities (such as neuroblasts in the segmental ganglia and innervation of the eye: Osorio et al. 1995) between particular crustaceans and Hexapoda. In any event, notions of Crustacea as a non-monophyletic group are soundly rejected in these studies.

The interpretation of Cambrian crustaceaform fossils remains an area for more work. This is highlighted by substantial differences in the phylogenetic positioning of such fossils as canadaspidids by Bergström and Hou and Walossek and Müller (who interpret them as non-crustaceans, near the base of the euarthropods) and Wills et al. (who resolve such Cambrian fossils as scattered through the Crustacea). Schram and Hof's position is intermediate in many respects, finding the majority of Cambrian crustaceaform taxa to nest together at the base of the crustacean clade. These differences must partly reflect methodological differences between the respective studies. The interpretation of particular characters differs profoundly whether a key characters/ground pattern reconstruction method is employed or whether computer-assisted parsimony methods form the basis for inferring phylogeny. Methodological issues aside, the need to clarify the actual anatomy of some of the fossils themselves, e.g., details of mouthparts in some of the Burgess Shale taxa, ensures that the Cambrian crustaceaform taxa will remain at the forefront of research.

## Significance of the Arachnomorpha

Wills et al. aptly observe that the greatest loss of phylogenetic diversity to affect the arthropods is the pruning (through extinction) of most of the lineage diversity of Arachnomorpha. Several contributions in the book emphasize the important role the early arachnomorphs — stem-group chelicerates — play in understanding character evolution across the major clades of arthropods. The extant chelicerates, as the sole post-Paleozoic survivors in Arachnomorpha, are sufficiently modified so that in many cases homologies with extant Crustacea are unclear (the conundrum of pycnogonid affinities serves as an example; Wheeler, Wills et al., and Selden and Dunlop all cite the position of pycnogonids as ambiguous). Structures in Cambrian arachnomorphs and stem-group crustaceans are more readily homologized, most notably the components of the biramous limb (Walossek and Müller). The establishment of these homologies is certainly a factor in accounting for the "TCC" resolution when fossils are included. The precise sequence of phylogenetic branchings and character transformations within the early arachnomorphs is of fundamental importance in pan-arthropod comparisons and requires more detailed study.

## Toward Synthesis

Studies of some classical morphological characters are emerging in new guises, with fossils playing a major role in posing the questions. For example, hexapod fossils have contributed to a theory that the arthropod limb base is fundamentally composed of subcoxal, coxal, and trochanteral segments (Kukalová-Peck 1992). Studies on much earlier schizoramian arthropods (Bergström and Hou; Walossek and Müller) dispute these claims, indicating instead that the common plan of stem-group crustacean and stem-group chelicerate limbs involves only a basipod; new proximal elements were later differentiated in the Crustacea. However, this question — the composition of the "coxopodite" — suggests itself as a prime target for gene expression studies. It is such instances of reciprocity between extinct and extant taxa that offer the best prospects for synthesis in arthropod phylogeny.

### References

Averof, M. and M. Akam. 1995. Insect-crustacean relationships: Insights from comparative developmental and molecular studies. *Philosophical Transactions of the Royal Society of London*, series B, 347:293–303.

Fortey, R. A., D. E. G. Briggs, and M. A. Wills. 1996. The Cambrian evolutionary "explosion": Decoupling cladogenesis from morphological disparity. *Biological Journal of the Linnean Society* 57:13–33.

Gauthier, J., A. G. Kluge, and T. Rowe. 1988. Amniote phylogeny and the importance of fossils. *Cladistics* 4:105–209.

Huelsenbeck, J. P. 1991. When are fossils better than extant taxa in phylogenetic analysis? *Systematic Zoology* 40:458–469.

Kukalová-Peck, J. 1992. The "Uniramia" do not exist: The ground plan of the Pterygota as revealed by Permian Diaphanopterodea from Russia (Insecta: Paleodictyopteroidea). *Canadian Journal of Zoology* 70:236–255.

Novacek, M. J. 1992. Fossils as critical data for phylogeny. In M. J. Novacek and Q. D. Wheeler, eds., *Extinction and Phylogeny*, pp. 46–88. New York: Columbia University Press.

Osorio, D., M. Averof, and J. P. Bacon. 1995. Arthropod evolution: Great brains, beautiful bodies. *Trends in Ecology and Evolution* 10:449–454.

Panganiban, G., A. Sebring, L. Nagy, and S. Carroll. 1995. The development of crustacean limbs and the evolution of arthropods. *Science* 270:1363–1366.

Patterson, C. 1981. Significance of fossils in determining evolutionary relationships. *Annual Reviews of Ecology and Systematics* 12:195–223.

Popadić, A., D. Rusch, M. Peterson, B. T. Rogers, and T. C. Kaufman. 1996. Origin of the arthropod mandible. *Nature* 380:395.

Rouse, G. W. and K. Fauchald. 1995. The articulation of annelids. *Zoologica Scripta* 24:269–301.

Schmidt-Ott, U., M. González-Gaitán, H. Jäckle, and G. M. Technau. 1994. Number, identity, and sequence of the *Drosophila* head segments as revealed by neural elements and their deletion patterns in mutants. *Proceedings of the National Academy of Sciences, USA.* 91:8363–8367.

Scholtz, G. 1995. Head segmentation in Crustacea: An immunocytochemical study. *ZACS Zoology* 98:104–114.

Wägele, J. W. 1993. Rejection of the "Uniramia" hypothesis and implications of the Mandibulata concept. *Zoologische Jahrbücher: Abteilung für Systematik, Ökologie, und Geographie der Tiere* 120:253–288.

Wiley, E. O., D. Siegel-Causey, D. R. Brooks, and V. A. Funk. 1991. *The Complete Cladist: A Primer of Phylogenetic Procedures.* University of Kansas Museum of Natural History, Special Publication 19:1–158.

# CHAPTER 1

## Molecular Systematics and Arthropods

*Ward Wheeler*

### Abstract

Nucleic acids have offered a wealth of novel information to systematic analyses of arthropod taxa. Recent molecular accounts of the systematics of extant arthropods have been marred, however, by three defects. The first of these is the lack of attention to potentially important groups such as pycnogonids and tardigrades, second is the nonuse of existing information, and third is the nongenerality of assumptions. Molecular and morphological data sets created for arthropod phylogeny have been treated *de novo* with little if any attempt to integrate existing work in these "novel" hypotheses, and these results are dependent on specific, unexamined assumptions. Only through the integration of research efforts, intelligent sampling, and the probing of our assumptions can coherent hypotheses be erected and tested.

The past decade has presented us with nearly annual molecular reanalyses of Arthropoda. Although there has been some consensus among these investigations, results have been highly dependent on the specifics of the study (Field et al. 1988; Abele et al. 1989; Lake 1990; Turbeville at al. 1991; Ballard et al. 1992; Wheeler et al. 1993; Boore et al. 1995; Friedrich and Tautz 1995; Garey et al. 1996; Giribet et al. 1996).

The initial molecular forays were limited to nuclear small ribosomal subunit (18S rRNA) sequences — usually RNA based. Field et al. (1988) were primarily interested in metazoan relationships; however, within their sample were several representatives of the major extant arthropod lineages. The scheme of relationships they proposed has a monophyletic Arthropoda with the basal dichotomy between myriapods (*Spirobolus*) and (chelicerates + (crustaceans + hexapods)) represented by *Limu-*

*lus*, *Artemia*, and *Drosophila* (fig. 1.1a). The authors note that this is at variance with traditional ideas of arthropod evolution and ascribe this difference to "fast-clock" taxa. In adding a pentastomid sequence to the Field et al. data, Abele et al. (1989) supported a monophyletic arthropod clade with Mandibulata (Crustacea + Myriapoda + Hexapoda) appearing for the first time (fig. 1.1b). Lake (1990) reanalyzed these data deriving a novel result — that the arthropods were neither monophyletic nor polyphyletic, but paraphyletic with respect to roundworms and mollusks (fig. 1.1c). This truly original idea had no previous complement in morphological or developmental work. Interestingly, when Lake's tree is rerooted such that arthropods are monophyletic, the internal arrangements are highly consistent with previous

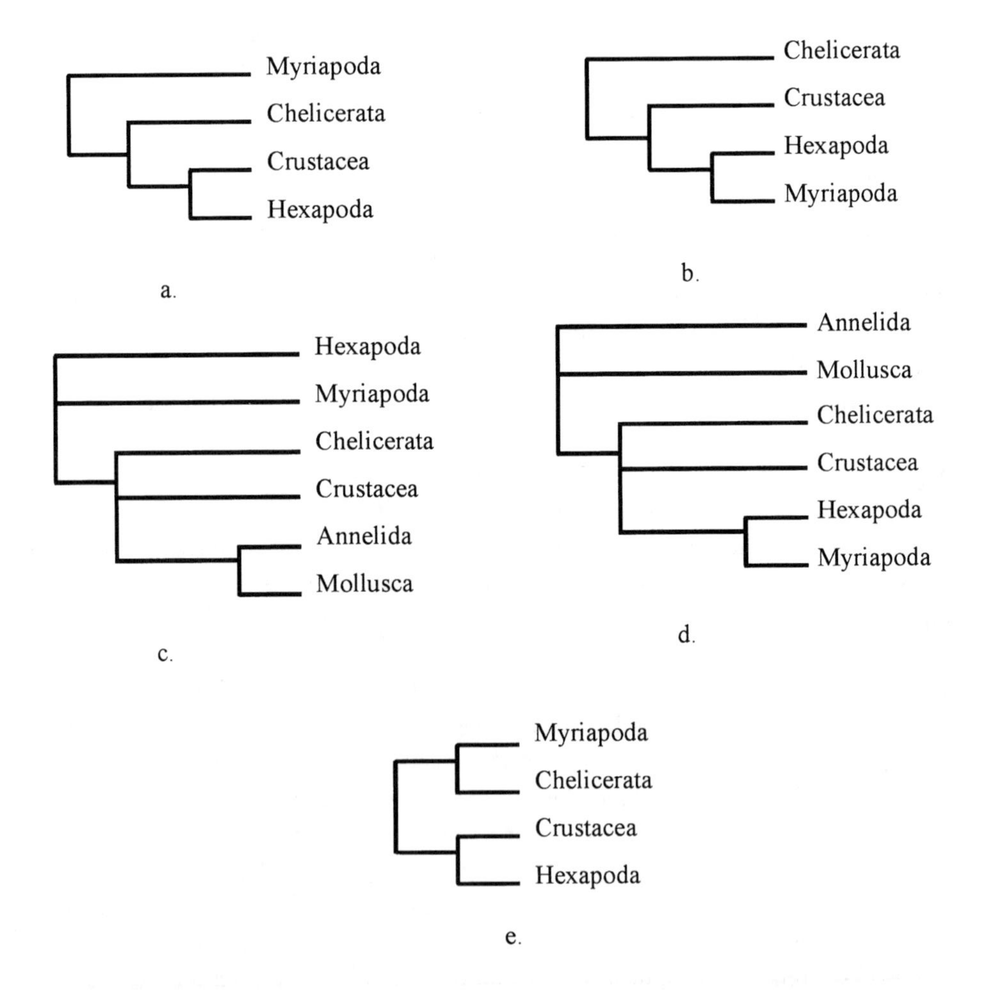


FIGURE 1.1
Small subunit ribosomal RNA-based arthropod phylogenies: *(a)* after Field et al. (1988); *(b)* after Abele et al. (1989); *(c)* after Lake (1990); *(d)* Lake (1990) rerooted with arthropods monophyletic; and *(e)* after Turbeville et al. (1991).

non-molecular analyses (Snodgrass 1938; Weygoldt 1986) (fig. 1.1d). Turbeville et al. (1991) added two new rRNA sequences (and a single DNA-based sequence) and jostled the remainder, straining to remove "fast-clock" taxa. They removed *Drosophila* and *Artemia* from analysis, substituting *Tenebrio* and *Procambarus*. They also supported a crustacean/hexapod clade with the myriapods now sister to the chelicerates (fig 1.1e).

In a departure from nuclear rRNA, Ballard et al. (1992) sequenced the 12S rDNA. This molecule is the mitochondrial version of the nuclear 18S sequenced so frequently. Ballard et al. displayed a tree for forty arthropod (and related) taxa, with good resolution for all the major lineages. The tree presented by the authors does not support a monophyletic Arthropoda, placing Myriapoda outside (Onychophora + (Chelicerata + (Crustacea + Hexapoda))) (fig 1.2a). Unfortunately, the authors mentioned but did not present the most parsimonious cladograms for the taxa (seven steps shorter out of 1418), and the data are considerably less informative (Carpenter et al. in press) (fig. 1.2b). Although their presentation lacked transparency, Ballard et al. (1992) laudably did sequence the crucial Onychophora.

Polyubiquitin was added to the mix by Wheeler et al. (1993). In addition to extending molecular analysis to a new locus, the authors attempted to explicitly integrate the character-based morphological data generated by other investigators. Although the ubiquitin sequences on their own seemed to offer little, they did aid in the resolution of arthropod groups when combined with the 18S rDNA data.

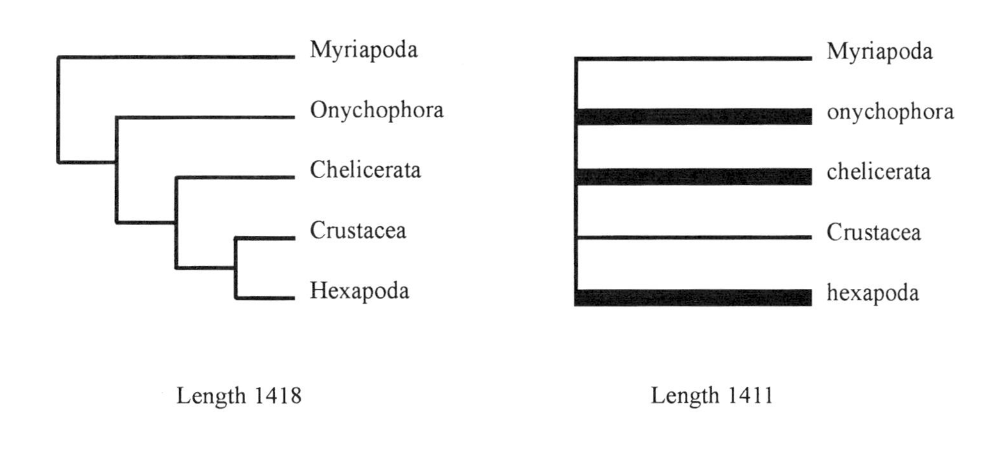


FIGURE 1.2

Mitochondrial small subunit-based arthropod phylogenies: (*a*) presented by Ballard et al. (1992) at length 1418 steps; (*b*) strict consensus cladogram (from 54 trees) at length 1411 steps reported but not presented. The taxa in lower case (and with thicker lines) are not monophyletic in the most parsimonious reconstructions.

Interestingly, the molecular data alone linked Crustacea with Hexapoda within Mandibulata. The main conclusions of this study came from the total evidence (Kluge 1989) analysis with morphological and molecular data yielding a phylogeny identical to that proposed by Snodgrass (1938) (fig. 1.3). Although presented (graphically) by the authors, this point was elegantly restated by Kraus and Kraus (1994).

The entirety of molecular data to this point has been sequence data. Boore et al. (1995) presented gene order data. Through studying the variation in the order of tRNA and other genes in the mitochondrial genome of arthropods and their relatives (including Onychophora), Boore et al. presented seven new characters that support a monophyletic Arthropoda and Mandibulata.

Friedrich and Tautz (1995) presented a somewhat retrograde analysis of arthropod taxa. The authors added a new genetic system to the analysis, the large ribosomal (28S) subunit. Their analysis included data from both the 18S and 28S loci, but did not include morphological data or analysis of crucial taxa such as the Onychophora. Additionally, the criteria for the exclusion of data and taxa seem ad hoc (removal of "fast-clock" taxa and "hard-to-align" areas, etc.). Although Manton's (e.g., 1964, 1973) Uniramia hypothesis cannot be directly tested with these data (because onychophorans were neglected), the authors proposed groupings of Hexapoda + Crustacea and Chelicerata + Myriapoda. This topology is identical to that shown in figure 1.1e.

The glaring hole in all of these analyses is the absence of the Tardigrada. This taxon is potentially the sister taxon of arthropods to the exclusion of Onychophora (reviewed in Brusca and Brusca 1990) but had never been sequenced. This was

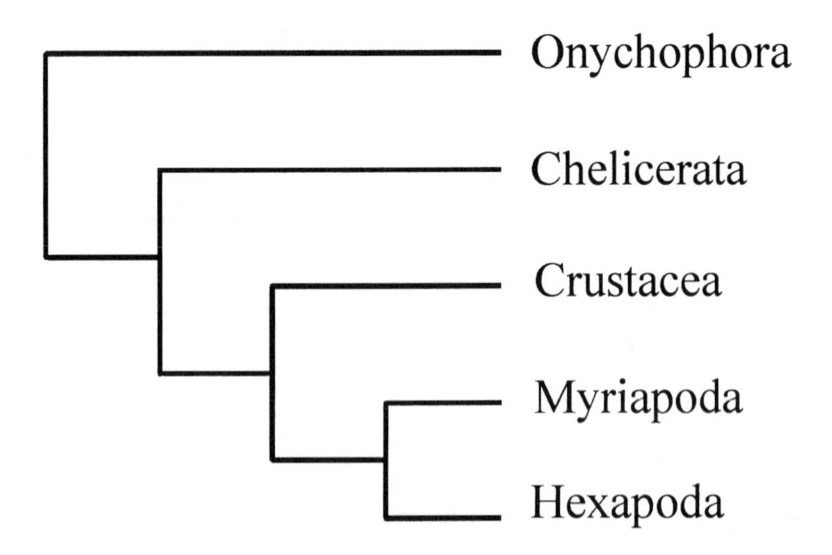


FIGURE 1.3

Total evidence phylogeny of Arthropoda supported by Wheeler et al. (1993). This is largely equivalent to the phylogenetic tree of Snodgrass (1938).

due, no doubt, to the technical difficulties in gathering pure samples of these near-microscopic creatures. Giribet et al. (1996) determined the 18S rDNA sequence for *Macrobiotus hufelandi*, one of the larger tardigrades. Their analysis confirmed the placement of the Tardigrada as sister to the arthropods, but could not place them relative to the Onychophora (no data). In line with other molecular work, the myriapod representative was sister to the arachnids, with crustaceans and hexapods interdigitated (fig. 1.4).

Garey et al. (1996) also analyzed tardigrade sequence (the same genus as Giribet et al. 1996) in their metazoan level sample. Their analysis did not contain any myriapods, pycnogonids, or Onychophora. A distance analysis (bootstrapped Neighbor-Joining) "supports" a tardigrade-arthropod grouping, and the parsimony analysis they performed shows a similar "bootstrap" support of 50%. It is difficult to interpret the analysis of Garey et al. since the authors do not present any of the most parsimonious cladograms or their consensus. Although laudable for the inclusion of tardigrade sequence, the impact of Garey et al. (1996) is unclear given their depauperate sampling and analytical fog.

## Common Themes

Although the molecular analyses to date have been polymorphous in the extreme, several common themes surface. The first is that with few exceptions investigators have generated data independently. Although integrating these studies, by adding data to an ever-growing pool of information, seems an obvious step, this was rarely

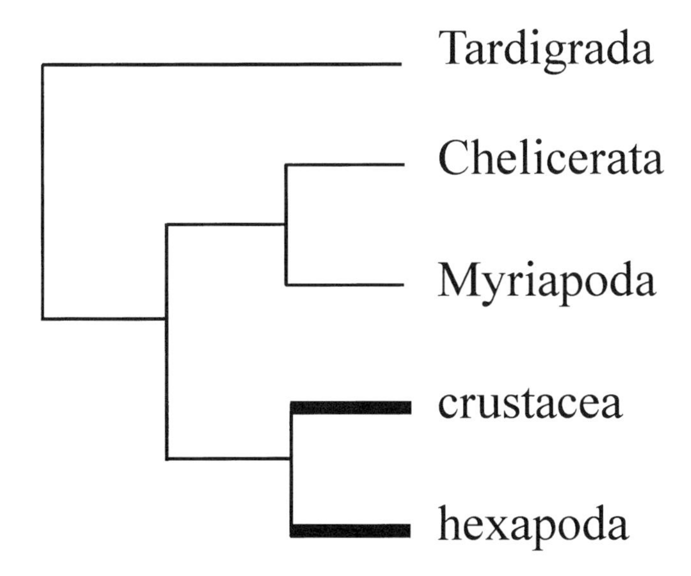


FIGURE 1.4
Small subunit ribosomal DNA-based arthropod phylogeny of Giribet et al. (1996). The taxa in lower case (and with thicker lines) are not monophyletic.

done. Phylogenetically, there are three groupings that are common to almost all the molecular work. They are: (1) monophyletic Arthropoda; (2) chelicerates + myriapods; and (3) crustaceans + hexapods. The second and third results are at great variance with morphological studies, which usually support Crustacea + (Hexapoda + Myriapoda).

## Yet Another Reanalysis

To address the problems mentioned above and to summarize current information on the extant arthropods, an analysis was performed including 18S and 28S rDNA, ubiquitin, and morphological information. This analysis examined thirty taxa (table 1.1), including both Onychophora and Tardigrada. Where sequences were not available (e.g., tardigrade 28S and ubiquitin), they were treated as missing data. The inclusion of all data sources was attempted with the caveat that 50% of the taxa must have been analyzed for a given data set to be included. For this reason the Ballard et al. (1992) data were not included — there was almost no overlap between that and other analyses. An exception to this were the Boore et al. (1995) data. These data were coded somewhat generously (see table 1.2) to surpass the 50% criterion. The nonsequence data (table 1.2; appendix 1.1) consisted of 117 discrete characters. They are based on the 100 characters Wheeler et al. (1993) garnered from the literature, to which were added the 7 gene arrangement characters of Boore et al. (1995), 1 feature of the Crustacea from Boxshall (1996), and 9 characters relating to myriapod-hexapod relationships from Kraus and Kraus (1994) and not included in the previous codings.

The sequence data are derived from several previous studies augmented by several generated for this analysis (table 1.1). There were approximately 1100 bases of the 18S rDNA (defined by the 5' and 3' end points of Wheeler et al. 1993), 350 bases of 28S rDNA, and 228 bases of ubiquitin. The entire contiguous 18S and 28S sequences were employed (as defined by primers, not the complete loci: Whiting et al. [1997]) with the exception of three regions. These areas (the E21–2 to E21–4 region of the 18S and two in the "D3" region of 28S: Hendriks et al. 1988; Gutell and Fox 1988) contained large inserts that were unique to some terminal taxa. They correspond to unique secondary-structure "loops" that had no homologue in other taxa. They are, in effect, autapomorphies for terminal taxa.

These data were analyzed separately and in combination. In the molecular and the combined (total evidence) analyses, two analytical parameters were varied: insertion-deletion cost and transition-transversion ratio (as in Wheeler 1995). When nonsequence data were included, they were weighted equal to the insertion-deletion cost, and when the transition-transversion ratio was set other than unity, the insertion-deletion cost was set according to the cost of transversions. In total, nine combinations of analysis parameters were employed for each of the molecules separately, the molecular data as an ensemble, and in combination with the nonsequence

TABLE 1.1
Taxa Used in the Study

| Higher Group | Taxon | 18S rDNA | 28S rDNA | Ubiquitin |
|---|---|---|---|---|
| ***Mollusca*** | | | | |
| Cephalopoda | *Loligo pealei* | Wheeler | ND | Wheeler |
| Polyplacophora | *Lepidochiton cavernae* | Wheeler | ND | Wheeler |
| ***Annelida*** | | | | |
| Polycheata | *Glycera* sp. | Wheeler | ND | Wheeler |
| Oligocheata | *Lumbricus terrestris* | Wheeler | ND | Wheeler |
| | *Tubifex* sp. | Friedrich | Friedrich | ND |
| Hirudinea | *Haemopis marmorata* | Wheeler | ND | Wheeler |
| ***Onychophora*** | | | | |
| Peripatoidae | *Peripatus trinitatis* | Wheeler | ND | Wheeler |
| Peripatopsidae | *Peripatoides novozealandia* | Wheeler | ND | Wheeler |
| ***Tardigrada*** | | | | |
| | *Macrobiotus hufelandi* | Giribet | ND | ND |
| ***Chelicerata*** | | | | |
| Pycnogonida | *Anoplodactylus portus* | Wheeler | Here | Wheeler |
| Xiphosura | *Limulus polyphemus* | Wheeler | Here | Wheeler |
| Scorpiones | *Centruroides hentzii* | Wheeler | Here | Wheeler |
| Uropygi | *Mastigoproctus giganteus* | Wheeler | Here | Wheeler |
| Araneae | *Peucetia viridans* | Wheeler | Here | Wheeler |
| ***Crustacea*** | | | | |
| Cirrepedia | *Balanus* sp. | Wheeler | Here | Wheeler |
| Malacostraca | *Callinectes* sp. | Wheeler | Here | Wheeler |
| Phyllopoda | *Artemia salina* | Nelles | Friedrich | ND |
| ***Myriapoda*** | | | | |
| Chilopoda | *Scutigera coleoptrata* | Wheeler | Here | Wheeler |
| | *Lithobius forficatus* | Friedrich | Friedrich | ND |
| Diplopoda | *Spirobolus* sp. | Wheeler | Here | Wheeler |
| | *Polyxenus lagurus* | Friedrich | Friedrich | ND |
| | *Megaphyllum* sp. | Friedrich | Friedrich | ND |
| ***Hexapoda*** | | | | |
| Collembola | *Pseudachorutes* sp. | Friedrich | Friedrich | ND |

TABLE 1.1 (*continued*)

| Higher Group Taxon | | 18S rDNA | 28S rDNA | Ubiquitin |
|---|---|---|---|---|
| Archeognatha | *Petrobius brevistylis* | Friedrich | Friedrich | ND |
| Odonata | *Libellula pulchella* | Wheeler | Whiting | Wheeler |
| Dictyoptera | *Mantis religiosa* | Wheeler | Whiting | Wheeler |
| Heteroptera | *Saldula pallipes* | Wheeler | Whiting | ND |
| Lepidoptera | *Papilio* sp. | Wheeler | Whiting | Wheeler |
| Diptera | *Drosophila melanogaster* | Tautz | Tautz | Lee |
| Coleoptera | *Tenebrio molitor* | Hendriks | Whiting | ND |

Giribet = Giribet et al. (1996); Hendriks = Hendriks et al. (1988); Friedrich = Friedrich and Tautz (1995); Nelles = Nelles et al. (1984); Lee = Lee et al. (1988); Tautz = Tautz et al. (1988); Wheeler = Wheeler et al. (1993); Whiting = Whiting et al. (1997); ND = no data; Here = this study.

data (morphology and other character data), for a total of forty-five phylogenetic analyses. These analyses were performed to assess the effect of variation in unmeasurable factors (such as insertion-deletion cost) on systematic conclusions. Phylogenetic analysis was performed using the direct character optimization method of Wheeler (1996), which directly diagnoses phylogenetic cost without an intervening multiple-alignment step. In each case, eleven addition sequences were employed (one based on proximity to the outgroup taxa and ten random) with TBR-type branch swapping using the program MALIGN, version 2.7 (Wheeler and Gladstein 1992, 1994). The values reported are all parsimony tree lengths weighted by the relative costs of gaps, transversions, transitions, and nonsequence character transformation costs.

In each case, the character incongruence values (Mickevich and Farris 1981) were calculated. This value was used to choose among these various analyses in order to present those schemes of relationship that best represent all the data.

## Results

The morphological data yielded two cladograms at 141 steps with a consistency index of 0.88 and a retention index of 0.97. The first cladogram shows a monophyletic Myriapoda and the second a paraphyletic Myriapoda with a sister group relationship between the diplopods and hexapods to the exclusion of the centipedes (as in Kraus and Kraus 1994). The strict consensus of the two leads to a trichotomy among Chilopoda, Diplopoda, and Hexapoda (fig. 1.5).

For the combined analyses, the phylogenetic tree lengths (weighted parsimony costs) are shown in table 1.3. For each of the three insertion-deletion costs and three transition-transversion ratios, five phylogenetic reconstructions were performed. These consisted of total evidence including nonsequence data (morphological and other) and sequence data, molecular sequence information (18S, 28S, and ubiquitin), and 18S rDNA, 28S rDNA, and ubiquitin independently. Character-based incongruity was calculated for the entire data, among the sequence data sets, and between

TABLE 1.2

Nonsequence character matrix

| | | | | | |
|---|---|---|---|---|---|
| *Lepidochiton* | 1111110000 | 0000000000 | 0000?00000 | 0?00?000?0 | 0000000000 |
| | 0000000000 | 0000000000 | 0000000000 | 0000000?00 | 0000?????0 |
| | 0000110??? | ??????? | | | |
| *Loligo* | 1111110000 | 0000000000 | 0000?00000 | 0?00?000?0 | 0000000000 |
| | 0000000000 | 0000000000 | 0000000000 | 0000000?00 | 0000?????0 |
| | 0000110??? | ??????? | | | |
| *Glycera* | 0000001111 | 1000000000 | 0000?00000 | 0?00?000?0 | 0000000001 |
| | 0000100111 | 1111000000 | 0000000000 | 0000000?00 | 0000?????0 |
| | 0000110??? | ??????? | | | |
| *Haemopis* | 0000001111 | 1111100000 | 0000?00000 | 0?00?000?0 | 0000000000 |
| | 0000100111 | 1111000000 | 0000000000 | 0000000?00 | 0000?????0 |
| | 0000110??? | ??????? | | | |
| *Lumbricus* | 0000001111 | 1111100000 | 0000?00000 | 0?00?000?0 | 0000000000 |
| | 0000100111 | 1111000000 | 0000000000 | 0000000?00 | 0000?????0 |
| | 0000110??? | ??????? | | | |
| *Tubifex* | 0000001111 | 1111100000 | 0000?00000 | 0?00?000?0 | 0000000000 |
| | 0000100111 | 1111000000 | 0000000000 | 0000000?00 | 0000?????0 |
| | 00001100?? | ??????? | | | |
| *Peripatoides* | 0000000000 | 0000011111 | 1110000000 | 0?00?000?0 | 0000000001 |
| | 10?1100111 | 1111111111 | 1110000000 | 0000000?00 | 0000??00?0 |
| | 000000001? | ??????? | | | |
| *Peripatus* | 0000000000 | 0000011111 | 1110000000 | 0?00?000?0 | 0000000001 |
| | 10?1100111 | 1111111111 | 1110000000 | 0000000?00 | 0000??00?0 |
| | 000000001? | ??????? | | | |
| *Macrobiotus* | 0000000000 | 0000000000 | 0000?00000 | 0?00?000?0 | 0000000000 |
| | 1000?00??1 | 111?11111? | ???100?00? | 1000000?01 | 000000000? |
| | ???????0?? | ??????? | | | |
| *Anoplodactylus* | 0000000000 | 0000000000 | 0000001111 | 1100000000 | 0000000000 |
| | 0100000111 | 1111211111 | 1111111111 | 1110000?00 | 0?01001001 |
| | ???????000 | 0000000 | | | |
| *Limulus* | 0000000000 | 0000000000 | 0000001110 | 0011100000 | 0000000000 |
| | 2200011111 | 1111211111 | 1111111111 | 1110000?00 | 1111001101 |
| | 1111000000 | 0000000 | | | |
| *Centruroides* | 0000000000 | 0000000000 | 0000001110 | 0011211110 | 0000000000 |
| | 2200001111 | 1111211111 | 1111111111 | 1110000?10 | 0101000001 |
| | 111100?000 | 0000000 | | | |
| *Mastigoproctus* | 00000000000 | 0000000000 | 0000011100 | 0112111100 | 0000000002 |
| | 200001111 | 1111211111 | 1111111111 | 1110000?10 | 0101000001 |
| | 111100?000 | 0000000 | | | |
| *Peucetia* | 0000000000 | 0000000000 | 0000001110 | 0011211110 | 0000000000 |
| | 2200001111 | 1111211111 | 1111111111 | 1110000?10 | 0101000001 |
| | 111100?000 | 0000000 | | | |
| *Artemia* | 0000000000 | 0000000000 | 0001110000 | 0000000000 | 0000000002 |
| | 3100011111 | 1111211111 | 1111111111 | 1111111000 | 0000111001 |
| | 1111001100 | 0000000 | | | |

TABLE 1.2 (*continued*)

| | | | | | |
|---|---|---|---|---|---|
| *Callinectes* | 0000000000 | 0000000000 | 0001110000 | 0000000000 | 0000000002 |
| | 3100011111 | 1111211111 | 1111111111 | 1111111000 | 0000111001 |
| | 1111001100 | 0000000 | | | |
| *Balanus* | 0000000000 | 0000000000 | 0001110000 | 0000000000 | 0000000002 |
| | 3100011111 | 1111211111 | 1111111111 | 1111111000 | 0000111001 |
| | 111100?100 | 0000000 | | | |
| *Scutigera* | 0000000000 | 0000000000 | 0000200000 | 0?000000?1 | 0000000002 |
| | 2011100111 | 1111211111 | 1111111111 | 1111111121 | 0000110011 |
| | 1111001010 | 0111000 | | | |
| *Lithobius* | 0000000000 | 0000000000 | 0000200000 | 0?000000?1 | 0000000002 |
| | 2011100111 | 1111211111 | 1111111111 | 1111111121 | 0000110011 |
| | 1111001010 | 0111000 | | | |
| *Spirobolus* | 0000000000 | 0000000000 | 0000200000 | 0?000000?1 | 0000000002 |
| | 2011100111 | 1111211111 | 1111111111 | 1111111120 | 0000110011 |
| | 1111001011 | 1000111 | | | |
| *Polyxenus* | 0000000000 | 0000000000 | 0000200000 | 0?000000?1 | 0000000002 |
| | 2011100111 | 1111211111 | 1111111111 | 1111111120 | 0000110011 |
| | 1111001011 | 1000111 | | | |
| *Megaphyllum* | 0000000000 | 0000000000 | 0000200000 | 0?000000?1 | 0000000002 |
| | 2011100111 | 1111211111 | 1111111111 | 1111111120 | 0000110011 |
| | 1111001011 | 1000111 | | | |
| *Pseudachorutes* | 0000000000 | 0000000000 | 0000200000 | 0000000000 | 1111111112 |
| | 2111100111 | 1111211111 | 1111111111 | 1111111121 | 0000110011 |
| | 1111001011 | 1000000 | | | |
| *Petrobius* | 0000000000 | 0000000000 | 0000200000 | 0000000000 | 1111111112 |
| | 2111100111 | 1111211111 | 1111111111 | 1111111121 | 0000110011 |
| | 1111001011 | 1000000 | | | |
| *Saldula* | 0000000000 | 0000000000 | 0000200000 | 0000000000 | 1111111112 |
| | 2111100111 | 1111211111 | 1111111111 | 1111111121 | 0000110011 |
| | 1111001011 | 1000000 | | | |
| *Tenebrio* | 0000000000 | 0000000000 | 0000200000 | 0000000000 | 1111111112 |
| | 2111100111 | 1111211111 | 1111111111 | 1111111121 | 0000110011 |
| | 1111001011 | 1000000 | | | |
| *Libellula* | 0000000000 | 0000000000 | 0000200000 | 0000000000 | 1111111112 |
| | 2111100111 | 1111211111 | 1111111111 | 1111111121 | 0000110011 |
| | 1111001011 | 1000000 | | | |
| *Mantis* | 0000000000 | 0000000000 | 0000200000 | 0000000000 | 1111111112 |
| | 2111100111 | 1111211111 | 1111111111 | 1111111121 | 0000110011 |
| | 1111001011 | 1000000 | | | |
| *Papilio* | 0000000000 | 0000000000 | 0000200000 | 0000000000 | 1111111112 |
| | 2111100111 | 1111211111 | 1111111111 | 1111111121 | 0000110011 |
| | 1111001011 | 1000000 | | | |
| *Drosophila* | 0000000000 | 0000000000 | 0000200000 | 0000000000 | 1111111112 |
| | 2111100111 | 1111211111 | 1111111111 | 1111111121 | 0000110011 |
| | 1111001011 | 1000000 | | | |

Characters 25, 50, 52, and 89 are nonadditive, the remaining multistates are ordered.

the nonsequence and sequence data (table 1.3). In no case was more than one most parsimonious result found. Hence, the polytomies in phylogenetic results are due to zero-branch-lengths rather than to incongruence. This is not as unusual as it sounds. Since the tree lengths are so long, there is ample opportunity for topologies to differ (however slightly) in length. Phylogenetic conclusions are fairly robust with most variation in the disposition of the pycnogonid *Anoplodactylus* (figs. 1.6–1.8). These achieved minima at two points (fig. 1.9). The overall and sequence-data incongruence were minimized (12.9% and 13.9%) when the insertion-deletion cost was four times that of a base substitution and transitions and transversions were equally weighted. The discordance between the nonsequence and sequence data (more or less morphological and molecular) was minimized when the insertion-deletion cost was twice that of base substitutions (0.996% — again with equal weighting of transversions and transitions). These results are in accord with those of Wheeler (1995), which is not terribly surprising given the similarity of data and problem.

## Discussion and Conclusions

The basic phylogenetic conclusions of these analyses are that Arthropoda, Mandibulata, Tracheata, and Myriapoda are all monophyletic. The data are less sanguine about the placement of the pycnogonids with the other chelicerates, but this appears to be more due to lack of evidence than to conflict among characters. Given

TABLE 1.3
Phylogenetic Tree Statistics

| | Insertion-Deletion Cost Ratio | | |
|---|---|---|---|
| | 2:1 | 4:1 | 8:1 |
| *Transversion-Transition Cost Ratio* | | | |
| 1:1 | 5221/4605/2762<br>374/672/564/<br>16.3%/17.3%/0.996% | 5910/4669/2992/<br>356/672/1128<br>12.9%/13.9%/1.91% | 7145/4531/2991/<br>118/672/2256<br>15.5%/15.1%/5.01% |
| 2:1 | 4040/3388/1998/<br>432/478/564<br>14.1%/14.2%/2.18% | 4832/3506/2174/<br>249/478/1128<br>16.6%/17.3%/4.10% | 5963/3465/2127/<br>86/478/2256<br>17.0%/22.3%/4.06% |
| 4:1 | 6673/5454/3158/<br>709/751/1128<br>13.9%/15.3%/1.36% | 8114/5335/3438/<br>352/751/2256<br>16.2%/14.9%/6.45% | 11132/6430/3961/<br>217/751/4512<br>15.2%/23.3%/1.71% |

In each cell: Total Evidence/Sequence Data/18S rDNA/28S rDNA/Ubiquitin/Nonsequence tree lengths/Mickevich-Farris (1981) Character Incongruence Overall/Among Sequence Data/Nonsequence vs. Sequence

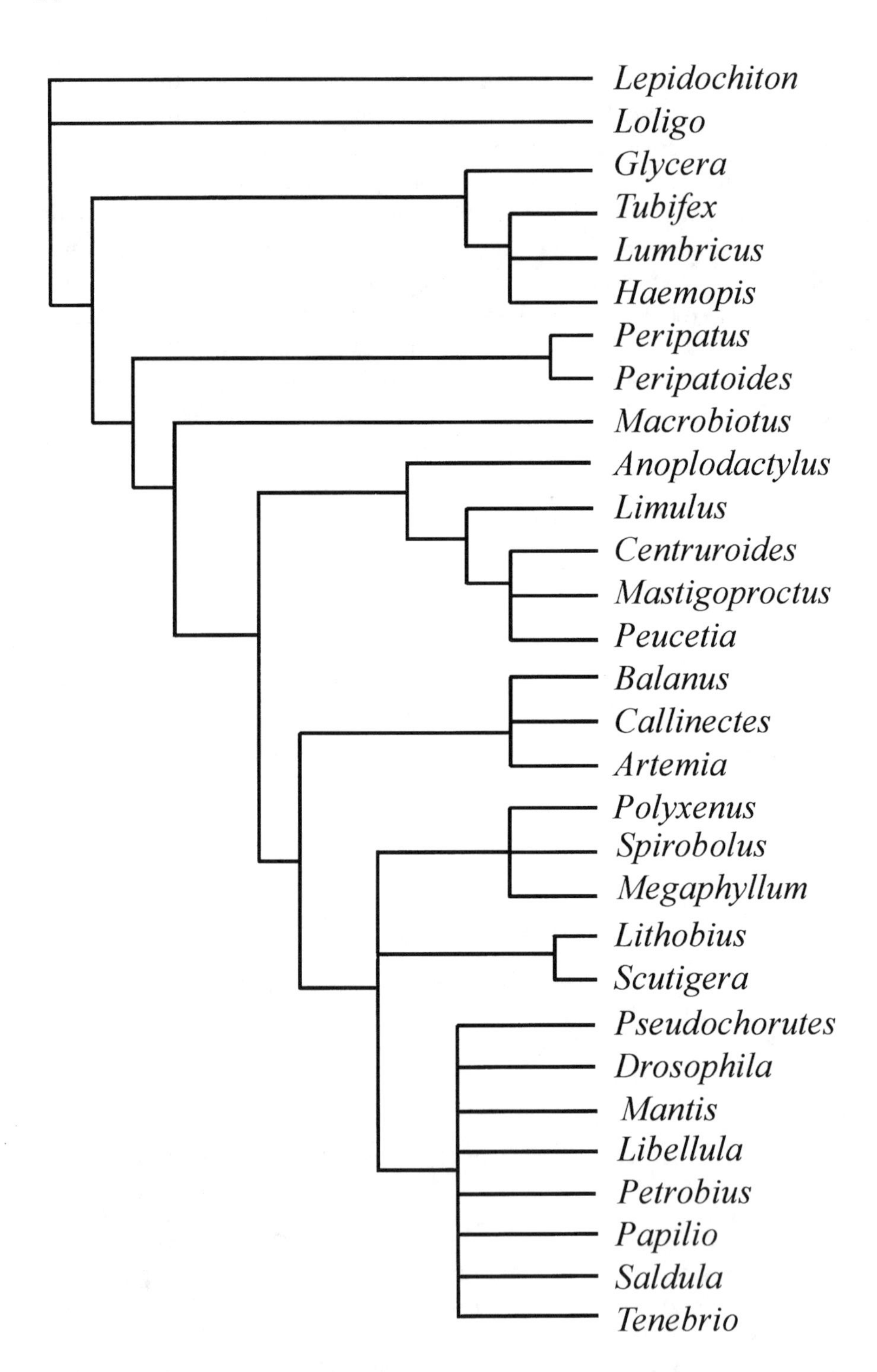


FIGURE 1.5

Nonsequence (morphology and other character data such as gene rearrangements) data-based phylogeny of the arthropod taxa in this study. There were two trees of length 141 with CI of 0.88 and RI of 0.97. This is the strict consensus of the two.

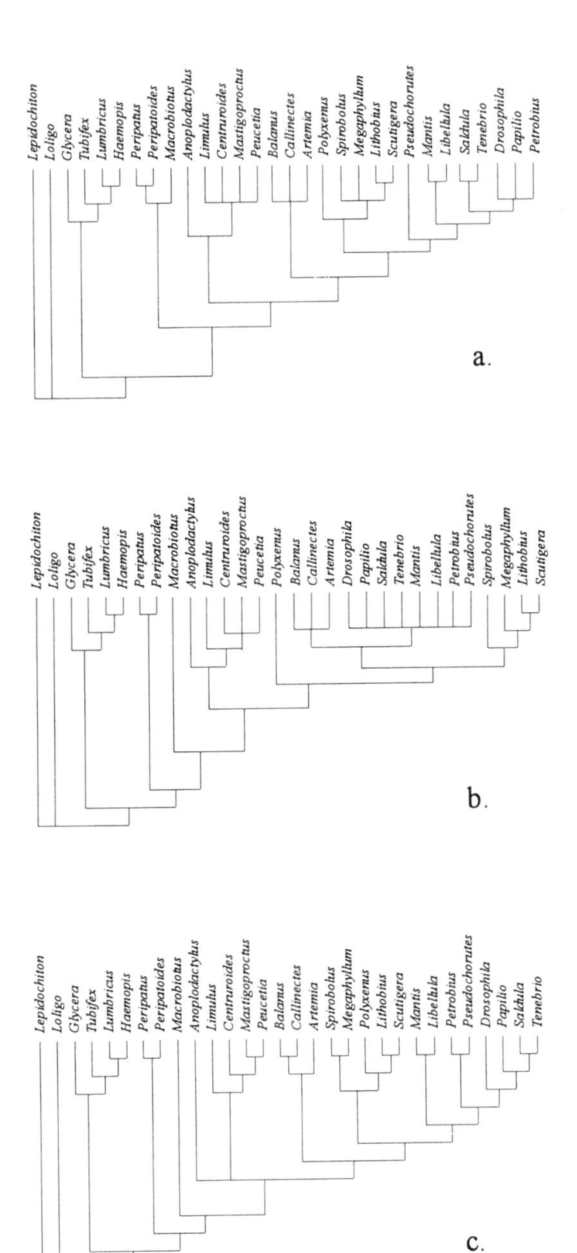


FIGURE 1.6

Total-evidence-derived phylogenies of arthropod relationships. These three cladograms and those in figures 1.7 and 1.8 are derived from nonsequence (morphological and other), small subunit, large subunit, and ubiquitin sequence data. Parsimonious reconstructions of phylogeny were derived via the method of Wheeler (1996) varying analysis parameters. These three cladograms were derived using an insertion-deletion cost (gap ratio) of twice that of a base substitution: *(a)* transversions weighted four times transitions; *(b)* transversions weighted twice transitions; *(c)* equal weighting of transversions and transitions. In each case, the nonsequence characters were assigned a weight equal to that of a sequence gap.

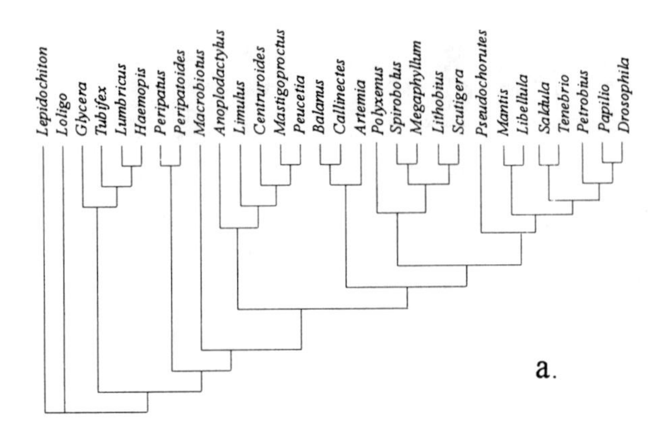


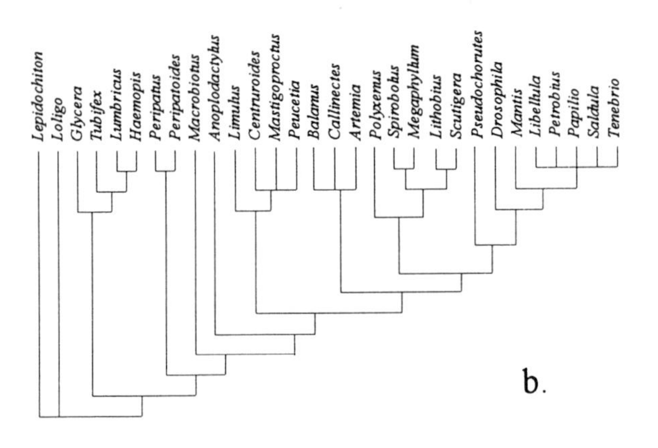


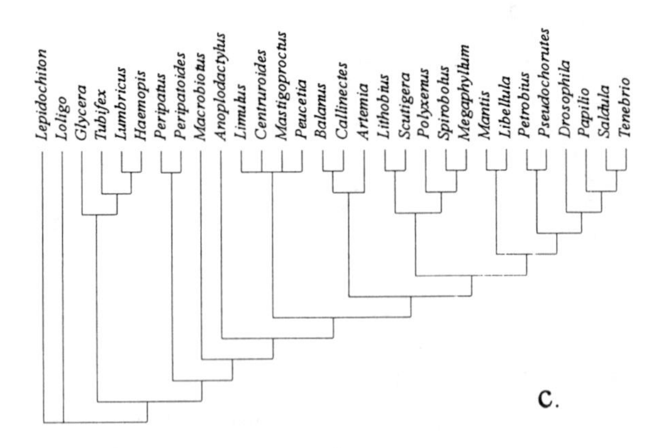


FIGURE 1.7

Total evidence cladograms derived using an insertion-deletion cost of four times the substitution cost (cf. figs. 1.6 and 1.8): *(a)* transversions weighted four times transitions; *(b)* transversions weighted twice transitions; *(c)* equal weighting of transversions and transitions.

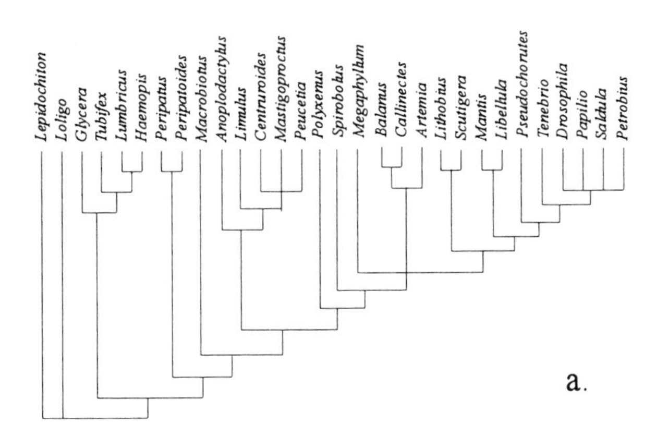


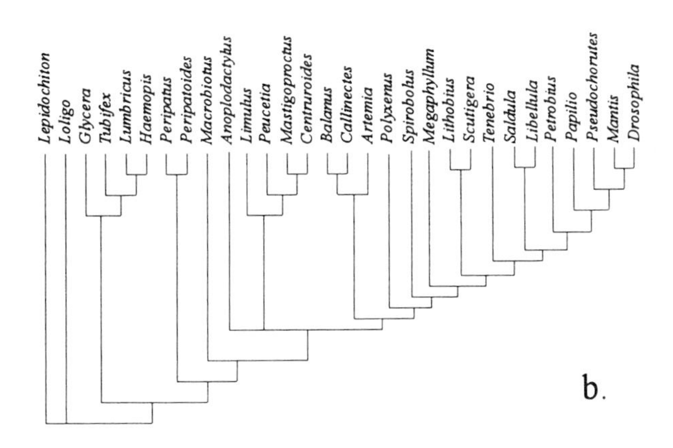


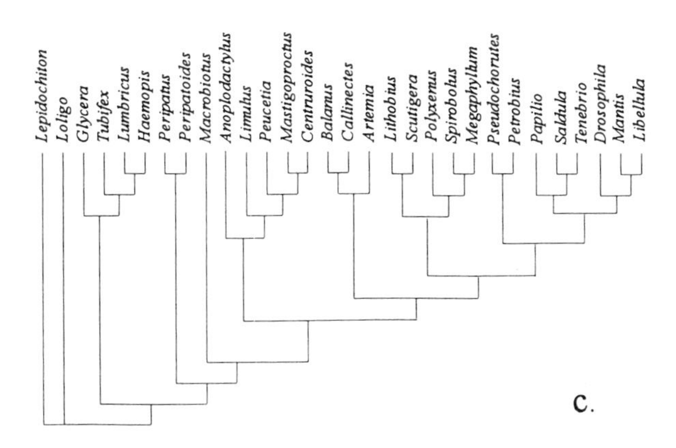


FIGURE 1.8

Total evidence cladograms derived using an insertion-deletion cost of eight times the substitution cost (cf. figs. 1.6 and 1.7): *(a)* transversions weighted four times transitions; *(b)* transversions weighted twice transitions; *(c)* equal weighting of transversions and transitions.

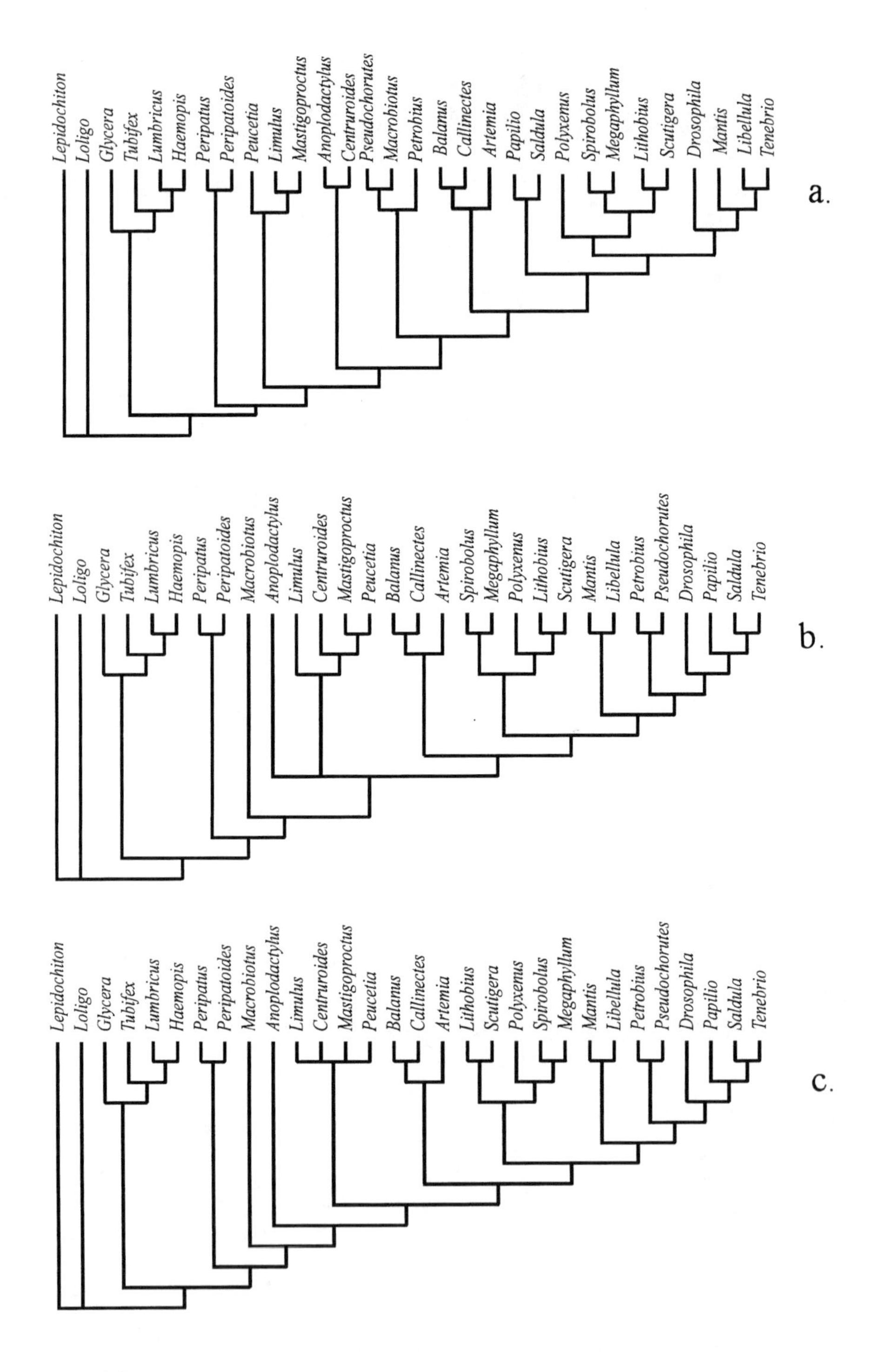


FIGURE 1.9

Arthropod phylogenies with minimal character-based incongruence as assayed by the Mickevich and Farris metric (1981): *(a)* optimal sequence-only cladogram; *(b)* optimal for nonsequence (morphological and other) versus sequence data; *(c)* optimal for all data.

the inherent uncertainties of weighting character information versus molecular sequence data, and transitions versus transversions versus insertion-deletion events, it is comforting to see the relative consistency among the results. In all cases (employing total evidence), the arthropods form a monophyletic group. In eight cases (of nine), Mandibulata is supported, and seven cases support Tracheata. Interestingly, the Crustacea + Hexapoda clade is present only with a transversion-to-transition cost ratio of two to one (the same analysis conditions and result of Friedrich and Tautz [1995]); under any other parameter regime this clade is not supported. This result highlights the importance of examining the robustness of phylogenetic conclusions for variation in these parameter values, which are essentially arbitrary assumptions (Wheeler 1995). Friedrich and Tautz reported a result (verified here) that is based on a particular (but not general) set of analysis conditions. It is dangerous to make strong phylogenetic claims without explicit analysis of sensitivity to parameter values.

Although the weight of nonsequence (e.g., morphological) data was kept the same as the insertion-deletion cost, the contribution of this weighted change information varied from approximately 10% of the total length of the cladogram (at an insertion-deletion cost ratio of 2:1 and transversion to transitions at equality) to over 40% where insertion-deletion events cost was eight times that of transversions and that of transversions four times that of transitions. This variation is due to the ability of the sequences to favor base substitution over insertion-deletion events while the nonsequence changes remain constant in type and number. Over this range, the combined data yielded remarkably homogeneous results. The main areas of inconsistency are in the placement of the pycnogonid *Anoplodactylus* and the status of the Myriapoda.

Roughly half the analyses support a monophyletic Chelicerata (*sensu* Weygoldt and Paulus 1979). Among the others, the pycnogonids are placed either agnostically, in an unresolved polytomy at the base of the euarthropods, or as sister to other euarthropods — a paraphyletic Chelicerata. In two-thirds of the cases examined, the myriapods are monophyletic (and in the least incongruous). Other situations show the myriapods as paraphyletic, but not in the sense of Kraus and Kraus (1994). These two areas may benefit disproportionately from increased sampling. The diversity of pycnogonids is not well circumscribed and major myriapod lineages (Pauropoda and Symphyla) are not sampled in these analyses.

The overall message of this analysis is simple. First, phylogenetically, these data support the Tardigrada as the sister taxon to the Euarthropoda to the exclusion of the Onychophora. Within the arthropods, Euchelicerata, Mandibulata, and Tracheata (= Atelocerata) are all generally supported. The placement of the pycnogonids is suggestive of the morphologically defined Chelicerata, but the evidence here is not strong. Similarly, some evidence supports the monophyly of the myriapods, but this is far from unequivocal.

Second, the combination of data — sequence and other — is certainly achievable and these results should be persuasive as to the importance of total evidence. Total evidence does not mean that more data will not be uncovered, but that responsible investigators must include all evidence within their grasp. Is it satisfactory to base conclusions on a minority of information? Is it scientific? And what of fossils (although none are included here)? Without the simultaneous analysis of sequence and anatomical information, there will be two worlds of systematics — molecular data from the extant taxa and anatomical from the extinct and extant. Synthesis will be impossible.

The third message concerns the robustness of phylogenetic conclusions. The results discussed above are generally supported. They are reasonably independent of perturbations in the tested assumptions (of course, the examinations performed here are simplistic — but salutary nonetheless). This need not have been true. Combining data and varying their relative importance could have caused chaotic, incoherent results. This may be true in other phylogenetic situations, but only by examining multiple parameter values (insertion-deletion costs and transition-transversion ratios) can this be determined.

Arthropods are too large and diverse a group to be allied based on a single shot in the dark, whether that be due to taxonomic, empirical, or epistemological myopia.

## ACKNOWLEDGMENTS

I would like to thank Norman Platnick, Amy Litt, and Michael Whiting for reading drafts of this manuscript.

## REFERENCES

Abele, L. G., W. Kim, and B. E. Felgenhauer. 1989. Molecular evidence for the inclusion of the phylum Pentastomida in the Crustacea. *Molecular Biology and Evolution* 6:685–691.

Anderson, D. T. 1979. Embryos, fate maps, and the phylogeny of arthropods. In A. P. Gupta, ed., *Arthropod Phylogeny*, pp. 59–106. New York: Van Nostrand Reinhold.

Ballard, J. W. O., G. J. Olsen, D. P. Faith, W. A. Odgers, D. M. Rowell, and P. W. Atkinson. 1992. Evidence from 12S ribosomal RNA sequences that onychophorans are modified arthropods. *Science* 258:1345–1348.

Bergström, J. 1979. Morphology of fossil arthropods as a guide to phylogenetic relationships. In A. P. Gupta, ed., *Arthropod Phylogeny*, pp. 3–58. New York: Van Nostrand Reinhold.

Boore, J. L., T. M. Collins, D. Stanton, L. L. Daehler, and W. M. Brown. 1995. Deducing the pattern of arthropod phylogeny from mitochondrial rearrangements. *Nature* 376:163–165.

Boudreaux, H. B. 1979. *Arthropod Phylogeny with Special Reference to Insects*. New York: John Wiley.

Boxshall, G. 1996. Limb segmentation patterns in copepods and other major crustacean groups. In *International Symposium on the Relationships of Major Arthropod Groups: Programme and Abstracts*, p. 5. London.

Brusca, R. C. and G. L. Brusca. 1990. *Invertebrates*. Sunderland, Mass.: Sinauer.

Carpenter, J. M., P. A. Goloboff, and J. S. Farris. In press. Randomization and confusion: A reply to Trueman. *Cladistics*.

Cisne, J. L. 1974. Trilobites and the origin of arthropods. *Science* 186:13–18.

Cutler, B. 1980. Arthropod cuticle features and arthropod monophyly. *Experientia* 36:953.

Eldredge, N. 1974. Revision of the suborder Synziphosurina (Chelicerata, Merostomata), with remarks on merostome phylogeny. *American Museum Novitates* 2543:1–41.

Field, K. G., G. J. Olsen, D. J. Lane, S. J. Giovannoni, M. T. Ghiselin, E. C. Raff, N. R. Pace, and R. A. Raff. 1988. Molecular phylogeny of the animal kingdom. *Science* 239:748–753.

Friedrich, M. and D. Tautz. 1995. Ribosomal DNA phylogeny of the major extant arthropod classes and the evolution of the myriapods. *Nature* 376:165–167.

Garey, J. R., M. Krotec, D. R. Nelson, and J. Brooks. 1996. Molecular analysis supports a tardigrade-arthropod association. *Invertebrate Biology* 112:79–88.

Giribet, G., S. Carranza, J. Baguñà, M. Riutort, and C. Ribera. 1996. First molecular evidence for the existence of a Tardigrada + Arthropoda clade. *Molecular Biology and Evolution* 13:76–84.

Gutell, R. and G. Fox. 1988. A compilation of large subunit RNA sequences presented in a structural format. *Nucleic Acids Research* 16 (supplement): r175–r269.

Hendriks, L., R. De Baere, C. Van Broeckhoven, and R. De Wachter. 1988. Primary and secondary structure of the 18S ribosomal RNA of the insect species *Tenebrio molitor. Federation of European Biochemical Societies* 232:115–120.

Hennig, W. 1981. *Insect Phylogeny*. New York: John Wiley.

Hyman, L. H. 1967. *The Invertebrates*. Vol. 6, *Mollusca I: Aplacophora, Polyplacophora, Monoplacophora, Gastropoda: The Coelomate Bilateria*. New York: McGraw-Hill.

Jamieson, B. G. M. 1988. On the phylogeny and higher classification of the Oligocheata. *Cladistics* 4:367–410.

King, P. E. 1973. *Pycnogonids*. London: Hutchinson.

Kluge, A. 1989. A concern for evidence and a phylogenetic hypothesis for relationships among *Epicrates* (Boidae, Serpentes). *Systematic Zoology* 38:1–25.

Kraus, O. and M. Kraus. 1994. Phylogenetic system of Tracheata (Mandibulata): on "Myriapoda" - Insecta interrelationships, phylogenetic age, and primary ecological niches. *Verhandlungen des naturwissenschaftlichen Vereins in Hamburg* (NF) 34:5–31.

Kristensen, N. P. 1975. The phylogeny of hexapod "orders": A critical review of recent accounts. *Zeitschrift für zoologische Systematik und Evolutionsforschung* 13:1–44.

Lake, J. A. 1990. Origin of the Metazoa. *Proceedings of the National Academy of Sciences USA* 87:763–766.

Lee, H. J., A. Simon, and J. T. Lis. 1988. Structure and expression of ubiquitin genes in *Drosophila melanogaster. Molecular and Cellular Biology* 8:4727–4735.

Manton, S. M. 1964. Mandibular mechanisms and the evolution of arthropods. *Philosophical Transactions of the Royal Society of London*, series B, 247:1–183.

Manton, S. M. 1973. Arthropod phylogeny: A modern synthesis. *Journal of Zoology* 171:111–130.

Manton, S. M. 1979. Functional morphology and the evolution of the hexapod classes. In A. P. Gupta ed., *Arthropod Phylogeny*, pp. 387–466. New York: Van Nostrand Reinhold.

Mickevich, M. F. and J. S. Farris. 1981. The implications of congruence in *Menidia. Systematic Zoology* 30:351–370.

Nelles, L., B.-L. Fang, G. Volckaert, A. Vandenberghe, and R. De Wachter. 1984. Nucleotide sequence of a crustacean 18s ribosomal RNA gene and secondary structure of eukaryotic small subunit RNAs. *Nucleic Acids Research* 14:2345–2364.

Paulus, H. F. 1979. Eye structure and the monophyly of Arthropoda. In A. P. Gupta, ed., *Arthropod Phylogeny*, pp. 299–383. New York: Van Nostrand Reinhold.

Pocock, R. I. 1893. On the classification of the tracheate Arthropoda. *Zoologischer Anzeiger* 16:271–275.

Schram, F. R. 1978. Arthropods: A convergent phenomenon. *Fieldiana, Geology* 39:61–108.

Schram, F. R. 1986. *Crustacea*. New York: Oxford University Press.

Shultz, J. W. 1990. Evolutionary morphology and phylogeny of Arachnida. *Cladistics* 6:1–38.

Snodgrass, R. E. 1938. Evolution of the Annelida, Onychophora, and Arthropoda. *Smithsonian Miscellaneous Collections* 97:1–159.

Snodgrass, R. E. 1952. *A Textbook of Arthropod Anatomy*. Ithaca: Cornell University Press.

Tautz, D., J. M. Hancock, D. A. Webb, C. Tautz, and G. A. Dover. 1988. Complete sequence of the rRNA genes in *Drosophila melanogaster*. *Molecular Biology and Evolution* 5:366–376.

Tiegs, O. W. and S. M. Manton. 1958. The evolution of the Arthropoda. *Biological Reviews of the Cambridge Philosophical Society* 33:255–337.

Turbeville, J. M., D. M. Pfeifer, K. G. Field, and R. A. Raff. 1991. The phylogenetic status of the arthropods, as inferred from 18S rRNA sequences. *Molecular Biology and Evolution* 8:669–702.

Weygoldt, P. 1986. Arthropod interrelationships: The phylogenetic-systematic approach. *Zeitschrift für zoologische Systematik und Evolutionsforschung* 24:19–35.

Weygoldt, P. and H. F. Paulus. 1979. Untersuchungen zur Morphologie, Taxonomie, und Phylogenie der Chelicerata, Teil 2: Cladogramme und die Entfaltung der Chelicerata. *Zeitschrift für zoologische Systematik und Evolutionsforschung* 17:117–200.

Wheeler, W. C. 1995. Sequence alignment, parameter sensitivity, and the phylogenetic analysis of molecular data. *Systematic Biology* 44:321–332.

Wheeler, W. C. 1996. Optimization Alignment: The end of multiple sequence alignment in phylogenetics? *Cladistics 12:1–9*.

Wheeler, W. C., P. Cartwright, and C. Y. Hayashi. 1993. Arthropod phylogeny: A combined approach. *Cladistics* 9:1–39.

Wheeler, W. C. and D. S. Gladstein. 1992. *MALIGN version 2.7: Program and Documentation*. New York.

Wheeler, W. C. and D. S. Gladstein. 1994. MALIGN: A multiple sequence alignment program. *Journal of Heredity* 85:417.

Whiting, M. F., J. C. Carpenter, Q. D. Wheeler, and W. C. Wheeler. 1997. The Strepsiptera problem: phylogeny of the holometabolous insect orders inferred from 18S and 28S ribosomal DNA sequences and morphology. *Systematic Biology* 46:1–68.

## APPENDIX 1.1

**Morphological characters. Literature sources refer to all characters listed since previous citation.**

1. Reduction of coelom and development of an open hemocoelic circulatory system
2. Mantle from closed body wall
3. Mantle shell gland produces spicules (and shell)

4. Ventral body wall muscles developed into a muscular foot
5. Radula
6. Chambered heart with separate atria and ventricles (Hyman 1967; Brusca and Brusca 1990)
7. Annelid head
8. Epidermal paired setae (or bundles)
9. Longitudinal muscles broken into bands instead of sheets (Brusca and Brusca 1990; Fitzhugh, pers. comm.)
10. Annelid nephridial system
11. Cuticle with collagen but no chitin except in setae and stomodaeum (Boudreaux 1979)
12. Clitellum
13. Hermaphroditism
14. Direct development without intervening larval stages
15. Cerebral ganglion moved into anteriormost trunk segment (Jamieson 1988; Brusca and Brusca 1990)
16. Suppression of external segmentation
17. Oblique muscle layer in body wall
18. Subcutaneous hemal channels
19. Oral papillae
20. Body papillae and scales
21. Slime glands
22. Non-migratory gastrulation
23. Lobopods with pads and claws (Boudreaux 1979; Brusca and Brusca 1990)
24. Two pairs of antennae
25. Third head metamere: 0 = unspecialized appendages; 1 = biramous antenna; 2 = intercalary (appendages absent) (Kraus and Kraus 1994)
26. Nauplius or egg-nauplius stage in ontogeny (Schram 1986)
27. Tagmosis into prosoma and opisthosoma without distinct head
28. First appendages chelicerae (or cheliphores) of three articles
29. "Typically" four pairs of walking legs (Weygoldt and Paulus 1979; Weygoldt 1986)
30. Opisthosoma reduced
31. Proboscis (King 1973; Weygoldt and Paulus 1979)
32. Inverse retina in four median eyes (Paulus 1979)
33. Prosoma a carapace-like shield
34. First or second opisthosomal segment modified into a genital somite
35. Opisthosomal respiratorial lamellae: absent (0); as book gills (1); enclosed to form book lungs (2)
36. Extraintestinal digestion
37. Five simple lateral eyes
38. Slit sensillae (Weygoldt and Paulus 1979)
39. Vitreous body present in median eyes (Paulus 1979) Characters 36 and 37 are questioned by Shultz (1990).
40. Palps on first and second maxillae absent (Brusca and Brusca 1990) Other features mentioned as myriapod synapomorphies may be more broadly distributed (such as the organ of Tömösváry) or have unclear homology relationships (stink glands).
41. Thorax divided into three segments, each with a pair of limbs

42. Locomotory limbs six-segmented
43. Abdomen with twelve segments (except for Collembola)
44. Distinct thorax and abdomen
45. "Knee" as joint versus segment
46. Labium
47. Hexapod-type cephalization
48. Abdominal cerci
49. Two primary pigment cells in ommatidia (Kristensen 1975; Boudreaux 1979; Paulus 1979; Hennig 1981)
50. Antennae and palps: 0 = nothing; 1 = prostomial/pre-oral palps; 2 = post-oral antennae
51. Lateral eyes: 0 = absent; 1 = simple; 2 = compound; 3 = stalked compound (ordered because 1, 2, and 3 all formed from a homogeneous secretion from a subcornegenous cell layer (Paulus 1979)
52. Median eyes 0 = none; 1 = four; 2 = two (unordered) (Weygoldt and Paulus 1979)
53. Tracheae. Onychophora are scored as unclear due to the structural dissimilarities between their tracheae and those in myriapods and hexapods. This feature also occurs in some araneamorph spiders, but again the structures, though superficially similar, appear quite different in detail.
54. Whole limb feeding structure. Although this homology statement is almost surely erroneous (Boudreaux 1979; and many others), this character is included to force a rigorous test of arthropod monophyly (Manton 1979).
55. Ordering of fate map tissues (anterior-stomodaeum-midgut-mesoderm-posterior) v. anterior-midgut-mesoderm-stomodaeum-posterior) (Pycnogonid from Schram 1978) (Anderson 1979)
56. Fundamentally biramous post-antennal appendages (Tiegs and Manton 1958)
57. Digestive diverticula
58. Segment origination in caudal elongation or proliferation zone (Weygoldt 1986)
59. Schizocoelous metamerism between pre-oral acron (prostomium) and the non-metameric telson (periproct)
60. Acronal protocerebrum serving the eyes and containing an association center connected with pedunculate bodies
61. Double ventral somatic nerve cord
62. Dorsal and ventral longitudinal muscles
63. Coelomoducts, their vestiges and derivatives
64. Dorsal blood vessel with forward-going peristalsis (Boudreaux 1979)
65. Loss of ectodermal cilia: 0 = many tissues ciliate; 1 = no cilia except in photoreceptor and sperm cells; 2 = no cilia except in sperm (although some sensillae may be cilia-derived) (Paulus 1979)
66. Elongated dorsal gonads
67. Development of ventrolateral appendages
68. Reduction of coelom-hemocoel and circulating system with dorsal blood vessel with paired ostia and pericardial sinus
69. Ecdysis
70. Cuticle of alpha chitin (as opposed to collagen in annelids) and protein (Cutler 1980)

71. Resilin protein present
72. Superficial blastoderm formation (Boudreaux 1979; Weygoldt 1986; Brusca and Brusca 1990)
73. Rhabdomeric retinular structure in eye facets (Paulus 1979)
74. Nephridia in, at most, first four cephalic and first two post cephalic segments (Weygoldt 1986)
75. Hard exoskeleton
76. Articulating jointed appendage with arthrodial membrane
77. Fully segmented sclerites
78. Cephalic (at least anterior) ecdysis glands
79. Cephalon with one pair of pre-oral and four pair post-oral appendages (Weygoldt 1986; Brusca and Brusca 1990)
80. All muscles striated
81. Suppression of all circular body wall muscle
82. Similar intersegmental tendon system (Boudreaux 1979)
83. Nephridia with sacculi (Weygoldt 1986)
84. Specialized ommatidial structures: two corneagene cells, four Semper cells, and a cone with four parts, retinula with eight cells (versus variable and higher number of subunits) (Paulus 1979)
85. Tripartite brain
86. Mandibles (main feeding appendage) with strong coxal endites on third post-acronal head segment (Weygoldt 1986)
87. Two pairs of maxillae on segments 4 and 5 (Brusca and Brusca 1990)
88. Loss of mandibular palpus (inactive recoded in crustacean area)
89. Malpighian tubules: 0 = absent; 1 = endodermally derived; 2 = ectodermally derived (Weygoldt 1986)
90. Posterior gonopore (terminal in Chilopoda and only the Ellipura among the hexapods) (Pocock 1893)
91. Exoskeleton with hard and strong dorsal side and soft ventral
92. Trilobation
93. Widened and broadened front end (Eldredge 1974; Weygoldt and Paulus 1979; Weygoldt 1986)
94. Embryos with four gangliar post-oral segments
95. Separation of cephalic and locomotory functions onto different tagma
96. Coxal endites on appendage of second somite (primitively on all) (Boudreaux 1979)
97. Anterior (0) v. posterior (1) directed mouth (posterior in TCC) (Cisne 1974)
98. Lamellar spines on appendages (Bergström 1979)
99. Pretarsal segment of leg (dactylopodite) has only a single muscle (Snodgrass 1952)
100. Single, anatomically pre-oral limb-bearing segment in larva (if not adult)
101. $tRNA^{C}$ is between $tRNA^{W}$ and $tRNA^{Y}$
102. $tRNA^{Y}$ is inverted with respect to $tRNA^{W}$
103. $tRNA^{M}$ is between $tRNA^{Q}$ and ND2
104. $tRNA^{S(AGN)}$ is between $tRNA^{N}$ and $tRNA^{E}$
105. l-rRNA / $tRNA^{L(CUN)}$ / $tRNA^{L(UUR)}$ / ND1
106. COI / $tRNA^{L(UUR)}$ / COII

107. l-rRNA / tRNA$^{L(CUN)}$ / ND1 (Boore et al. 1995—somewhat charitably coded)
108. Unique crustacean limb segmentation pattern based on a shared coxa-basis muscular arrangement (Boxshall 1996)
109. Tarsal claws (paired?) (Hennig 1981)
110. Coxal vesicles, styli
111. Maxillary plate (mouth cavity bordered by second maxillae)
112. Appendages of first post-cephalic segment transformed into maxillipedes
113. Specialization in the ventral border of the mouth cavity
114. Stemmata
115. Diplosegments
116. Antennae with four sensory cones in distal segment
117. First post-cephalic segment into collum (Kraus and Kraus 1994)

Characters 25, 50, 52, and 89 are nonadditive, the remaining multistates are ordered.

# CHAPTER 2

## An Arthropod Phylogeny Based on Fossil and Recent Taxa

*Matthew A. Wills, Derek E. G. Briggs,*
*Richard A. Fortey, Mark Wilkinson, and Peter H. A. Sneath*

## Abstract

Morphological data from problematic fossil taxa offer important new insights into perennial questions of deeper branching in arthropod phylogeny. A comprehensive new database of ninety-seven morphological characters coded for twenty-five Recent, twenty-eight Cambrian, and nine other key fossil genera is presented. Parsimony analysis reveals much robust structure, although reduced cladistic consensus techniques demonstrate that resolution over large portions of the tree is frustrated by the mobility of a small number of rogue taxa. Crustaceans (including a number of crustaceanomorph fossil taxa) form a clade in opposition to the Arachnomorpha (including the Trilobita, Chelicerata, and a large number of problematica from the Burgess Shale). The Marrellomorpha (including both Cambrian and Devonian representatives) oppose this grouping and represent an extinct but high-ranking arthropod bodyplan. The deepest bifurcation within the euarthropods is that between biramous groups (Schizoramia) and the Atelocerata (Hexapoda and Myriapoda). Three basal clades in pectinate succession herald the appearance of all "euarthropod" characters; tardigrades, lobopodians, and a group comprising *Opabinia* and *Anomalocaris*. A similar gross topology results from successive approximations weighting of the data, and the major clades also appear as phenetic clusters. Omitting data from all fossil or all Recent genera results in cladistic topologies with many elements in common with the trees for all taxa. The lower-level relationships of several problematic groups are discussed, and alternative placements proposed in the literature are considered.

The arthropods have attracted considerable attention recently in new attempts to unravel their evolution and phylogeny (Wheeler et al. 1993; Wills et al. 1994, 1995;

Averof and Akam 1995). Historically, much controversy has surrounded the relationships of the four major arthropod groups (Crustacea, Chelicerata, Uniramia, and the extinct Trilobita), and the question of whether they all share an ancestor that was itself an arthropod. Both morphological and embryological research have been the classical tools, but paleontological and, more recently, molecular lines of evidence have also been brought to bear on the problem.

Most comparative anatomists have concentrated on just one or two organ systems or aspects of morphology (see contributions in Gupta 1979), and consequently they put tremendous weight on a limited subset of the available characters. Unfortunately, while the four major groups are relatively clearly defined both morphologically and taxonomically (with the possible exception of the Crustacea; Schram 1982), very few shared derived characters exist to identify links between them. Moreover, these characters conflict rather badly, resulting in numerous, alternative hypotheses of relationships.

Molecular approaches also have yet to converge on a definitive solution. They have been hampered by the sampling of only small numbers of taxa and particularly by the inclusion of lone exemplars of (taxonomically) major groups. Taxa have frequently been excluded for reasons that are unclear. Several studies, for example, have included a small number of arthropod taxa in wider examinations of protostome phylogeny. Wheeler et al. (1993) presented an analysis of arthropods using 18S rDNA and polyubiquitin sequence data, to which they added morphological and developmental characters. Each of the data sets in isolation supported arthropod monophyly, as did the "total" analysis (Kluge 1989). The principal strength of this work is that both morphological and molecular data were utilized. A considerable drawback is the paucity of information from fossils, the trilobites being the only extinct group included. Wheeler et al. (1993) stressed the importance of fossils for correct interpretations of character transition sequences but did not consider the possibility that they might overturn phylogenies.

Data from problematic arthropod fossils have seldom been factored into objective phylogenetic studies (Bergström 1992; Wilson 1992). The vast majority of problematica date from the Lower Paleozoic, their greatest abundance being in the Cambrian (Wills and Sepkoski 1993). Attempts to interpret problematic arthropods from the Burgess Shale (Whittington 1985) have prompted a debate regarding levels of bodyplan disparity in Cambrian and Recent metazoans (Gould 1989, 1991, 1993; Briggs et al. 1992a,b, 1993; Foote and Gould 1992; Lee 1992; McShea 1993; Ridley 1993; Wills et al. 1994). Most workers have preferred either to assign these fossils to taxonomic groups based solely on the Recent fauna (e.g., the "shoehorning" of Walcott 1908: see Gould 1989), consign them to "wastebasket" taxa (e.g., Trilobitomorpha: Størmer 1944), or grant them each a distinctness worthy of class or subphylum status (Gould 1989). None of these approaches has proved satisfactory. The discovery that many Burgess problematica are phenetically and phylogenetically intermediate between established groups (Wills et al. 1994) highlights the fact that fossils often represent novel combinations of character states and as such are essential to a valid phylogenetic analysis (Bergström 1979; Doyle and Donoghue 1987; Gauthier et al. 1988).

The database discussed here provides the basis for an analysis of relationships among living and fossil taxa. Preliminary results have been outlined elsewhere (see Wills et al. 1995), but several additional analyses are presented here together with a detailed discussion of the characters used.

## Principal Issues in Arthropod Evolution

Excellent syntheses of the controversies in arthropod evolution can be found in Gupta (1979), Wheeler et al. (1993), Wägele (1993), and Averof and Akam (1995). The following questions inform most of the discussions (see Wills et al. 1995):

1. Whether the crustaceans and tracheates form a clade (Mandibulata) (Snodgrass 1938; Sharov 1966; Boudreaux 1979; Paulus 1979; Weygoldt 1979) or are not closely related (Tiegs 1947; Tiegs and Manton 1958; Cisne 1974, 1982; Hessler and Newman 1975).
2. Whether the trilobites constitute a sister group to the chelicerates alone (Sharov 1966) or to the chelicerates plus the crustaceans (Cisne 1974; Hessler and Newman 1975).
3. Where the Onychophora and Tardigrada lie with respect to the tracheates and to the rest of the euarthropods in general (Dzik and Krumbiegel 1989; Brusca and Brusca 1990; Robison 1990; Monge-Nájera 1995; Nielsen 1995).
4. Whether the euarthropods arose once from a single soft-bodied ancestor that was not itself an arthropod (monophyly: Sharov 1966; Baccetti 1979; Boudreaux 1979; Callahan 1979; Clarke 1979; Gupta 1979; Paulus 1979; Weygoldt 1979), or whether two (diphyly: Tiegs 1947; Tiegs and Manton 1958; Hessler and Newman 1975; Cisne 1974, 1982) or more (polyphyly: Manton 1977) such events occurred.
5. Whether the tracheates (Hexapoda and Myriapoda) are monophyletic (see Tiegs and Manton 1958), or the myriapods branched off lower in the phylogeny (Ballard et al. 1992; Turbeville et al. 1991; Averof and Akam 1995).
6. Whether the Crustacea are monophyletic (Cisne 1982; Schram 1986) or paraphyletic (Lauterbach 1983; Averof and Akam 1995). The group is not easily defined on morphological grounds (Schram 1986), and it is not clear how far its acceptance as a clade has been an assumption.

## Models Based on Molecular Data

The present study utilizes data from comparative morphology and paleontology but includes no molecular information. The insights provided by molecular sequences are limited by the small number of taxa for which such data are presently available.

Field et al. (1988) used distance-based methods (Fitch and Margoliash 1967) to analyze 18S rDNA data for *Artemia*, *Drosophila*, *Spirobolus*, and *Limulus* among other protostome taxa. In this instance the arthropods failed to form a clade, but a polyphyletic genesis for the group could not be supported. Several augmentations and reanalyses of these data have failed to yield consistent results. Lake (1990), for example, analyzed just a subset of the sequences and, using his own method of invariants (Lake 1987), obtained a paraphyletic ladder of arthropods between non-arthropodan protostomes, worms, and deuterostomes. Abele et al. (1989) briefly examined 18S sequences of arthropods in an attempt to establish the position of the pentastomids within the crustaceans. The most parsimonious cladogram of the taxa analyzed revealed the arthropods as monophyletic. Addition of a few chelicerate sequences to the data of Field et al. (1988) and removal of some "fast-clock" taxa also yielded a monophyletic Arthropoda (Turbeville et al. 1991). Three principal reasons can be invoked to explain why these different approaches yielded incompatible results based on what are essentially the same data: (1) Only cursory attention was paid to sequence alignment in any of these analyses. This is particularly important in that RNA can only be sequenced in one direction and accordingly has a high error rate in comparison with DNA. (2) Certain taxa were omitted from some analyses. The taxa that were analyzed were represented only by single species. Other taxa, notably the Onychophora, were never included at all. (3) Different methods of analysis were used by different workers.

As noted above, Wheeler et al. (1993) utilized 18S rDNA data, to which were added sequences from the 188 base pair polyubiquitin locus. The 18S rDNA data are present in hundreds of copies and not translated, while the polyubiquitin locus is present in relatively few (4–10) copies and codes for protein. The two sequences therefore fulfill markedly different functions, and any pattern common to these two disparate systems is more likely to represent common descent rather than immediate adaptation to an internal or external environment. Wheeler et al. (1993) also coded a number of morphological and developmental characters. They analyzed both sets of data in isolation and in combination (Kluge 1989). The resulting phylogeny was consistent with that of Snodgrass (1938), except that he resolved the Myriapoda as paraphyletic. Wheeler et al. (1993) placed the Onychophora basal to two opposing clades — the Arachnata (Trilobita and Chelicerata) and the Mandibulata (Crustacea and Atelocerata).

## The Importance of Fossils

There is considerable divergence of opinion among cladists on the role of fossils. Nelson (1978) regarded them as "data in need of interpretation," preferring to interpolate fossils into a well-supported phylogeny for living taxa (Patterson 1981; Ax 1987). By contrast, Doyle and Donoghue (1987) and Gauthier et al. (1988) argued that as units of analysis, fossils are no different from living taxa and should therefore be included in analyses from the outset. Among arthropod workers opinion is divided between those who believe that the fossils do not provide information in addition to that contained in the living fauna (Brusca and Brusca 1990) and those who regard them as essential to any phylogeny reconstruction (e.g., Bergström 1979).

Cladistic principles are only applicable to monophyletic groups, e.g., biological groups or clades including (only) a species and all its descendants (Hennig 1966). Clearly, this definition incorporates any fossil members of a clade (Bergström 1979). Cambrian problematica are taxonomic anomalies, in many cases because they possess unfamiliar combinations of features. Taxa with novel character distributions may radically influence clade topology (Donoghue et al. 1989), which is a strong argument in favor of their inclusion. The diversity of the modern arthropod fauna is largely the product of the Early Paleozoic radiation (Gould 1989; Briggs et al. 1992a,b), which also produced the morphological intermediates that have been lost through extinction. The inclusion of fossils allows all the products of this radiation to be embraced.

Several pivotal fossil groups are discussed below. These represent taxa previously omitted from detailed discussion and analysis by the authors (see Wills et al. 1995 for a summary). They serve to illustrate the diversity of interpretation.

## Some Enigmatic Fossil Groups

### Anomalocaridids, *Opabinia*, and *Kerygmachela*

Anomalocaridids were originally described from the Stephen Formation of Mount Stephen, but they are also known in some detail from the Burgess quarry, the Lower Cambrian Chengjiang fauna of Yunnan Province, southern China, and the Lower Cambrian of Poland (*Cassubia*). Despite their relatively large size (Chinese anomalocaridids may have attained lengths of two meters), the cuticle of these predators was sclerotized but not mineralized (Whittington and Briggs 1985). Recent descriptions of anomalocaridid morphology are available in Chen et al. (1994). The dorsal cuticle of the trunk was almost certainly divided into tergites. A cephalon of four somites was well-demarcated from a trunk bearing eleven weakly sclerotized and flaplike appendages that are presumed to have facilitated a form of metachronal, underwater flight (Briggs 1994). A pair of stoutly stalked (presumably compound) eyes were located dorsolaterally on the head shield. The massive anteriormost appendages, unlike most of the rest of the animal, were strongly sclerotized and jointed, flexing ventrally and posteriorly to bring their tips just below a ventral mouth that was bounded by a circlet of thirty-two narrow, radiating plates. Though there is no clear homology with the appendages of other taxa, the form of these anterior limbs has prompted speculation that anomalocaridids lie immediately basal to the clade of euarthropods (Collins 1996). The other possibility is that there is a less intimate association, with independent derivation of the jointed appendages. Anomalocaridids are united by an impressive number of synapomorphies (Chen et al. 1994), and the validity of their grouping is not questioned here.

*Opabinia* was a more modestly proportioned Burgess Shale form (up to 70 mm excluding the anterior appendages) originally described by Walcott (1912). Various interpretations of this animal have been advanced. Walcott (1912) and Hutchinson (1930) believed it to be a branchiopod crustacean (see also Fedotov 1925; Richter 1932; Linder 1946; Tiegs and Manton 1958; cf. Raymond 1935); Størmer (1944) placed it in his Trilobitomorpha, and Simonetta (1970) reconstructed it as an arthropod of unknown affinities. The most thorough analysis was made by Whittington (1975), who determined that the cuticle was not biomineralized but sufficiently rigid to maintain the shape of lateral lobes down the body and a tail fan composed of several blades. The body was divided into a head bearing a single pair of fused appendages (the clawed "proboscis") and a segmented trunk. The mouth opened backward in the ventral posterior portion of the cephalon, leading anteriorly into a U-shaped portion of the gut. Five large eyes were arranged on the dorsal cephalon, all supported by broad, fleshy stalks. The sclerotized trunk was divided into fifteen segments plus a terminal division. Each of the trunk segments bore a pair of lobelike appendages, the posterior margin of each lying over the anterior margin of the next one down the trunk (imbricating in the opposite direction to those of the anomalocaridids). All but the first were divided into two finely separated layers — a lower, smooth lobe of tissue overlain by a narrower sheet or gill composed of imbricating lamellae lying parallel to the main axis of the body. The terminal division of the trunk bore three rounded, dorsolaterally directed tail-lobes, plus a small, terminal pair of curved spines.

*Kerygmachela*, a soft-bodied taxon from the Lower Cambrian Sirius Passet fauna of north Greenland, was about 20 cm long (Budd 1993). Budd assigned this form to the Superphylum Lobopodia but considered it to possess biramous appendages and therefore to be pivotal to an understanding of early arthropod and lobopod evolution. The animal was divided into a well-defined but poorly developed cephalic region bearing a single pair of stout, uniramous appendages, and a trunk of eleven divisions, each with a pair of purportedly biramous appendages. Terminally, there was a pair of segmented furcal rami, prominent but slender and almost equal in length to the rest of the animal. The frontal appendages were composed of a thick, fleshy, basal portion with transverse wrinkling, tapering slightly distally to four long, fine, terminal spines. The inner surface of the basal portion of each appendage was armed with a row of smaller spines. The appendages were directed anteriorly in an apparently raptorial or "grasping" attitude, diverging from each other at an angle of about 45 degrees. Wrinkled structures on the dorsal surface of each trunk lobe have been interpreted as gills (Budd 1993). Subtriangular lobes projecting from beneath the lateral lobes in two specimens are thought to represent lobopodal walking limbs.

Budd (1993) raised the possibility that ventral structures reconstructed as gut diverticulae in *Opabinia* (Whittington 1975) might be reinterpreted as lobopodous limbs similar to those in *Kerygmachela*. However, this contention remains unsubstantiated (Chen et al. 1994). Budd also cited the presence of wrinkled and spinose frontal appendages as a probable synapomorphy uniting *Kerygmachela* and *Opabinia*. On this basis he proposed a clade comprising these two taxa, and probably *Anomalocaris*, in opposition to the biramous euarthropods. This arrangement implied an independent derivation of the Schizoramia and Uniramia (*sensu* Harvey and Yen 1989), and therefore a diphyletic origin of the euarthropods. The anterior trunk of *Opabinia* is interpreted here as a pair of highly modified, fused appendages, but their morphology is considered too far removed from the frontal appendages of *Kerygmachela* to treat the two types as homologous.

Chen et al. (1994) considered *Kerygmachela* to be the closest relative of the anomalocaridid clade, with *Opabinia* in opposition to this grouping. Both *Kerygmachela* and *Anomalocaris* have a pair of stout, movable anterior appendages, but these are only very lightly sclerotized in the former and not composed of articulating podomeres as in *Anomalocaris*. Both taxa possessed a central body of eleven segments, but the number of cephalic appendages differed considerably (four in *Anomalocaris*, one in *Kerygmachela*). Hence, neither the total number of segments nor the number of segments incorporated into the cephalon coincide. Both taxa possessed a bilateral series of imbricating flaps down the trunk, but the morphology of these flaps is different.

## *Marrella*, *Mimetaster*, and *Vachonisia*

*Marrella*, the "Lace Crab" (Walcott 1912), is the commonest non-trilobite Burgess Shale arthropod. The body was divided into a cephalon bearing two pairs of appendages, plus a trunk of twenty-four to twenty-six somites bearing biramous limbs, and a telson (Whittington 1971). The cephalic shield bore two pairs of long, stout, dorsoventrally flattened spines. There were no eyes or other sense organs. The first appendage was a uniramous, multisegmented antenna. The second, which inserted immediately above and behind it, was composed of a massive, long, basal podomere, followed by five much shorter podomeres fringed with dense setae.

Behind the cephalic shield, the body was subcircular in cross section, tapering toward the posterior. All somites lacked tergites, but each bore a pair of biramous appendages that decreased in size posteriorly. These appendages were superficially trilobite-like (Størmer 1944), comprising a segmented inner ramus and a filament-bearing, multisegmented outer ramus. However, the precise number of segments in the endopod, the absence of gnathobases, endopodal spines or claws, and the presence of annulations on the outer ramus of *Marrella* are all detailed points of difference. The inner ramus of the first eight or nine trunk appendages was relatively long and functioned as a walking leg (see Briggs and Whittington 1985). Whittington

(1971) concluded that *Marrella* appeared to belong to a group separate from trilobites and thought that the presence of two pairs of antennae *could* suggest some crustacean affinities. Stürmer and Bergström (1976), by contrast, believed that *Marrella* was probably closely related to the Devonian *Mimetaster* and that both were primitive arachnomorphs.

The most detailed study of *Mimetaster* from the Hunsrück Slate was made by Stürmer and Bergström (1976). The head of the animal was strongly demarcated from the body. It was composed of a relatively large cephalic shield with a massive, ventral labrum, from which arose three pairs of uniramous appendages. The first was a filiform, segmented antenna. The second and third were massive walking limbs, the second almost disproportionately large (longer than the entire body). The most similar appendage in any other arthropod is probably the second cephalic appendage of *Marrella*. However, despite their similar size and attitude relative to the body, these appendages differed in the arrangement and structure of their component podomeres and are not considered to be homologous in the present analysis. The main, central, oval body of the cephalic shield (tergal shield of Stürmer and Bergström 1976) was only about one fifth of its entire length (up to 45 mm). Six very long and prominent spines extended laterally from the margin of the shield. A pair of stalked, lateral (presumably compound) eyes arose from the dorsal surface of the carapace.

The trunk of *Mimetaster* was composed of at least thirty somites, decreasing in size and tapering toward the posterior, as in *Marrella*. Each somite bore a pair of biramous appendages, which decreased in length posteriorly. Gnathobases were absent. The outer ramus was composed of numerous short, broad podomeres, narrowing distally and each bearing a long, flattened spine. Most of the outer rami on the posterior of the trunk could not have reached the substrate when *Mimetaster* was walking.

Birenheide (1971) and Whittington (1971) remarked on the similarities between *Mimetaster* and *Marrella*, citing the head shield spines, the tapering body without epimeres, the labrum, and the similar form of the trunk appendages. Stürmer and Bergström (1976) considered the lamellate spines on the outer rami to be closer to the condition in chelicerates and trilobites since nothing similar is known from crustaceans.

The rare Hunsrück Slate fossil *Vachonisia* was first described by Lehmann (1955) and later by Stürmer and Bergström (1976). A tergal shield covered the entire length of the body. A median notch, enclosing a rostrum-like structure, divided the anterior of the shield into two winglike lobes. The lateral margins were almost parallel, converging only slightly posteriorly. There was a V-shaped indentation along the posterior margin. Ventrally, a flattened doublure flanked a central, heart-shaped concavity (apex posterior), which accommodated the appendages (much as in the Recent xiphosurans). There was a large, lobate labrum, but no eyes. The first appendage was a short, uniramous antenna. The second to fourth pairs were by far the largest, probably ambulatory, and demarcated from those that followed by a short gap. Pairs two and three appear to have been armed with spinelike, medially directed endites, terminating just behind the labrum.

*Vachonisia* possessed approximately eighty pairs of homonomous trunk appendages (Stürmer and Bergström 1976), much smaller than those of the cephalon, and decreasing in size very regularly along the length of the body. Each comprised a quadrangular basal podomere, lacking gnathobases, plus two rami. The inner ramus was relatively stout, curved medially and composed of five or six anteroposteriorly flattened podomeres. The outer ramus was filiform and composed of numerous podomeres each bearing a long "spine," very similar to the comblike exopods of *Mimetaster.*

The dorsal shield of *Vachonisia* is superficially similar to that of *Lepidurus*, prompting speculation (in the absence of a detailed description of its appendages) that it might be a branchiopod crustacean (Lehmann 1955; Tasch 1969). However, the shield in *Vachonisia* is fused to the entire body, unlike the cephalic shield of crustaceans. The trunk appendages are very similar to those of *Mimetaster* and *Marrella*, which also share the regular decrease in the size of the trunk appendages and the relatively large cephalic limbs. Stürmer and Bergström (1976) believed *Vachonisia* to be a sister taxon to *Mimetaster* and *Marrella*.

## Euthycarcinoids

The euthycarcinoids are a group of just eight species in six genera known from the Late Silurian of Western Australia, the Late Carboniferous of France and the United States, and the Middle Triassic of France and eastern Australia. Specimens range in length from about 20 mm to 120 mm. Their affinities have been the subject of considerable controversy, various interpretations of their morphology placing them within the Crustacea, Chelicerata, and Uniramia.

Euthycarcinoids were originally thought to possess mandibles and biramous trunk appendages. Handlirsch (1914) speculated on copepod affinities for the first-known species, *Euthycarcinus kessleri* (cf. Gall and Grauvogel 1964). Following his original description of *Synaustrus brookvalensis*, Riek (1964) speculated that the euthycarcinoids were more properly regarded as merostomoid trilobitoids. Riek (1968) claimed the presence of two pre-oral antennae and two pairs of much reduced maxillae, prompting a reevaluation of the group as branchiopods. Schram (1971) cast serious doubt on the supposed crustacean affinities of the group. However, only when Bergström (1980) reexamined material of *Synaustrus* and *Kottixerxes* were euthycarcinoid appendages interpreted as uniramous, a view corroborated by descriptions of new material from the United States (Schram and Rolfe 1982), France (Heyler 1981; Rolfe et al. 1982; Rolfe 1985), and later Australia (McNamara and Trewin 1993). Furthermore, neither the presence of two pairs of antennae (Riek 1964, 1968) nor uropods could be substantiated (McNamara and

Trewin 1993). Many segments in euthycarcinoids are composed of two somites (diplosegments), a feature normally associated with myriapodous uniramians (Diplopoda and Pauropoda). Bergström (1980) concluded that the euthycarcinoids probably merited a taxon of equivalent rank to the Myriapoda and Hexapoda. Schram and Rolfe (1982) formalized this concept by placing the Euthycarcinoidea as a separate subphylum within the atelocerates. McNamara and Trewin (1993) extended Harvey and Yen's (1989) concept of the Uniramia as a phylum exclusive of the onychophorans to include the Euthycarcinoidea.

The uniramian hypothesis has been challenged in recent years. Starobogatov (1988) proposed a classification of the euthycarcinoids that incorporated the Aglaspidae, a Cambrian family originally regarded as chelicerates (Størmer 1955; Bergström 1971; cf. Briggs et al. 1979). McNamara and Trewin (1993) rejected such an assignment.

Considerable confusion has existed over the homology of the cephalic structures of euthycarcinoids and the number of somites incorporated into the cephalon (Gall and Grauvogel 1964; Riek 1964, 1968; Bergström 1980; Schram and Rolfe 1982; Schram and Emerson 1991). McNamara and Trewin (1993) were able to homologize all the somites of a representative euthycarcinoid (*Kalbarria*) with a representative hexapod (*Drosophila*). According to their scheme, the anus is located on the twenty-first somite in both taxa, but the euthycarcinoid has an additional post-anal extension. The first somite in both taxa bears an antenna, while the second lacks appendages of any kind. McNamara and Trewin (1993) believed the third and fourth to bear the labrum and mouth respectively. Hence, the mandibular somites of both taxa are also homologous, while the maxillary somite of hexapods corresponds to the first pre-abdominal somite of euthycarcinoids.

## *Cheloniellon*

*Cheloniellon calmani* from the Devonian Hunsrück Slate was first described by Broili (1933) and later in more detail by Stürmer and Bergström (1978). The dorsal aspect was oval, comprising ten large, tuberculate tergites followed by a much smaller cylindrical sclerite and the telson. The first tergite carried a pair of anterolaterally directed, sessile eyes. The dorsum could be divided into an axis and broad, lateral pleurae although there was no marked furrow between these regions. Ventrally, the tergites were fringed by a narrow doublure. A median oval, tuberculate plate, with a spinose posterior margin, was interpreted by Stürmer and Bergström (1978) as the labrum. There was a total of fifteen pairs of appendages. The first six were located beneath the first two tergites (five beneath the first and one beneath the second), interpreted as forming the cephalon. The first appendage was a uniramous, multisegmented, filiform antenna. The second was located anterolateral to the labrum in a pre-oral position. Appendages three to six were post-oral, basically homonomous and arranged around the mouth. Each was uniramous, comprising a massive gnathobase and endopod. The first eight trunk appendages were biramous and lacked a gnathobase. Each comprised an endopod of five podomeres and an exopod of (probably) two proximal podomeres with a platelike, distal podomere fringed with long setae along its posterior margin. The last appendages were very long, uniramous, unjointed "furcal rami."

The nature of the trunk appendages, dorsal trilobation, and filiform antennae prompted Broili (1933) to compare *Cheloniellon* with the trilobites. Stürmer and Bergström (1978), however, made comparisons with the morphology of chelicerates, highlighting the presence of four gnathobasic limbs in *Cheloniellon*. They regarded the exopodal lamellar spines as morphologically intermediate between the narrow lamellar spines of trilobites and the platelike lamellar gills of xiphosurids. The second appendage of *Cheloniellon*, although not chelate, was considered to be homologous with a chelicera. Stürmer and Bergström (1978) thus argued that the first appendage (equivalent to the antenna of trilobites and *Cheloniellon*) was lost during the evolution of the chelicerates. This is not, however, the stance taken here; antennae and chelicerae are homologized as appendages of the first post-acronal or deutocerebral somite (Brusca and Brusca 1990). Stürmer and Bergström (1978) concluded that *Cheloniellon* is not a chelicerate and retained it in the Trilobitomorpha.

## The Database

Twenty-three Recent euarthropods were selected to represent modern class- or subclass-level taxonomic diversity. A modern onychophoran (*Peripatoides*) and a tardigrade (*Echiniscus*) were also included. Pycnogonids were omitted. There is considerable difficulty in homologizing regions of the pycnogonid body, and their position in preliminary analyses was somewhat variable. Typically the pycnogonids branched off near the base of the cladogram, a position considered to be an artifact of unsatisfactory character coding. Twenty-five Cambrian (those preserved in sufficient detail to permit credible comparisons with modern forms), one anomalous Silurian, and five Devonian euarthropod genera were added to these living taxa. The Cambrian problematica *Anomalocaris*, *Aysheaia*, *Kerygmachela*, and *Opabinia* were also included. The data are presented in table 2.1.

### Missing Data

Several characters in the data matrix (table 2.1) incorporate missing data entries ("?"). These are included for one of two reasons (Platnick et al. 1991):

1. For some taxa, information is simply unknown. This often affects the coding of soft-part or developmental characters for fossils but may also be a problem where a Recent organism has been insufficiently studied. The

TABLE 2.1
The data matrix used in the analyses. The numbered characters and codes are explained in the text.

| | | | | | |
|---|---|---|---|---|---|
| *Acerentomon* | 1000000000 | 0000000000 | 010000?050 | a00??02002 | k111100100 |
| | c10????000 | 0000000000 | 1111011111 | 211101011? | ?101010 |
| *Aglaspis* | 1011000111 | 1010101000 | ?100110040 | d10?010000 | g011100100 |
| | c00????010 | 0000021001 | ??0?01???? | ????0????? | ??????? |
| *Agnostus* | 1101000011 | 1110000000 | 111000?040 | a010210000 | b010100101 |
| | f010100011 | 0000010000 | ??0?01???? | ????1????? | ????11? |
| *Alalcomenaeus* | 1001000101 | 0000000000 | 1102110040 | g110010000 | g010100101 |
| | h011010010 | 0000012010 | ??0?01???? | ????0????? | ??????? |
| *Alima* | 1001200101 | 0000000001 | 1112111051 | a111111000 | k110100100 |
| | c00????100 | 0000110000 | 1101011011 | 0111010110 | 1111011 |
| *Androctonus* | 1001000000 | 0000100000 | 00?1100060 | c10?410000 | j1111000?? |
| | ?00????000 | 0000021000 | 1101011011 | 1111110110 | 0111110 |
| *Anomalocaris* | 1001000000 | 0000000000 | 00?0110040 | h010000000 | g0110000?? |
| | ??10010000 | 0001000100 | ??0?00???? | ?????????? | ??????? |
| *Argulus* | 1001000000 | 0000000000 | 1112111050 | a10?111000 | c111100100 |
| | a010100000 | 0000000000 | 1101011011 | 0111010110 | 1111011 |
| *Artemia* | 1000000000 | 0000000000 | 1102110050 | b10?111000 | o110100100 |
| | a010010110 | 0001000000 | 1101011011 | 0111110110 | 1111011 |
| *Aysheaia* | 0000000000 | 0000000010 | 00?000?010 | a10?000000 | e011000111 |
| | ??0????000 | 0001000000 | ??0?00???? | ????0????? | ??????? |
| *Baltoeurypterus* | 1101000101 | 1000100000 | 1111110060 | c10?010000 | j1111000?? |
| | ??0????000 | 0000021000 | ??0?01???? | ????0????? | ??????? |
| *Branchiocaris* | 1002011000 | 0000000000 | 011000?020 | a10?300000 | q0101110?? |
| | ??10110000 | 0000000000 | ??0?11???? | ????0????? | ??????? |
| *Bredocaris* | 1001000000 | 0000000000 | 1100111050 | a110011000 | e110100101 |
| | c010000011 | 0001000000 | ??0?11???? | ?????????? | ??????1 |
| *Burgessia* | 1001000000 | 1000000000 | 10?000?040 | a110000000 | d011100100 |
| | e010010000 | 0000021000 | ??0?01???? | ????1????? | ??????? |
| *Calanus* | 1001000000 | 0000000000 | 111200?050 | a110111000 | g110100100 |
| | b010100000 | 0001000000 | 1101011011 | 0111010110 | 1111011 |
| *Campodea* | 1000000000 | 0000000000 | 010000?050 | a00??02002 | j111100100 |
| | c10????000 | 0001000000 | 1111011111 | 211101011? | ?101010 |
| *Canadaspis* | 1002010000 | 0000000000 | 1110111050 | b10?111000 | i110100101 |
| | i010010010 | 0000000000 | ??0?11???? | ????0????? | ??????? |
| *Cheloniellon* | 1011000111 | 0000001000 | ?110110060 | a10?010000 | g011100100 |
| | d011100000 | 00000?0000 | ??0?01???? | ????0????? | ??????? |
| *Corynothrix* | 1000000000 | 0000000000 | 0103110050 | a00??02002 | e110100100 |
| | c10????000 | 0001000000 | 1101011111 | 2111010111 | 1101010 |
| *Cypridina* | 1002001001 | 0000000000 | 1110111050 | a110111000 | g0101000?? |
| | ??0????000 | 0001000000 | 1101111011 | 0111010110 | 1111011 |
| *Derocheilocaris* | 1001000000 | 0000000000 | 111300?050 | a110011000 | g110100100 |
| | a00????000 | 0001000000 | 1101011011 | 0111010110 | 1111011 |
| *Echiniscus* | 1000000000 | 0000000000 | 00?0100010 | k10?000000 | a011000110 |
| | ??0????000 | 0000010000 | 1100010?10 | 0101000000 | 0001010 |
| *Emeraldella* | 1001000101 | 0000001000 | 110000?060 | a10?010000 | k111100101 |
| | e011011010 | 00000?1001 | ??0?01???? | ????0????? | ??????? |

TABLE 2.1 (*continued*)

| | | | | | |
|---|---|---|---|---|---|
| *Galathea* | 1001100101 | 0000000001 | 1112111051 | a111111000 | k111100100 |
| | d00????100 | 0000112000 | 1101011011 | 0111110110 | 1111011 |
| *Habelia* | 1011000110 | 1000000000 | ???000?030 | a1100?0000 | g010100101 |
| | ?0110100?0 | 0000011010 | ??0?01???? | ????0????? | ??????? |
| *Julus* | 1000000100 | 0000000000 | 0100100140 | a00??0212? | q011110100 |
| | e00????000 | 1?10000000 | 1111011111 | 2111011110 | 1101010 |
| *Kalbarria* | 1100000100 | 0010000000 | ?100110030 | a00??00000 | a011000110 |
| | ??0????000 | 0000011000 | ????01???? | ????0????? | ??????? |
| *Kerygmachela* | 0010000000 | 0000000010 | 0??000?010 | j110000000 | d011000100 |
| | ??10010000 | 00010?0000 | ??0?00???? | ????0????? | ??????? |
| *Leanchoilia* | 1001000110 | 0000001000 | 010000?030 | g110000000 | f010100100 |
| | g011010000 | 0000021010 | ??0?01???? | ????0????? | ??????? |
| *Lepas* | 1002001000 | 0000000000 | 111200?050 | a00??11000 | d110100100 |
| | j010100000 | 0000000000 | 1101111011 | 0111110110 | 1111011 |
| *Lepidocaris* | 1001000000 | 0000000000 | 110000?050 | b110111000 | m110100100 |
| | a010000010 | 0001000000 | ??0?01???? | ????0????? | ??????1 |
| *Lepidurus* | 1101000001 | 0010000000 | 1103110050 | b10?111000 | q111111100 |
| | a010010110 | 0001010000 | 1101111011 | 0111110110 | 1111011 |
| *Lepisma* | 1000000000 | 0000000000 | 0102110050 | a00??02002 | k111100100 |
| | d10????000 | 0001000000 | 1111011111 | 2111010111 | 1101010 |
| *Lithobius* | 1000000000 | 0000000000 | 0100110150 | a00??02111 | n011100100 |
| | d00????000 | 0100000000 | 1111011111 | 2111011110 | 1101010 |
| *Marrella* | 1001000001 | 0000000100 | ?10000?020 | a10?500000 | q011101100 |
| | d011100000 | 0000000000 | ??0?01???? | ????0????? | ??????? |
| *Martinssonia* | 1001000000 | 0000000000 | 110000?050 | a110010000 | d1101000?? |
| | ??0????001 | 0000010000 | ??0?11???? | ????0????? | ??????0 |
| *Mimetaster* | 1001000000 | 0000000100 | ?101110030 | a10?700000 | q011101100 |
| | f011100000 | 0000000000 | ??0?01???? | ????0????? | ??????? |
| *Molaria* | 1001000111 | 1000001000 | 110000?040 | a110010000 | d010100101 |
| | e011010010 | 0000021000 | ??0?01???? | ????1????? | ??????? |
| *Nahecaris* | 1002201000 | 0000000000 | 1110111051 | a110111000 | l111100100 |
| | f0100000?0 | 0001011000 | ??0?11???? | ????0????? | ??????? |
| *Naraoia* | 1001000111 | 1100000000 | 110000?040 | a110010000 | 1011100101 |
| | e011011010 | 0000010000 | ??0?01???? | ????1????? | ????11? |
| *Nebalia* | 1002201100 | 0000000000 | 1110111051 | a10?111000 | 1110100100 |
| | b010010100 | 0001000000 | 1101111011 | 0111010110 | 1111011 |
| *Odaraia* | 1002011000 | 0000000000 | 1112111050 | ?10?111000 | q010111100 |
| | d010010010 | 0000000000 | ??0?11???? | ????0????? | ??????? |
| *Olenoides* | 1211000111 | 1111011000 | 1103110040 | a110010000 | i111100101 |
| | e011001010 | 0000010000 | ??0?01???? | ????1????? | ????11? |
| *Opabinia* | 0000000000 | 0000000000 | 1102110010 | i010000000 | h0110000?? |
| | ??10010000 | 0000000100 | ??0?00???? | ????1????? | ??????? |
| *Pauropus* | 1000000000 | 0000000000 | 010000?140 | a00??0212? | g011110100 |
| | f00????000 | 1?10000000 | 1101011111 | 211101111? | ?101010 |
| *Peripatoides* | 0000000000 | 0000000010 | 00?0100030 | a10?603000 | l011000110 |
| | ??0????000 | 0001010000 | 1110000?11 | 0100000110 | 0101000 |
| *Periplaneta* | 1000000000 | 0000000000 | 0102110050 | a00??02002 | k111100100 |
| | c10????000 | 0000000000 | 1111011111 | 2111110111 | 1101010 |

TABLE 2.1 (*continued*)

| | | | | | |
|---|---|---|---|---|---|
| *Perspicaris* | 1002011000 | 0000000000 | ?110111050 | b10?1??000 | l1101000?? |
| | ??100100?0 | 0001000000 | ??0??1???? | ????0????? | ??????? |
| *Rehbachiella* | 1001000000 | 0000000000 | 1100111050 | a110111000 | j111101100 |
| | c010000010 | 0001000000 | ??0111???? | ?????????? | ??????1 |
| *Sanctacaris* | 1001000111 | 1000100000 | ?100111061 | e110010000 | i111100101 |
| | g010010010 | 0000022010 | ??0?01???? | ????0????? | ??????? |
| *Sandersiella* | 1001000101 | 0000000000 | 110001?050 | a110111000 | o110100100 |
| | e010100110 | 0001000000 | 1101111011 | 0111110110 | 1111011 |
| *Sarotrocercus* | 1001000111 | 0000001000 | ?100111020 | f010000000 | d0101000?? |
| | ??11010000 | 00000?1000 | ??0?01???? | ????0????? | ??????? |
| *Scutigerella* | 1000000000 | 0000000000 | 010000?150 | a00??02112 | k011100100 |
| | c00????000 | 0100000000 | 1111011111 | 211101111? | ?101010 |
| *Sidneyia* | 1101000101 | 0000001000 | 10?0111010 | a110010000 | e111100101 |
| | g011010010 | 0000112000 | ??0?01???? | ????0????? | ??????? |
| *Skara* | 1001000001 | 0000000000 | 110000?050 | a110011000 | h1101000?? |
| | ??0????001 | 0001000000 | ??0?01???? | ????0????? | ??????1 |
| *Speleonectes* | 1001000101 | 0000000000 | 110000?051 | a110111000 | p011100100 |
| | c011100000 | 0001000000 | 1101011011 | 011101011? | ?11101? |
| *Tachypleus* | 1101000011 | 1010100000 | 1111110060 | c10?410000 | f011100100 |
| | c010010000 | 0000021000 | 1101011011 | 0111010110 | 0111110 |
| *Triarthrus* | 1201000111 | 1111011000 | 1113110040 | a110010000 | q111100100 |
| | e011000010 | 0000010000 | ??0101???? | ????1????? | ????11? |
| *Vachonisia* | 1?01000001 | 0000000000 | ?10000?040 | b10?010000 | q011101100 |
| | d011100000 | 0000000000 | ??0?01???? | ?????????? | ??????? |
| *Waptia* | 1002000000 | 0000000000 | ?110111050 | a10?1????? | l110100100 |
| | ??0????0?0 | 0001000000 | ??0?11???? | ????0????? | ??????? |
| *Weinbergina* | 1111000111 | 0001100000 | 1110110070 | c10?010000 | i0111000?? |
| | ??11010000 | 0000021000 | ??0?01???? | ?????????? | ??????? |
| *Yohoia* | 1001000111 | 0010000000 | 0100111040 | g10?000000 | i1101000?? |
| | ??10010000 | 0000012010 | ??0?01???? | ????0????? | ??????? |
| Annelida | 0000000000 | 0000000010 | 00?000?000 | ?00??0???? | ?00????0?? |
| | ??0????00? | 00000?0000 | 0000000?11 | 000000001? | ?000000 |
| Mollusca | 0000000000 | 0000000000 | 00?000?000 | ?00??0???? | ?00????0?? |
| | ??0????00? | 0000000000 | 0000000?00 | 0000?0000? | ?010000 |

inclusion of incomplete information on state distribution is preferable to the wholesale exclusion of a character.

2. The coding of some characters may be contingent on the presence of other structures. For example, the condition of the second appendage cannot be meaningfully coded in taxa that lack the limb altogether.

At present, parsimony programs make no distinction between these two sources of missing data. Importantly, however, neither influences resolution. According to Swofford (1990:17), "only those characters that have non-missing values will affect the location of any taxon on the tree." In retrospect, however, the placement of taxa on the tree, as determined by informative characters, may influence the parsimonious reconstructions of the values of missing entries (Wilkinson 1992). DELTRAN optimization, however, favors convergences near the terminals and therefore delays character transitions so as to support fewer taxa for which the state is unknown. This avoids the problem of spurious support at nodes.

A concentration of missing entries may mean that there are not sufficient characters with a positive influence on the placement of a particular taxon to provide an unambiguous hypothesis of its relationships (i.e., the taxon is underdetermined). Including such taxa may also lead to a loss of information regarding the relationships of taxa that are not underdetermined (Crepet and Nixon 1989; Nixon and Wheeler 1992). Even taxa without missing data may be underdetermined where incongruous characters result in exactly balanced support for alternative groupings (Wilkinson 1992). Such problems are most appropriately addressed by using sensitive consensus methods (e.g., reduced cladistic consensus, RCC) rather than at the stage of coding or analysis. Taxa that include no unique character-state combinations, apart from missing entries, can safely be omitted from analyses.

## Rooting

Morphologically, the arthropods are more similar to the annelids than to any other major protostome group (Schram 1978), and their union within the Articulata has a long history (e.g., Cuvier 1817; Haeckel 1866; Kozloff 1990). A less widespread view is that mollusks and arthropods are more closely related to each other than either is to the annelids (Lemche 1959a,b; Fretter and Graham 1962). A third possibility is that the annelids, mollusks, and certain other, smaller phyla (Eutrochozoa: Ghiselin 1988) share a more Recent common ancestor than any does with the arthropods. This finds some support both from morphology (e.g., Pelseneer 1906; Naef 1913, 1924) and more recently from 18S rRNA data (e.g., Field et al. 1988; Raff et al. 1989; Lake 1990).

Since the derivation of the arthropods from soft-bodied protostomes is widely accepted (Weygoldt 1979; Brusca and Brusca 1990), and all models require a common protostome ancestor, the broader questions are most appropriately addressed by rooting a cladistic analysis with a hypothetical, generalized, soft-bodied protostome. In practice, details of coding differ little whether an annelid or mollusk form is used.

## The Interpretation of Characters

A problem inherent in analyses such as this is the variation in the interpretation of characters at the outset. There are few features widely accepted as good homologies even within the four major arthropod groups, and there is considerable debate about the polarity of others. Diametrically opposed interpretations often derive from similarly painstaking anatomical studies (e.g., Manton 1977 v. Boudreaux

1979). Characters that appear to provide good homologies for a subset of arthropods may also occur in a limited number of (phenetically) very dissimilar forms. Several characters supporting major branches have high retention indices (Farris 1989) but relatively low consistency indices because of reversals and parallel gains in or near terminals. Each character is discussed at length below, but some specific points of interpretation are treated first.

Uniramous versus Biramous Limbs. Biramous trunk appendages have long been quoted as a synapomorphy uniting the crustaceans and trilobites (e.g., Vandel 1949). The extant chelicerates and atelocerates, by contrast, have uniramous trunk limbs (Manton 1977). Studies of exopod morphology in crustaceans and trilobites (e.g., Størmer 1944; Snodgrass 1952) have revealed differences assumed by some workers to preclude homology. The crustacean exopod arises from the second protopodal segment (the basis), whereas in trilobites an undivided protopod bears both rami. However, a biramous ancestor may have undergone fusion of the protopodal elements or, alternatively, division of a single-element protopod. While the evidence for secondary uniramy in the chelicerates is fairly convincing (Schram and Emerson 1991), this has not been forced on the analysis here. Loss of the outer ramus is not recognized as a separate character state.

Number and Homology of Cephalic Segments and Appendages. The subject of the homology of cephalic appendages is particularly controversial. The scheme used here was also adopted by Brusca and Brusca (1990), although it should be noted that rather different interpretations were offered by Manton (1977) and Schram (1978).

The presegmental acron does not bear appendages in any known arthropod; it is associated with the anteriormost cerebral ganglion, or protocerebrum. The acron is absent from the onychophorans (but cf. Manton 1977). The first appendages of all living classes derive from the second (deutocerebral or first post-acronal) somite. In the uniramians, crustaceans, and some arachnomorphs, this appendage is an antenna (composed of several to many relatively unspecialized podomeres and performing a sensory function). Cheliceriformes have a chelicera or chelifore in the place of an antenna and lack a deutocerebral ganglion (Schram 1978).

In all groups for which early ontogeny is documented, with the exception of the pycnogonids, the mouth develops embryonically as part of the third somite (tritocerebral or second post-acronal). In the Pycnogonida the mouth appears to be situated embryonically just posterior to the acron. In crustacean ontogeny the migration of the second appendage to a pre-oral position also suggests that the location of the mouth is not a critical character. Tritocerebral appendages are antennae in the Crustacea, pedipalps (arachnids) or first walking legs (xiphosurans) in the chelicerates, and first walking legs in the trilobites. The atelocerates all secondarily lack appendages on this somite, although transitory limb-buds are present in an embryonic stage of the symphylan *Hanseniella*. The third post-acronal somite bears mandibles in the Atelocerata and Crustacea. The trilobites and chelicerates have walking legs in this location.

Cephalization may have occurred on several occasions, in which case the arrangement of appendages in the cephalon of different classes may not necessarily be homologous. Numbers of cephalic somites are treated no differently here than any other character; free reversals and transitions between states are allowed.

Use of the Eighth Appendage in Coding Post-cephalic Characters. A specific appendage has been coded to avoid anomalies caused by variation in the position of the cephalon/trunk border in different groups. The eighth appendage is located behind the cephalon in all taxa. This rationale also serves to code the morphology of trunk exopods where these structures are present, minimizing the use of missing data. In a few genera (e.g., Derocheilocaris) the more posterior trunk appendages lack exopods, but the anteriormost, including the eighth, are biramous. Conversely, the more anterior trunk appendages in forms such as Tachypleus lack an exopod, but exopods are present on the eighth and following ones. Use of the eighth appendage also effectively codes the presence of gnathobases in only those taxa with a long series of trophic appendages. A number of crustacean taxa have gnathobasic maxillipedes on the anterior trunk somites but lack protopodal endites on the following limbs. Moreover, all taxa with a gnathobasic eighth limb also have gnathobases on the more anterior trunk limbs.

## Arthropod Pattern Theory (APT)

Schram and Emerson (1991) hypothesized that biramous arthropod appendages resulted from the fusion of pairs of originally uniramous limbs on successive segments. The most primitive arrangement of appendages and segments was considered to occur in the hexapods and some myriapods, where each segment (monomere or monosegment) bears a single pair of appendages. In many myriapods and in the extinct euthycarcinoids, the dorsal tergites of successive pairs of monomeres are fused while the ventral sternites remain detached, each bearing a pair of uniramous appendages. These diplomeres or diplosegments correspond to two hexapodan monomeres. Schram and Emerson (1991) considered that fusion of the sternites in some diplopodous arthropodan ancestor yielded a condition with two pairs of uniramous appendages borne on a single segment, or duplomere. The basal crustacean *Tesnusocaris* was reconstructed with two pairs of uniramous appendages on each trunk segment — an inner, medial endopede probably used for sculling and an outer exopede apparently used to row (Emerson and Schram 1990). The trunk appendages of *Branchiocaris* were similarly interpreted with the outer flaplike exopede separate from the inner flipper-like, segmented endopede (rather than the latter simply forming a strengthening ridge on the proximal anterior margin of the lobate appendages as argued by Briggs [1976]). The biramous appendages in most arthropods were considered to have developed by the basal fusion of duplopodous, uniramous limbs. Thus each trunk segment (duplosegment) was considered to correspond to two segments in the hexapods and some myriapods. Secondary uniramy (effected by the loss of one or the other ramus, usually the outer) has probably occurred on a number of occasions in distantly related groups (e.g., Isopoda, Arachnida).

Schram and Emerson (1991) argued that the principles of bodyplan construction implicit in APT are fundamental to any consideration of arthropod phylogeny. A series of nodes and fields were recognized in all Recent and fossil arthropod groups. Tagma boundaries, gonopores, and body terminations tended to occur within nodes. Here we prefer to use a much more comprehensive list of characters and less restrictive coding. This list naturally includes a number of characters relevant to APT, but recoding this subset according to the tenets of Schram and Emerson (1991) would have few implications for the present analysis.

## Description of Character States

### Hard-Part Characters

#### *Cuticle and Specializations*

1. Cuticle sclerotization: *(0) cuticle unsclerotized, or very weakly sclerotized but not forming articulating plates (sclerites); (1) cuticle sclerotized and forming sclerites.*

A thin, unsclerotized, and highly permeable cuticle is found in the annelids and lobopods. A sclerotized cuticle composed of articulating plates (sclerites) is classically a diagnostic feature for the euarthropods. The question of whether this character evolved once or on more than one occasion in a number of independent lineages is central to the monophyly/polyphyly debate on the origins of the phylum. Some species of tardigrades (e.g., *Echiniscus*) have a series of symmetrically arranged dorsal and lateral sclerotized plates, which are thought to be homologous with the sclerites of euarthropods (Brusca and Brusca 1990; Kinchin 1994).

2. Ecdysial sutures: *(0) not visible; (1) visible and marginal; (2) visible and dorsal.*

Visible marginal sutures are limited to a small number of arachnomorph taxa and some notostracan branchiopods. In this database only the trilobites *Olenoides* and *Triarthrus* have prominent dorsal sutures. The Onychophora molt often (as frequently as every two weeks), but the thin cuticle has no visible suture lines (Ruhberg 1985). Tardigrades have a slightly thicker cuticle, molted periodically but with no visible sutures (Baccetti and Rosati 1971). The annelids do not molt.

3. Cuticle sculpture: *(0) smooth; (1) strongly tuberculate.*

A strongly tuberculate cuticle is a feature of the trilobite *Olenoides*, the chelicerate *Weinbergina*, and a limited number of taxa that have frequently been allied to these two groups (e.g., *Aglaspis* and *Cheloniellon*) (Størmer 1944; Stürmer and Bergström 1978).

4. Cephalic shield: *(0) absent or developed only as a simple head capsule; (1) simple head shield, entire carapace as a single unit, or followed by a variable number of tergites in association with it; (2) bivalved carapace.*

The homology of carapaces is a contentious issue within specific groups (e.g., Crustacea) as well as within the arthropods as a whole (Schram 1986). Secretan (1980) considered that all crustaceans (except anostracan branchiopods) have a carapace to a greater or lesser degree. A head shield was assumed to be a prerequisite for the development of a complete carapace. The crustacean carapace is usually considered to derive from the maxillary segment (see Schram 1986). However, work by Dahl (1991) demonstrated its derivation from the maxillipedal or first trunk somite in many groups (see also Boxshall 1983; Müller 1983), with a variety of subsequent developmental patterns (Dahl 1983, 1991). Schram (1986) considered that various carapace forms within the Crustacea may not be homologous.

In those arthropod taxa with no cephalic shield, the head is (at most) encased in a capsule of cuticle, and the appendages articulate directly with this (e.g., lobopods, atelocerates, *Artemia*). In all other carapace types the cephalic shield gives the cephalon a dorsoventral division, and the appendages emerge from beneath it.

In a small number of taxa the shield is well developed but only as a single element. This state encompasses a range of morphologies from the condition in *Calanus* and *Galathea* to that in *Argulus*, although it does tend to be a feature of crustaceans. A similar condition is seen in many chelicerates, arachnomorphs, and some other crustaceans, where the anteriormost shield element articulates with one or more subsequent plates of cuticle, resulting in a cephalic shield with tergites (e.g., *Emeraldella*: Bruton and Whittington [1983]). The distinction is subjective, and the two variants of the entire carapace are therefore scored as a single character state.

Bivalved carapaces occur predominantly within the crustaceans. The shield usually extends back over a large portion of the trunk and is divided sagittally by a hinge or pronounced fold (as in *Nebalia*).

5. Anterior margin of carapace: *(0) rounded or invaginated; (1) produced into a fixed rostral spine; (2) produced into an articulating rostral spine.*

Articulating rostral spines are typical of many malacostracans (Rolfe 1969). In *Nebalia* this takes the form of a large plate that functions to restrict the flow of fluid through the carapace (Bergström et al. 1987). The eumalacostracan *Galathea* has a fixed rostral spine (unique among the taxa in the database).

6. Posterior margin of carapace: *(0) straight or rounded; (1) with a sharp, V-shaped invagination, diverging at an angle of less than 100 degrees from the apex.*

A V-shaped posterior margin is always associated with long, bivalved carapaces that have straight hinges in lateral aspect. Taxa with this posterior invagination include the fossils *Branchiocaris*, *Canadaspis* (Briggs 1978), *Odaraia*, and *Perspicaris*. Many other bivalved taxa, by contrast, have an almost straight posterior border to the carapace, often with an indistinct or curved hinge line.

7. Ventral extent of the carapace: *(0) carapace covering the anterior dorsum and possibly the sides of the animal, but leaving the appendages to protrude more than half of their length beneath it; (1) carapace all-enveloping.*

All-enveloping carapaces are found in the living ostracodes (*Cypridina*), barnacles (*Lepas*), and phyllocarids (*Nebalia*), and in the fossils *Perspicaris*, *Branchiocaris* (Briggs 1976), and *Odaraia* (Briggs 1981). The ventral extension is also associated with a tendency for the ventral margins to curve under the animal, forming, at the extreme, the tubelike carapace in *Odaraia*. This ventral extension is not equivalent to a doublure (character 10), where the margin of the carapace is distinctly reflexed or at least bent sharply through an angle.

8. Style of post-cephalic articulation: *(0) without overlapping pleurae; (1) with overlapping pleurae.*

A number of arthropod groups that lack a fully sclerotized cuticle have developed isolated, non-articulating cuticular plates, commonly on the dorsum. Recent work on Cambrian lobopods (Ramsköld 1992a,b; Hou and Bergström 1995) suggests that this was once a more widespread phenomenon (e.g., *Hallucigenia*, *Microdictyon*). These cuticular plates probably provided some degree of protection for the animal (many are armed with spines) and afforded a substrate for the attachment of underlying limb muscles. Abuttal or overlapping of these plates provides the simplest means of covering the dorsum, retaining regions of lightly sclerotized cuticle in between (as the simplest form of articulation).

More intimate association between plates is achieved by broadly overlapping their lateral extensions as in many arachnomorphs and some Crustacea (e.g., *Sandersiella*, *Galathea*). Such pleurae typically have rounded margins in lateral aspect and sweep back a short distance over the anterior margins of the distal portions of the succeeding tergites (e.g., Hessler and Sanders 1973).

9. Extent of the anteroposterior division of the dorsal cuticle: *(0) not trilobed; (1) trilobed.*

The trilobites and a number of other arachnomorph taxa have the dorsal body surface divided longitudinally by a pair of anteroposterior furrows or depressions, forming a medial region, flanked on either side by lateral lobes (Harrington 1959; Cisne 1974). Trilobation of the cuticle in this manner does not correspond to the presence of overlapping pleurae since either condition is possible without the other.

10. Reflexion of the more heavily sclerotized dorsal cuticle of the cephalic shield to form a doublure: *(0) not reflexed; (1) reflexed.*

In trilobites the cephalic exoskeleton is continued on the ventral side of the animal as a reflexed rim or doublure of variable width. In many trilobites this is a broad band extending a considerable distance adaxially. A similar reflexion occurs beneath the pygidium. It is particularly pronounced in the Illaenidae, where the doublure covers more than 70% of the ventral side of the pygidial region (Harrington 1959; Bergström 1973). The anterior doublure is usually marked by terrace lines running subparallel to the margin of the cephalon. A soft, ventral membrane was attached to the inner edge of the reflexed margin in life and ran to the bases of the appendages. A number of arachnomorph taxa show a similar development of the marginal cephalic shield. *Sidneyia*, for example, had a massive sclerotized plate beneath the carapace, extending further back than the head shield in dorsal and lateral aspect. A small number of crustacean genera have reflexed carapace margins, notably the cephalocarids and remipedes, taxa regarded by Hessler (1969) and Schram (1986), respectively, as particularly primitive within the group. The ventral carapace of stomatopods and some decapods projects medially to meet an epistomal plate lying between the bases of the antennae and the mandibles.

The head capsules of atelocerates lack any dorsoventral division, and the simple continuation of the dorsal cuticle beneath the head is not considered homologous with the presence of a doublure in other groups. A form of reflexion does occur along the trunk of some Thysanura, where the dorsal tergites double back along a considerable portion of their length before joining with the bases of the thoracic appendages (Manton 1977). However, the character under discussion applies only to the doublure of the cephalic shield.

11. Marginal rim: *(0) absent; (1) present.*

A marginal rim takes the form of a shallow furrow or groove running around the periphery of the cephalic shield and is visible in lateral aspect. It is distinct from the doublure and occurs in taxa with and without reflexion of the carapace margin. The rim is particularly well developed in the chelicerate *Tachypleus*, but it also occurs in the trilobites and in a number of arachnomorphs (Briggs and Fortey 1989).

12. Pygidium: *(0) absent; (1) present.*

The terminal tagma in trilobites consists of a number of fused or partially budded body segments, plus a post-segmental telson (Harrington 1959; Whittington 1992). Though somewhat variable in form, its morphology is distinct from anything found in other arthropod taxa.

13. Genal spines: *(0) absent; (1) present.*

Genal spines take the form of prominent lateral extensions of the posteriormost corners of the cephalic shield. They occur in a number of trilobites (Harrington 1959; Whittington 1992), chelicerates (Størmer 1955), and other arachnomorphs (notably the aglaspidids).

14. Eye ridges: *(0) absent; (1) present.*

Eye ridges are distinctive features of some trilobites and take the form of raised, generally narrow bands running across the fixigenae from the vicinity of the antero-lateral corners of the glabella to the anterior extremities of the palpebral lobes (Harrington 1959). The presence of eye ridges does not necessarily imply the presence of eyes, although in this analysis all taxa with ridges also have compound eyes on the carapace.

15. Cardiac lobe: *(0) absent; (1) present.*

The cardiac lobe (a prominent site of muscle attachment) is defined either by a pair of sharply emplaced cardiac furrows on the dorsal prosoma (as in most xiphosurans, chasmataspids, aglaspids, eurypterids, and primitive arachnids) or by a pair of colored bands (as in *Sanctacaris*) (Eldredge 1974).

16. Articulating half-rings: *(0) absent; (1) present.*

The mesotergites of some trilobites are composed of two elements: an axial ring (forming the bulk of the mesotergite) and an articulating half-ring. The axial ring is a comparatively short, wide band that is visible whether the animal is outstretched or enrolled. The articulating half-ring is a crescentic extension from the anterior portion of the mesotergite, separated from the axial ring by a transverse furrow (Harrington 1959). This tonguelike flange inserted beneath the posterior margin of the preceding mesotergite. The posterior border of each ring is also connected to the anterior margin of the one succeeding it by a narrow band of flexible cuticle (e.g., *Olenoides*, *Triarthrus*). For *Agnostus* this character has been scored as absent (0) because it lacks the articulating half-ring on the anterior thoracic segment. The prozonites of diplopods are not considered homologous with articulating half-rings, since they are not separated from the main body of the tergite (metazonite) by a furrow (Hoffman 1969).

17. Outline of posterior trunk tergites in dorsal aspect: *(0) straight or slightly curved; (1) strongly curved or semicircular.*

In most taxa where trunk tergites are distinct, their posterior margins are either approximately straight or very slightly concave in dorsal aspect. In a number of arachnomorph taxa, including some trilobites (but not the arachnids themselves), the tergites on the posterior portion (typically half) of the trunk show a marked curvature. This increases toward the posterior of the trunk, such that the terminal tergites are U-shaped or semicircular. This character is exemplified by *Leanchoilia* (Bruton and Whittington 1983) in which the posterior margins of the last tergite recurve so strongly that lateral movement of the articulated terminal spine is restricted. Less pronounced, but remarkably similar, conditions occur in *Emeraldella*, *Sarotrocercus*, *Molaria*, and *Aglaspis*. Similarly the curvature of the broad tergites in *Sidneyia* increases markedly toward the posterior.

18. Prominent, thick, and armlike lateral extensions from the carapace, approaching the length of the body itself: *(0) absent; (1) present.*

The detailed morphology of these extensions differs between the Cambrian genus *Marrella* and the Lower Devonian *Mimetaster* (Birenheide 1971; Stürmer and Bergström 1976). However, they are massive and very distinctive structures that occur in no other arthropod taxon. They cannot be considered homologous with the genal spines of arachnomorphs.

19. Transverse wrinkling of the dorsal cuticle: *(0) absent; (1) present.*

Transverse wrinkling of the cuticle may be an important character uniting the lobopods (Budd 1993). It probably reflects the presence of transverse vascular channels, with circular musculature deflected around them. These wrinkles are distinct from transverse annulations (Hou and Bergström 1995).

20. Well-developed epistome: *(0) absent; (1) present.*

The epistome is present as a large, sclerotized plate beneath the anterior of the carapace of some eumalacostracans and stomatopods (*Galathea* and *Alima* in this study). In *Galathea* the epistome lies between the antennae and the mandibles, extending anteriorly between the antennal bases and abutting the anterior margin of the labrum posteriorly. In *Alima* it occupies a similar position but is less extensive anteriorly and is continuous posteriorly with the labrum. The plate gives the anterior cephalon a very rigid, boxlike structure. The rostral plate in trilobites is bounded by sutures and comprises part of the doublure; it is not regarded as homologous with the epistome.

*Mouth Region*

21. Orientation of anterior esophagus: *(0) anterior to posterior or ventral to dorsal (mouth terminal or ventral); (1) posterior to anterior (mouth facing backward).*

In a substantial minority of the arthropod taxa coded, the alimentary canal either extends directly backward from the mouth or extends dorsally before recurving. This condition pertains in all possible outgroups (e.g., annelids and mollusks exhibit both anteriorly and ventrally facing mouths) and is probably the primitive state for the arthropods. The mouth of the Cambrian lobopod *Aysheaia* is situated terminally (the gut running straight back along the body) while that of modern lobopods is ventral. In atelocerates ventral orientation of the mouth, which is associated with herbivorous and bloodsucking modes of feeding (hypognathous condition in insects, e.g., *Locusta*), is considered more primitive, while an anteriorly facing mouth is probably a secondary specialization associated with raptorial feeding (prognathous condition in insects, e.g., Staphylinidae [Schaefer and Leschen 1993]). These taxa contrast with the condition in the chelicerates, trilobites, and almost all crustaceans (Cisne 1974) where the mouth faces posteriorly, the anterior region of the esophagus leading anteriorly into a U-shaped loop and subsequently reflexing to pass back along the trunk above the mouth.

22. Labrum: *(0) absent; (1) present.*

In the earliest articulates the gut was probably a simple, straight tube opening forward at the extreme anterior of the animal. In *Aysheaia* the anteriorly situated mouth is surrounded by a ring of thin oral papillae, while the first true appendages (which are considered to be incorporated into the cephalon) probably served only a limited, if any, trophic function (Whittington 1978). The process of directing food into the mouth was probably relatively simple since there were no highly specialized mouthparts as in most atelocerates and crustaceans, nor a series of strongly gnathobasic limbs as in chelicerates and trilobites. Such modifications are associated with a ventral migration of the mouth, which is often accompanied by a curvature of the anterior part of the gut so that the mouth faces backward into a ventral food groove. The oral opening is usually covered in such taxa with a ventral "labral plate" (labrum or hypostome in trilobites). Fortey (1990) discussed the probable symplesiomorphy of this structure in all euarthropod taxa.

In a small number of taxa (e.g., *Androctonus*, *Burgessia*) the labrum has been lost. No distinction is made here between those taxa believed to lack the labrum primitively and those thought to have lost it secondarily.

23. Condition of labrum: *(0) attached; (1) natant.*

Fortey (1990) recognized three basic conditions of association between the hypostome and cephalic shield in trilobites. The primary conterminant hypostome condition corresponds to the attached state recognized here. The natant condition is that referred to as "detached" by Whittington (1988). A third state in trilobites (not represented in any of the taxa considered here) is the secondary conterminance or "reattachment" of the labrum. Attachment of the labral plate to the dorsal cuticle is considered to be the primitive condition. Taxa in which a labrum is not present or in which it has not been recognized are scored as missing data with respect to this character.

*Eyes/Sense Organs*

24. Number of median eyes: *(0) none; (1) two; (2) three; (3) four.*

All extant arthropods, with the exception of the Myriapoda, have median eyes or frontal ocelli at some stage in their development. In crustaceans these "nauplius eyes" persist into the adult but serve as photoreceptors only in the larvae. In insects these organs are referred to as frontal eyes or ocelli. The homology of median eyes within the chelicerates, crustaceans, and hexapods is shown by the innervation of all these organs from the middle region of the protocerebrum (Paulus 1979).

*Tachypleus* has a single pair of externally visible, median eyes (Fahrenbach 1971; Jones et al. 1971). Just beneath the median eyes are two photosensitive organs (Millecchia et al. 1966) sending axons along the nerves of the median eyes to the middle part of the protocerebrum. Paulus (1979) believed these to be a second pair of rudimentary median eyes and therefore speculated that the merostomes may have had ancestors with four external median eyes. With the exception of the pseudoscorpions and the totally blind Ricinulei and Palpigradi, all groups of arachnids have a single pair of median eyes and show no evidence of the rudimentary "endoparietal eyes" of *Tachypleus*. Here, the total number of externally visible median eyes has been coded. This is approximately equivalent to counting only well-developed median eyes, meaning that the character can be reliably coded for fossil as well as living taxa.

25. Lateral eyes: *(0) absent; (1) present.*

There are two distinct structural and functional types of lateral eyes, both innervated from lateral parts of the protocerebrum: (1) Simple eyes, composed of a single lens with a cup-shaped retina, and (2) faceted or compound eyes, composed of many ommatidia. The former type is restricted to the arachnids and forms an aberrant eye in some fleas, lice, and insect larvae. Faceted lateral eyes are the most common and occur in crustaceans, most adult insects, and most of the fossil arthropods considered. Both forms of eyes are considered homologous in that they represent photoreceptors innervated from the same region of the brain.

Lateral, simple eyes are identical in their structure to dorsal or median eyes but differ in their innervation. Light falls from a single, usually strongly biconvex lens onto different regions of the retina, depending on its angle of incidence.

All arachnids have lateral eyes of the simple type (up to five, relatively large, circular organs). The five lateral eyes of the Scorpiones (Scheuring 1913) and Uropygi can be interpreted as a dispersed *Tachypleus*-type. The only exclusively fossil group of chelicerates (the eurypterids) included in this analysis had a pair of faceted lateral eyes, apparently similar to those of modern Xiphosura (e.g., *Tachypleus*).

26. Compound eyes: *(0) absent; (1) present.*

In compound eyes, an array of lenses (each composed of a cornea and crystalline cone) may form separate portions of the entire image on distinct retinulas or rhabdoms, or it may partially combine these images optically (Kunze 1979). The distinction between apposition and superposition eyes is not considered useful because transitions between these states occur at too fine a taxonomic level. As noted above, not all lateral eyes are compound (e.g., Onychophora, Chelicerata), and similarly not all compound eyes are lateral (*Opabinia*). The number of compound eyes is not a useful character. The usual number is two, and only *Opabinia* possesses more.

Ironically, the complex structure of the compound eye has been used to argue both for and against a monophyletic Euarthropoda. Tiegs and Manton (1958) stressed differences in ultrastructure between compound eyes in the Chelicerata, Crustacea, Myriapoda, and Insecta, and contended that the different eye types must have evolved independently (although, of course, this factor need not necessarily preclude monophyly for the euarthropods as a whole). Paulus (1979), by contrast, maintained that a primitive faceted eye is probably symplesiomorphic for the euarthropods and that detailed differences therefore represent secondary specializations.

According to Paulus (1979), the ancestors of arthropods probably had eyes similar to those of the Polychaeta (isolated lens eyes with no fixed number of elements). He argued that the fossil evidence of faceted eye distribution in the major groups (Størmer 1944) suggests that these organs are homologous within the arthropods. By contrast, Horridge (in Manton 1977:274–278) favored the view that the compound eye evolved many times within the arthropods, as did the eyes of annelids, mollusks, and coelenterates. The coding here assumes that lateral eyes are homologous (following Paulus 1979).

27. Position of lateral eyes: *(0) on dorsal cuticle of cephalon; (1) free or pedunculate beneath the carapace.*

Eyes fixed to the dorsal cuticle of the cephalon are typical of arachnomorphs (e.g., trilobites, scorpions) and atelocerates. Pedunculate eyes are typical of crustaceans; the notostracan branchiopods are exceptional in having their compound eyes located on the carapace. The eyes in a number of problematic fossil arthropods are also stalked. Stalks may emerge from beneath the anterior carapace margin (e.g., *Sarotrocercus*); such eyes are usually accommodated in a carapace groove (e.g., *Sidneyia*, *Galathea*). Eyes may also be located beneath the carapace but without the support of a stalk (e.g., *Bredocaris* and *Argulus*). This condition is coded in the same manner as stalked forms. The polarity of the character is uncertain, and taxa lacking lateral eyes are scored as missing data.

28. Organs of Tömösváry: *(0) absent; (1) present.*

The organs of Tömösváry are paired structures located at the bases of the antennae in myriapods. Each organ is composed of a disc with a central pore into which the endings of subcuticular sensory cells converge. They are thought to detect vibrations and may be auditory in function (Camatini 1980).

*Cephalic (or Anterior) Appendages*

29. Number of post-acronal somites incorporated into the cephalon: *(0) none; (1) one; (2) two; (3) three; (4) four; (5) five; (6) six.*

The presence of a well-differentiated head is a fundamental feature of the Euarthropoda. The ancestral stock from which the euarthropods arose probably lacked a clearly differentiated cephalon (e.g., Whittington 1981). Locomotory appendages, and therefore somites, were progressively recruited from the trunk and incorporated into the cephalon to serve sensory and trophic functions. Particular numbers of cephalic appendages may not have arisen independently, in that greater numbers are easily derived through a series of intermediates. This certainly appears to have occurred in the origin of the crustaceans from primitive precursors (Briggs 1983). Weygoldt and Paulus (1979) referred to the possibility that the chelicerate prosoma evolved by the fusion of a cephalon composed of four somites (as in the trilobites) with two additional trunk somites. Work on homeobox genes (Averof and Akam 1995) suggested that it may be possible for groups of limbs to be lost or gained together under some circumstances, and this is reflected in the unordered treatment of this character here.

30. Nature of first appendage: *(0) uniramous; (1) biramous.*

A uniramous first appendage is characteristic of lobopods, most hexapods and myriapods, the trilobites, many other arachnomorph taxa, and also many crustaceans. A biramous first appendage is present in a small number of crustaceans as well as in the Cambrian *Sanctacaris*. There is some uncertainty as to whether the biramous limbs of schizoramian post-antennulary appendages are strictly comparable to the biramous antennules of crustaceans. This probably does not represent a problem in the present context as the first appendage has been coded for uniramy or biramy in isolation.

31. Form of the first appendage: *(a) antennule or antenna; (b) vestigial antennule or antenna (unsegmented or single-segmented appendages, found solely in some crustaceans and the Protura); (c) chelicera; (d) relatively small, possibly chelate and grasping, but not homologous with the chelicera of arachnids* (Aglaspis: *Briggs et al. 1979); (e) biramous, raptorial, and forming an integral part of an anteriorly directed grasping basket* (Sanctacaris); *(f) as in* Sarotrocercus; *(g) "great appendage" composed of three massive proximal podomeres, with distal spines. In* Leanchoilia *and* Alalcomenaeus *the n-1 and n-2 podomeres are barlike with a flagellar extension (Whittington 1981; Bruton and Whittington 1983); (h) massive and sclerotized, bearing large spines on the inner margin, and capable of flexing beneath the animal*

*to the ventral mouth* (Anomalocaris: *Whittington and Briggs 1985); (i) fused and proboscis-like, with a terminal claw bearing serrations on the inner margins* (Opabinia: *Whittington 1975); (j) fleshy, unsclerotized, and unjointed "grasping" appendage, bearing spines on the inner margin* (Kerygmachela: *Budd 1993); (k) simple, unsclerotized lobopod with terminal hooks* (Echiniscus).

Chelicerae are relatively small chelate or subchelate grasping appendages, capable of swinging down toward the mouth and serving a trophic function. They were most elaborately developed in the eurypterids (where they were composed of three segments, the last two forming the chela) and greatly reduced in some of the arachnids (where the appendage is subchelate and has just two segments). Despite a considerable range of form, their homology is widely accepted (Shultz 1990).

32. Inner ramus of second appendage: *(0) absent; (1) present.*

Coding the second cephalic appendage as uniramous or biramous is an oversimplification. Some taxa have a uniramous second appendage because they lack the outer ramus, while others have no inner ramus.

33. Outer ramus of second appendage: *(0) absent; (1) present.*

34. Form of outer ramus of appendage of second post-acronal somite: *(0) filiform or filamentous; (1) developed as an antennal scale or scaphocerite.*

35. Form of the appendage of the second post-acronal somite: *(0) unspecialized and undifferentiated from those that follow; (1) antenniform; (2) as in* Agnostus *(autapomorphic state); (3) as in* Branchiocaris *(autapomorphic state); (4) pedipalps; (5) as in* Marrella *(autapomorphic state); (6) bladelike jaw as in* Peripatoides *(autapomorphic state); (7) as in* Mimetaster *(autapomorphic state).*

An antenna is considered to be an appendage composed of at least one segment or at most of a series of largely undifferentiated segments (in all practical instances differentiated from the appendages succeeding it). Importantly, antennae show no modifications for locomotion over a solid substrate.

36. Gnathobases on appendages of the third post-acronal somite: *(0) absent; (1) present.*

In many arachnomorphs (e.g., trilobites, chelicerates), most of the cephalic appendages (usually with the exception of the first antenna or chelicera) are armed with broad biting surfaces, little different from those of the trunk (where present). These appendages usually also retain a locomotory function. In crustaceans the mandible is usually more strongly gnathobasic than the other appendages, bearing a variously developed protopodal enditic process. The limb itself is most commonly reduced to a non-locomotory palp. Gnathobases may not meet medially.

This character distinguishes between taxa with and without cephalic gnathobases, but it is preferable to code it this way, without reference to the cephalon itself, which is not a rigidly defined structure. In six of the taxa considered here (*Sidneyia*, *Aysheaia*, *Opabinia*, *Branchiocaris*, *Marrella*, and *Sarotrocercus*) the cephalon comprises only one or two somites. In all but *Sidneyia* it would make no difference how the cephalon was defined since the succeeding trunk limbs, like those of the cephalon, lack gnathobases. However, all but the first limb of *Sidneyia* (which is an antenna) are massively gnathobasic. This coding accommodates taxa with many similar cephalic gnathobases (e.g., *Tachypleus*) and those with just a single, well-differentiated, gnathobasic mandible (e.g., *Odaraia*). If the presence of cephalic gnathobases is to be considered a homology, then the differentiation of these by form or degree must be a more derived condition and is codable as a separate character (below).

37. Form of third appendage: *(0) unspecialized; (1) gnathobasic mandible; (2) mandible biting with tip; (3) slime papillae (autapomorphic state).*

A gnathobasic mandible is defined here as an appendage specialized to perform a trophic function and differentiated from those preceding and succeeding it by the degree to which the enditic processes are developed. Mandibles perform no locomotory function (Manton 1977). The distinction between the gnathobasic mandibles of crustaceans and the whole-limb jaws of the atelocerates (biting with their tips) is well recorded (Manton 1977). However, Snodgrass (1938, 1950, 1958) considered both mandible types to be homologous in that they both represent appendages of the second post-oral somite. Brusca and Brusca (1990) considered that both forms could easily have been derived from a simple jointed appendage of the third post-acronal somite, and it is these specializations that are scored as homologies in this case.

38. Gnathal lobe of third appendage: *(0) absent; (1) present.*

The presence of a gnathal lobe is limited to myriapod taxa and is a development of the whole-limb jaw.

39. Degree of fusion of the appendage of the fourth post-acronal somite: *(0) not fused; (1) medially coalesced; (2) fused to form a gnathochilarium.*

The appendages of the fourth post-acronal somite are fused to various degrees in a number of atelocerates. The first maxillae are medially coalesced in the Chilopoda and Symphyla, whereas the same appendages are completely fused to form a flaplike gnathochilarium in the Diplopoda and Pauropoda (Blower 1974; Manton 1977).

40. Degree of fusion of the appendage of the fifth post-acronal somite: *(0) not fused; (1) medially coalesced; (2) fused to form a labium.*

In the Chilopoda the second pair of maxillae (like the first) is medially coalesced. In the Symphyla and Insecta the second pair of maxillae is fused to form a platelike labium or lower lip. The second maxilla is absent from the Diplopoda and Pauropoda (Blower 1974; Manton 1977).

*Trunk*

41. Total number of somites in cephalon and trunk, excluding the telson (not a true somite) or terminal division: *(a) 5; (b) 9; (c) 10; (d) 12; (e) 13; (f) 14; (g) 15; (h) 16; (i) 17; (j) 18; (k) 19; (l) 20; (m) 22; (n) 23; (o) 24; (p) 25; (q) > 25.*

42. Differentiation of trunk somites: *(0) no gross differentiation of somites; (1) trunk somites grouped into tagmata.*

Cephalization occurs from the anterior of the animal backward, while abdominalization is effected by the differentiation of posterior somites, typically by modification of their appendages. The presence of an abdomen in most crustaceans or uniramians would appear to be a reliable character, but it is not clear whether such abdomens are homologous, especially with the pygidium in trilobites or the opisthosoma in chelicerates. The problems are even greater where some fossil taxa are concerned. Nonetheless, although it is potentially misleading at a high level, this character is retained because it provides resolution within smaller groups.

43. Ventrolateral arthropod or lobopod appendages: *(0) absent; (1) present.*

The ventrolateral appendages in all arthropod and lobopod taxa are considered to be homologous (Manton 1977; Wheeler et al. 1993) but not with the parapodia of polychaetes, although they are primitively biramous. Although parapodia are promoted and remoted by extrinsic musculature, Mettam (1967) suggested that primitively the parapodium was operated by intrinsic musculature alone and that the extrinsic muscles are actually derivatives of part of the oblique musculature that braces the body wall.

44. Attitude of the trunk appendages relative to the body: *(0) pendent or located beneath the trunk; (1) laterally deflected.*

Lateral deflection of the limbs (Bergström 1992) is exemplified by many trilobites (e.g., *Olenoides*). Similar lateral deflection is seen in marrellomorphs, atelocerates, chelicerates, and a number of other arachnomorphs (e.g., Emeraldellida and Limulavida). For terrestrial arthropods the attitude of the limbs relative to the body may well be influenced by mechanical constraints associated with size and particular gait patterns. Lateral displacement is exhibited by almost all taxa walking over hard substrates without the support of water. Pendent limbs are more typical of crustaceans, and they also occur in *Leanchoilia* and *Yohoia* (note that Bergström [1992] regarded these fossils as allied to the crustaceans rather than to the arachnomorphs).

45. Gross morphology of trunk appendages: *(0) lobopodous or very lightly sclerotized and lamellate; (1) strongly sclerotized and rigid or composed of a number of articulating podomeres.*

Unsclerotized, lobopodous, ventrolateral appendages are present in the modern Onychophora and the Cambrian *Aysheaia*. Flaplike, unsclerotized, ventrolateral appendages are present in the Cambrian *Anomalocaris*, *Opabinia*, and *Kerygmachela*. Euarthropod taxa possess sclerotized and typically jointed, ventrolateral trunk appendages.

46. Correlation between trunk segments and appendages: *(0) segmentation distinct, with one pair of appendages per division, or segmentation indistinct (simple absence of visible intersomite boundaries) with one limb pair per embryonic somite; (1) segmentation distinct, with more pairs of appendages than divisions.*

The onychophoran lobopods, the Cambrian lobopod *Aysheaia*, and a number of other non-sclerotized forms lack any clear trunk segmentation. In *Aysheaia* the trunk is distinctly annulated, but these rings are not equivalent to somites. Whittington (1978) assumed that there was one pair of appendages per somite, except in the anterior trunk/cephalon. Most euarthropods have a clearly segmented trunk, with one limb pair per somite. In *Lepas* and *Cypridina*, all traces of trunk segmentation (except the presence of trunk cirri and two thoracopods respectively) are absent.

In *Vachonisia*, *Lepidurus*, *Branchiocaris*, and *Odaraia*, the trunk is distinctly segmented, but each segment can give rise to more than one pair of appendages. In the myriapods *Julus* and *Pauropus* all trunk divisions (diplosegments) give rise to two pairs of appendages. Polarity is uncertain. However, those taxa with more pairs of appendages than somites are usually primitive in a number of respects (large numbers of trunk somites with many homonomous appendages).

47. Relative length of limbs along the length of the trunk: *(0) more or less constant or increasing; (1) trunk limbs regularly diminishing in length.*

A regular decrease in the length of the trunk appendages posteriorly is often a feature of taxa with large numbers of homonomous and unspecialized limbs, exemplified in *Lepidurus*, *Marrella*, and *Vachonisia* (Stürmer and Bergström 1976).

48. Inner ramus of trunk (eighth) appendages: *(0) absent; (1) present.*

Many chelicerates (e.g., *Androctonus*) lack an inner ramus to their opisthosomal appendages. In addition, the broad, flattened trunk limbs of *Perspicaris*, *Sarotrocercus*, and *Yohoia* are considered to represent the outer rami only.

49. Inner ramus of trunk (eighth) appendages with a terminal group of reflexed, clawlike spines: *(0) not clawed; (1) clawed.*

Terminal, reflexed, clawlike spines are observed exclusively in weakly sclerotized taxa. In *Aysheaia* they take the form of a radially arranged group of up to seven sclerotized hooks, curving away from the blunt tips of the limbs (very similar to the claws in some tardigrades, e.g., *Echiniscus*). The claws of *Peripatoides* and other modern Onychophora are similarly positioned, although fewer in number (Ramsköld and Hou 1991).

50. Inner ramus of trunk (eighth) appendages: *(0) not spinose; (1) spinose.*

Spines are distinguished from bristles in that they do not articulate with the podomeres from which they arise. Although predominantly a feature of arachnomorph taxa, spines also occur in a limited number of crustaceans and the lobopod *Aysheaia*.

51. Number of podomeres in inner ramus of trunk (eighth) appendages: *(a) one; (b) three; (c) four; (d) five; (e) six; (f) seven; (g) eight; (h) 11; (i) 14; (j) > 14.*

52. Form of "knee" in inner ramus of trunk (eighth) appendages: *(0) segment; (1) joint.*

In the hexapods the major trunk appendage geniculation ("knee") takes the form of a joint (see Wheeler et al. 1993). In other taxa where a clear geniculation is present, it involves the insertion of a short podomere (Manton 1977).

53. Outer ramus of trunk (eighth) appendages: *(0) absent; (1) present.*

Absence of the outer trunk rami is characteristic of the atelocerates but also of a small number of crustaceans (e.g., some eumalacostracans) and arachnomorphs (*Androctonus*, *Aglaspis*). Taxa lacking an outer ramus were scored with missing data for characters 54 to 57.

54. Filaments on the outer ramus of the trunk (eighth) appendages: *(0) absent; (1) present.*

In many schizoramian taxa with numerous, largely homonomous trunk limbs (e.g., *Olenoides*, *Speleonectes*, *Marrella*) the outer rami bear numerous, regularly spaced filaments. Most crustaceans and a limited number of arachnomorphs lack these filaments.

55. Segmentation of the outer ramus of the trunk (eighth) appendages: *(0) unsegmented; (1) segmented.*

Segmentation of the outer ramus is usually a feature of crustaceans, although it also occurs in *Marrella*, *Branchiocaris*, and *Agnostus*. An entire and unsegmented outer ramus is characteristic of most trilobites and arachnomorphs.

56. Organization of the outer rami of the trunk (eighth) appendages: *(0) around a simple, longitudinal rhachis; (1) around a flattened lamellate or lobate region.*

Broad, lamellate outer rami are a characteristic of the chelicerate *Tachypleus*, *Sanctacaris* (interpreted as a chelicerate by Briggs and Collins [1988]) and the Burgess Shale problematica *Yohoia*, *Sarotrocercus*, and *Burgessia*.

57. Distal lobe on outer ramus of trunk (eighth) appendages: *(0) absent; (1) present.*

The distal lobe is characteristic of the arachnomorphs *Emeraldella* and the trilobites *Olenoides* and *Naraoia*.

58. Epipodite arising from trunk (eighth) appendages: *(0) absent; (1) distinctive.*

Epipodal lobes occur in a number of crustacean taxa, notably *Artemia*, *Lepidurus*, and *Nebalia* (Schram 1986).

59. Gnathobases or protopodal endites on trunk (eighth) appendages: *(0) absent; (1) present.*

60. "Orsten"-type appendages: *(0) absent; (1) present.*

"Orsten"-type appendages are defined as those in which at least the exopod is multisegmented and each podomere has a rigid, prominent, long seta or bar (e.g., Müller 1983; Müller and Walossek 1985).

61. Suppression of the sixth pair of appendages (anteriormost trunk appendages): *(0) not suppressed; (1) suppressed.*

Suppression of the sixth appendages is restricted to the Diplopoda and Pauropoda (Blower 1974; Manton 1977).

62. Sixth appendage (anteriormost trunk) modified as poison fangs: *(0) not modified; (1) modified.*

In the chilopods and symphylans the first trunk appendages are modified as raptorial poison fangs. These are much more prominent in the Chilopoda (Lewis 1981).

63. Collum: *(0) absent; (1) present.*

In the diplopods the first trunk somite is expanded into a conspicuous, heavily sclerotized collar (collum) between the head and the remainder of the trunk. This is much longer dorsally than ventrally (wedge-shaped in lateral aspect). A homologous structure is developed in the pauropods, although the expansion is much greater on the ventral surface (Blower 1974).

64. Mobile appendages on terminal division: *(0) absent; (1) present.*

Crustaceans and some atelocerates possess mobile appendages (caudal rami or cerci) on the terminal division of the trunk. The term "terminal division" rather than "telson" is used to refer to the last discrete segment or division of the body, regardless of whether it contains the anus. The telson is not a true somite, but rather a post-segmental remnant. Hence, the telson of crustaceans is regarded as being homologous with the telson or terminal spine of *Tachypleus*. See also discussion of character 66.

65. Condition of the appendages of the posteriormost trunk somite: *(0) absent or, if present, independent of the telson; (1) present as uropods incorporated into a tail fan.*

The incorporation of uropods from the pre-telsonic somite into a tail fan comprising these paddle-shaped organs and a lamellate telson is a highly distinctive modification. This occurs with astonishing similarity in *Galathea* and *Sidneyia*.

66. Location of the anus: *(0) terminal; (1) ventral in the terminal division of the body; (2) ventral in the penultimate division of the body.*

In the majority of protostome groups the anus is terminal. Primitively the gut is a simple tube, with an anterior and posterior opening at either extremity of the body. In some Crustacea and a few problematic Cambrian genera the anus has assumed a ventral position but remains within the terminal trunk division. In the chelicerates and a substantial number of extinct forms the anus lies at the base of the penultimate trunk division. Confusion over the location of the anus arises from the status given to the telson. Some definitions consider the telson to be the body segment bearing the anus, but this results in circularity. Attempts have been made to distinguish true telsons from anal segments even *within* the Crustacea (Bowman 1971). The telson is regarded as the terminal unit of the body, its mesoderm derived directly from the teloblasts after the cessation of metamere budding (Schram 1986). This definition implies that the telson is not a true segment (e.g., Kaestner 1970; McLaughlin 1980; but cf. Moore and McCormack 1969). The terminal trunk division of all euarthropod groups is regarded as homologous in that it is post-segmental (Brusca and Brusca 1990) or at least contains post-segmental remnants (Schram 1986).

67. Gross form of the terminal body division: *(0) rounded; (1) styliform; (2) paddle-shaped or platelike.*

A styliform terminal extension is a familiar feature of the living *Tachypleus* and scorpions. In the true scorpions (represented here by *Androctonus*) this spine is distally modified as a sting or aculeus. Terminal spines are also well-developed in some of the eurypterids (*Baltoeurypterus*) as well as in a number of problematic genera. A paddle-shaped terminal extension is a distinctive feature of the arachnomorphs *Sanctacaris*, *Sidneyia*, *Alalcomenaeus*, and *Yohoia*.

68. Terminal body division with three, dorsolaterally directed lobes, forming a tail fan: *(0) absent; (1) present.*

A similar, three-lobed tail fan occurs in the Cambrian genera *Anomalocaris* (Chen et al. 1994) and *Opabinia* (Whittington 1975).

69. Border of the posterior body division: *(0) smooth; (1) flanked with fine spines.*

The border of the posterior body division is spinose in the Cambrian arthropods *Sanctacaris*, *Alalcomenaeus*, and *Leanchoilia*.

70. Postventral plate: *(0) absent; (1) present.*

A postventral plate is present in the Cambrian taxa *Emeraldella* and *Aglaspis*. The plate in *Aglaspis* was first illustrated by Raasch (1939) but described in more detail by Briggs et al. (1979). The plate in *Emeraldella* was described by Bruton and Whittington (1983).

## Soft-Part Characters

71. Cuticle of alpha chitin and protein: *(0) no; (1) yes.*

The cuticle in onychophorans, tardigrades, and euarthropods is composed of alpha chitin rods in a protein matrix. The cuticle of fossil arthropods was probably of similar composition, but this cannot be known with certainty. In annelids, by contrast, the cuticle is composed of collagen (Cutler 1980).

72. Resilin: *(0) absent; (1) present.*

The presence of the protein resilin has been demonstrated in all the major groups of living arthropods (Anderson and Weis-Fogh 1964) as well as in the Onychophora. Resilin is a highly efficient natural rubber, particularly associated with storing energy at joints.

73. Tracheae: *(0) absent; (1) present.*

The arthropods have developed two mechanisms for breathing air: the tracheae of myriapods, most hexapods, and onychophorans, and the book lungs of arachnids — an autapomorphy not coded in the present study. Tracheae are branched, usually anastomosing invaginations of the external cuticle, opening both internally and externally. Internally, the branching tracheoles extend into reservoirs of tracheolar fluid (continuous with the hemolymph), and gas exchange is by simple diffusion.

74. Intersegmental tendon system: *(0) absent; (1) present.*

Boudreaux (1979) demonstrated that all arthropods possess a similar intersegmental tendon system, which is absent in annelids, onychophorans, and tardigrades.

75. Carapace adductor muscles: *(0) absent; (1) present.*

The presence of adductor muscles is associated with the development of a bivalved carapace, but the character is not entirely limited to bivalved taxa. The living notostracan branchiopods (e.g., *Lepidurus*) have transverse muscles within their relatively flimsy, univalved carapaces, as do the brachypodans (e.g., *Sandersiella*). Adductors are also present in the univalved carapaces of the "Orsten" forms *Martinssonia* and *Bredocaris*. The presence of adductor muscles in fossil taxa can be inferred from scars on the carapace.

76. Suppression of all circular body wall musculature: *(0) no; (1) yes.*

The presence of circular somitic musculature is often associated with a flexible, elastic cuticle and a hydrostatic skeleton. Contraction of circular muscles in annelids, for example, extends the body along its longitudinal axis. In oligochaetes especially, contraction of circular muscles may occur in peristaltic waves down the body as a component of burrowing activity. A body restricted by a rigid exoskeleton composed of articulating plates of cuticle is incapable of such longitudinal extension, and the circular muscle has become redundant.

77. Muscle: *(0) striated and smooth; (1) all striated.*

The muscle tissue of all modern arthropods is of the striated (fast) type; smooth (slow) muscle is lacking. Mollusks, annelids, and onychophorans possess both muscle types. Tardigrades were long thought to have only smooth muscle, and the absence of striated fibers was used as an argument against their close affinity with the arthropods. Kristensen (1978), however, demonstrated the presence of both smooth and striated fibers in tardigrades, the latter being more prominent in the most primitive species. Furthermore, the striated muscle in tardigrades is cross-striated, similar to that in arthropods, whereas that of the onychophorans is obliquely striated. Nothing can be deduced regarding the musculature of the fossils under consideration.

78. Pretarsal segment of inner ramus of trunk appendages with only a single muscle: *(0) no; (1) yes.*

The presence of a single muscle at this location is a character state restricted to the atelocerates.

79. Structure of brain: *(0) less than bipartite; (1) bipartite or tripartite.*

In annelids, onychophorans, tardigrades, and arthropods, the cerebral ganglion or brain is specialized into two or three regions — forebrain, midbrain (where present), and hindbrain (otherwise the protocerebrum, deutocerebrum, and tritocerebrum). The protocerebrum derives from the acron and innervates the eyes. Antennulary nerves run from the deutocerebrum to the first antenna (antennule in crustaceans). The tritocerebrum forms a pair of circumcentric connectives that extend around the esophagus to a subesophageal or subenteric ganglion and link the brain with the ventral nerve cord. The three brain regions are differentiated to varying degrees, being most distinct in the uniramians. A deutocerebrum is absent in the chelicerates. The cerebral ganglion of mollusks is not differentiated in this manner.

80. Acronal protocerebrum serving the eyes (where present) and containing an association center connected with pedunculate bodies: *(0) absent; (1) present.*

81. Malpighian tubules: *(0) absent; (1) endodermal; (2) ectodermal.*

Malpighian tubules are excretory structures that occur in many terrestrial arthropods. They take the form of between two and several hundred blind diverticulae arising from the gut wall and extending into the hemocoel. In atelocerates the tubules arise from the hindgut (ectodermal), typically near its junction with the midgut. In arachnids they arise from the posterior midgut (endodermal) and are not homologous with those in atelocerates. Malpighian tubules are absent from aquatic crustaceans and xiphosurans and also from the onychophorans and tardigrades.

82. Gonads: *(0) ventral; (1) dorsal.*

Elongate, dorsally situated gonads are a general character of arthropods, onychophorans, and tardigrades. In annelids the gonads are located more ventrally and are typically rather saclike.

83. Nephridia with sacculi: *(0) absent; (1) present.*

Nephridia have open nephrostomes in annelids and drain blood directly from an open hemocoel to the outside. With the evolution of a hemocoelic circulatory system in arthropods, this system became functionally untenable. Arthropods have evolved a variety of excretory structures but all are internally closed. The inner, blind end of the arthropod nephridium is a coelomic remnant called a sacculus.

84. Reduction in the number of nephridia (number of post-cephalic segments containing nephridia): *(0) more than four; (1) four or less.*

Most annelids possess serially arranged nephridia, one per segment, the nephridial pore lying in the segment posterior to the nephridium. Onychophora have a similar arrangement, with each pair of nephridia located in the ventrolateral hemal sinuses. In arthropods, nephridial elements are reduced in number and restricted to the cephalon and anterior somites of the trunk. In most adult crustaceans only a single pair of nephridia persists, and this is usually associated with either the second antenna or second maxilla (as antennary glands [= green glands] or maxillary glands). In chelicerates there may be as many as four pairs of nephridia (coxal glands), opening at the base of the walking legs.

85. Gut diverticulae: *(0) absent; (1) present.*

86. Cephalic (at least anterior) ecdysial glands: *(0) absent; (1) present.*

The hormone ecdysone initiates the molting cycle in arthropods. The pathway controlling its release may differ in insects and crustaceans. However, all euarthropods appear to share the presence of ecdysial glands in the cephalon. In crustaceans, these "Y-organs" are located either at the base of the antenna or near the mouthparts and are controlled by a complex neurosecretory apparatus within the eyestalks. Onychophora and Tardigrada appear to lack distinct ecdysial glands even though they molt.

87. Repugnatorial glands: *(0) absent; (1) present.*

Myriapods possess lateral repugnatorial glands on the trunk segments that are particularly well developed in the geophilomorph centipedes (Camatini 1980).

88. Reduction of the coelom (hemocoel and circulatory system with dorsal blood vessel, paired ostia, and pericardial sinus): *(0) no; (1) yes.*

The coelom functions as a hydrostatic skeleton in annelids, acted upon by the longitudinal and particularly the circular muscles of the body wall. A rigid arthropod exoskeleton entails modifications associated with the formation of an open circulatory system, filling the body cavity with blood and bathing the internal organs directly. The body wall musculature assists the dorsal blood vessel in moving body fluids in annelids. The absence of a flexible body wall in arthropods, coupled with their relatively large size, required the dorsal vessel to become a highly muscularized, specialized pumping structure. Blood returns to the heart down a pressure gradient via a non-coelomic pericardial sinus and perforations in the heart wall (ostia). A reduced coelom and a heart are also present in the onychophorans and tardigrades.

89. Dorsal blood vessel: *(0) absent; (1) present.*

The principal blood vessel in annelids, onychophorans, tardigrades, and arthropods is located dorsally, and blood is pumped toward the anterior. This contrasts with the situation in mollusks, where the heart is more centralized, arteries leave it anteriorly and posteriorly, and the vessels themselves do not provide the primary pumping action.

90. Two primary pigment cells in ommatidia: *(0) absent; (1) present.*

The cornea of each ommatidium is formed (like the rest of the cuticle) by epidermal cells. Each lens is the product of two epidermal corneagen cells. In the hexapods, after forming the cornea, these cells withdraw to the sides of the ommatidium where they form two primary pigment cells.

91. Specialized ommatidial structures: *(0) variable and higher number of components than state 1; (1) two corneagen cells, four Semper cells (usually producing a cone with four parts), and a retinula with eight cells (one or two of which may be modified as short basal cells).*

Each ommatidium has a specialized structure in atelocerates and crustaceans (Paulus 1979). Beneath the cornea are four Semper cells, which usually produce the crystalline cone. The cone itself is a hard, transparent, intracellular structure, bordered laterally by the primary pigment cells. Immediately behind the cone (where present) are elongate, sensory neurons or retinular cells.

92. Cilia: *(0) many tissues ciliate; (1) no cilia except in photoreceptors and/or sperm.*

*Developmental Characters*

93. Ordering of fate map tissues from anterior to posterior: *(0) anterior — stomodaeum — midgut — mesoderm — posterior; (1) anterior — midgut — mesoderm — stomodaeum — posterior.*

The myriapods, hexapods, onychophorans, and annelids share a similar ordering of three of their fate map tissues. The stomodaeum lies anterior to the midgut and mesoderm as opposed to posterior in crustaceans and chelicerates (Anderson 1973) and pycnogonids (Schram 1978).

94. Superficial blastoderm formation: *(0) no; (1) yes.*

Superficial blastoderm formation is a feature of all euarthropods and the Onychophora, but it is absent from the annelids and mollusks (Anderson 1973).

95. Embryos with four gangliar post-oral segments: *(0) no; (1) yes.*

The presence of four post-oral segments bearing ganglia is a feature of the embryos of the living chelicerates and apparently the trilobites (Manton 1977).

96. Number of anatomically pre-oral limb-bearing segments in the larva: *(0) two or more; (1) one or less.*

The hexapods, myriapods, crustaceans, and trilobites pass through a phase in their ontogeny with just one limb-bearing segment anterior to the mouth. In crustaceans the second antenna migrates pre-orally only *after* the nauplius or egg-nauplius phase. In the chelicerates the chelicera arises behind the mouth embryologically but usually migrates pre-orally in the adult. The chelifore of pycnogonids remains post-oral.

97. Nauplius or egg nauplius in development: *(0) absent; (1) present.*

All crustaceans either pass through a free nauplius larval stage (ostracodes, cirripedes, copepods, and some malacostracans) or show some evidence of a nauplius stage within the egg (Schram 1986). The nauplius is a phase in which the first three appendages (antennule, antenna, and mandible) are all used for locomotion.

## Methods

The present analyses represent an advance upon those of Briggs and Fortey (1989), Briggs et al. (1992a), and Wills et al. (1994) in the inclusion of additional, post-Cambrian arthropod genera (*Kalbarria*, *Mimetaster*, *Vachonisia*, *Baltoeurypterus*, *Weinbergina*, *Cheloniellon*, *Lepidocaris*, *Nahecaris*) and several pivotal taxa (*Anomalocaris*, *Opabinia*, *Kerygmachela*, Tardigrada). The list of cladistically informative characters is also considerably extended, with particular deference to recent morphological analyses of extant taxa (e.g., Wheeler et al. 1993).
Four cladistic analyses were run:

1. All taxa included. All characters unordered and equally weighted. This forms the basis for the bulk of the discussion.
2. All fossil taxa excluded. All characters unordered and equally weighted.
3. All living taxa excluded. All characters unordered and equally weighted.
4. All taxa included. All characters unordered and successively reweighted according to the maximum value of the rescaled consistency index.

Data were analyzed using PAUP 3.0 on a Macintosh Quadra 700. The number of taxa precluded the use of exact tree-building techniques (exhaustive and branch and bound), so heuristic searches were used throughout. No heuristic search method guarantees finding the most parsimonious tree or population of trees. Preliminary experimentation showed that none of the determinate addition sequences (e.g., as-is, simple, closest) found even a subset of the shortest trees available using stochastic techniques, even when a large number (up to 500) trees were saved at each step in the sequence. Searches were therefore made using several random additions of taxa, followed by TBR (tree bisection and reconnection) branch swapping (the most exhaustive algorithm available). Fifty random addition replicates were made, filtering and condensing the trees produced.

The data were also clustered phenetically. A Manhattan (simple character distance) matrix was produced for the ingroup. Taxa were then clustered with UPGMA (unpaired group mean average).

### Consensus Methods

Large numbers of most parsimonious trees (MPTs) were generated by some of the cladistic analyses, and the results are presented as consensus trees. Many authors have discussed the relative advantages and drawbacks of consensus methods

(Miyamoto 1985; Hillis 1987; Carpenter 1988; Swofford 1990). Three methods are used here: (1) 50% majority rule plus all compatible groupings; (2) Adams consensus; (3) reduced cladistic consensus.

The majority rule tree is fully resolved, but it merely presents the most common statements of cladistic relationship, many of which are not true of all the fundamentals. An Adams consensus consigns taxa of variable location to a polytomy at the lowest node they traverse, but such a representation is ambiguous with respect to the identity of the "mobile" taxa and also implies spurious relationships.

A collection of reduced cladistic consensus (RCC) trees (Wilkinson 1994) can be found with the useful property of graphically representing all positive cladistic information common to the set of fundamentals, but at the cost of pruning one or more taxa (the "basic" trees). A set of additional RCC trees (derivatives) can be constructed by combining basic RCC trees. These add no information that is not already present in the set of basic trees, but they may have greater resolution than any individual basic RCC tree. Moreover, the primary RCC trees (those with the largest number of nodes and secondarily the largest numbers of taxa) frequently are derivatives. First derivative RCC trees can be produced by the fusion of two basic RCC trees. Fusion of basic RCC trees need not necessarily produce a derivative RCC tree, since the resultant tree might simply be a subtree of one of the basic trees. Second derivative RCC trees are produced from the pairwise fusion of each first derivative RCC with those basic RCCs not used in their construction, discarding any redundant products. The process can be continued to *n*th order derivatives. These derivative trees incorporate *only* those taxa common to all the trees being fused, but they include *all* the relationships between them that are represented in any of the basic trees.

## Results and Discussion

### 1. All taxa, with all characters weighted equally

The analysis resulted in 5544 equally parsimonious fundamental trees (462 steps, CI = 0.33, RI = 0.67). The 50% majority rule (fig. 2.1), Adams (fig. 2.2), and primary derivative reduced cladistic consensus (Wilkinson 1994) (fig. 2.3) trees were all evaluated for this set.

Seven "basic" RCC trees were found from the 5544 fundamentals. The single most fully resolved RCC (the primary derivative RCC tree) was produced as a third order derivative from the forced fusion of four of these (fig. 2.3). This includes thirteen more nodes than the strict consensus (approximately a 40% increase) with the removal of just six taxa (*Branchiocaris*, *Derocheilocaris*, *Habelia*, *Kalbarria*, *Leanchoilia*, and *Sarotrocercus*).

#### *Gross Topology*

Several general points emerge from the majority rule consensus (fig. 2.1), and the primary derivative RCC tree (fig. 2.3). Fossil and Recent taxa interleave to a great

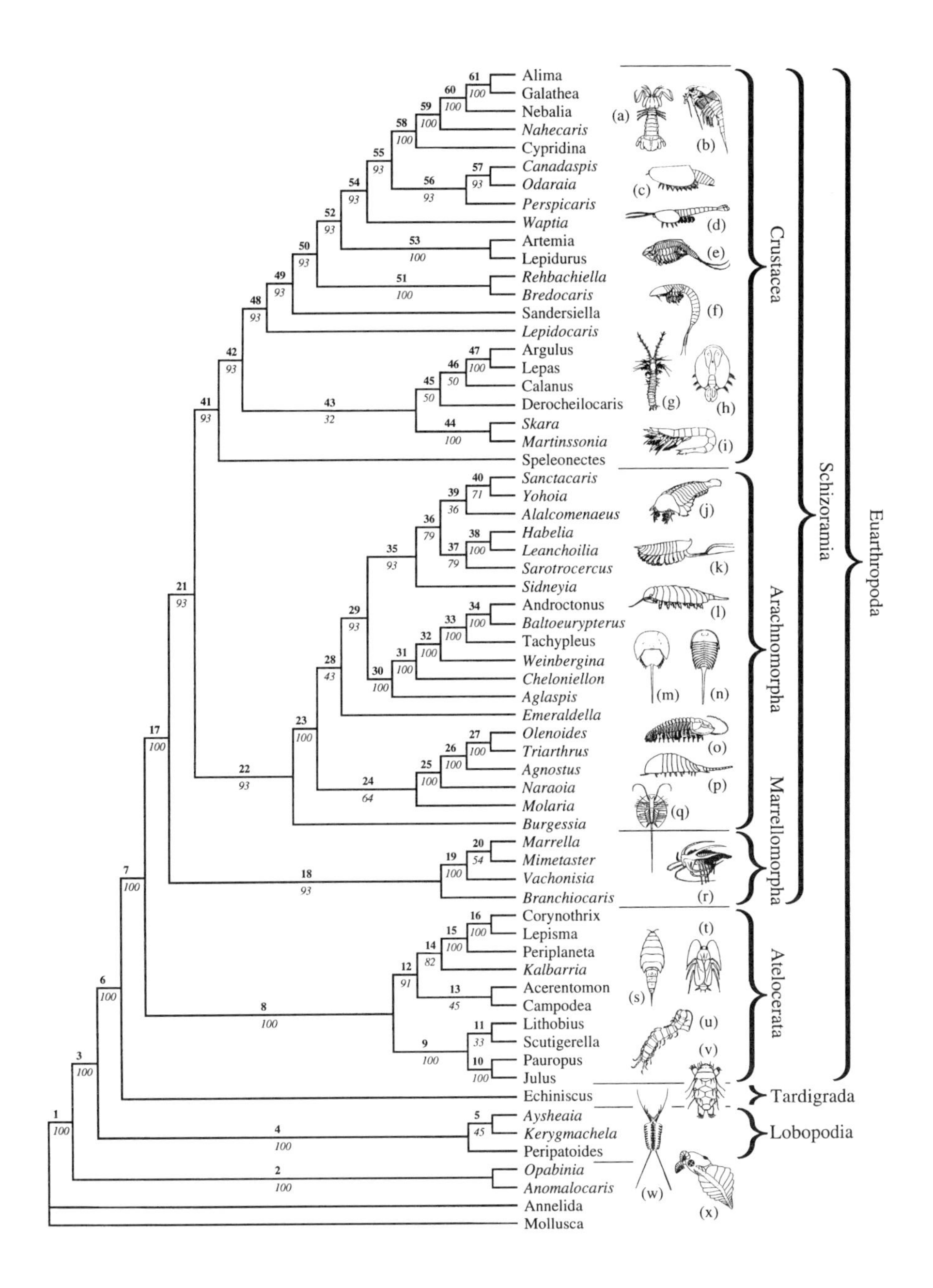


FIGURE 2.1

50% majority rule consensus of Recent and fossil arthropods (5544 fundamentals, 462 steps, CI = 0.333, RI = 0.667) plus all compatible groupings. All characters weighted equally. Nodes numbered above branches. Figures below branches indicate the percentage of fundamentals in which a grouping occurs. Genera illustrated are: *(a) Alima, (b) Nebalia, (c) Canadaspis, (d) Waptia, (e) Lepidurus, (f) Sandersiella, (g) Derocheilocaris, (h) Argulus, (i) Martinssonia, (j) Sanctacaris, (k) Leanchoilia, (l) Sidneyia, (m) Tachypleus, (n) Aglaspis, (o) Triarthrus, (p) Molaria, (q) Burgessia, (r) Marrella, (s) Kalbarria, (t) Periplaneta, (u) Lithobius, (v) Echiniscus, (w) Kerygmachela, (x) Anomalocaris.* See appendix 2.1 for apomorphies.

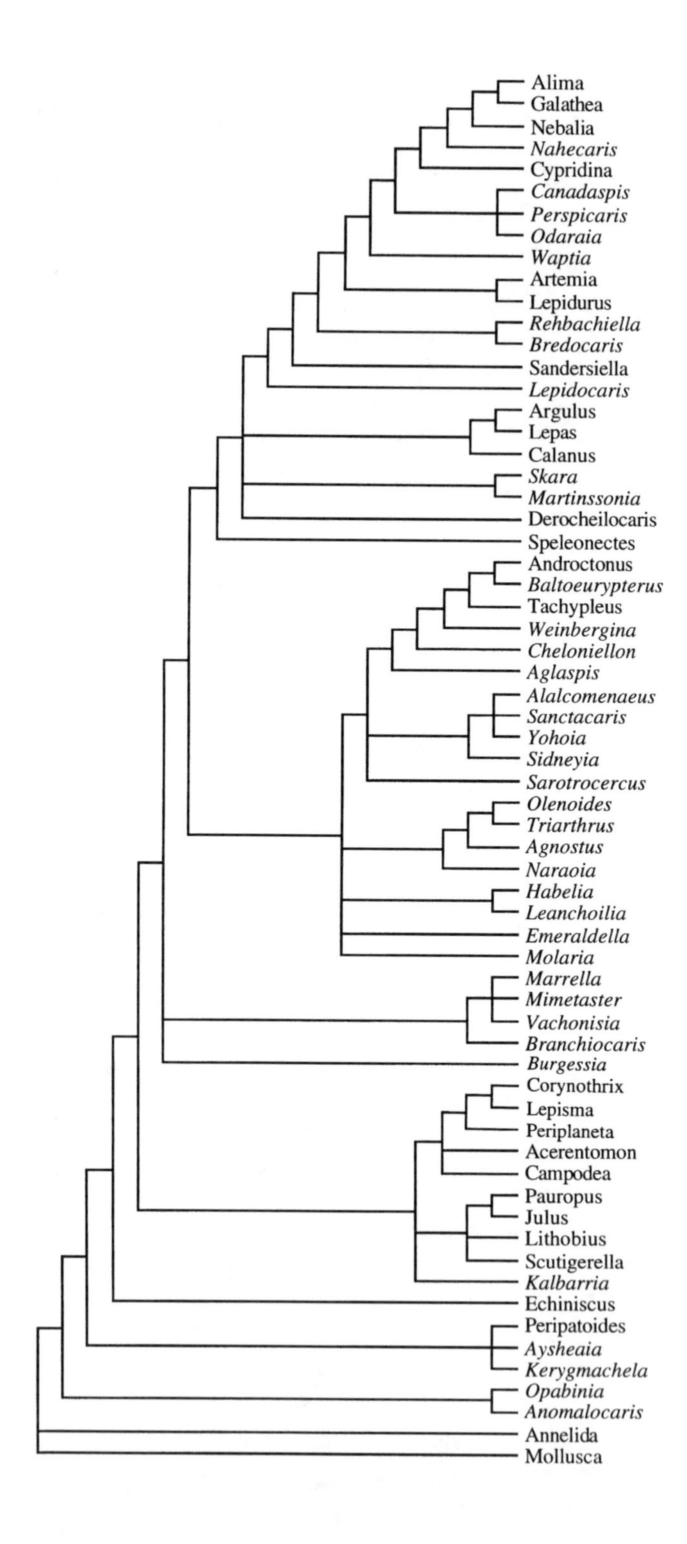


FIGURE 2.2

Adams consensus of Recent and fossil arthropods (5544 fundamentals). All characters weighted equally.

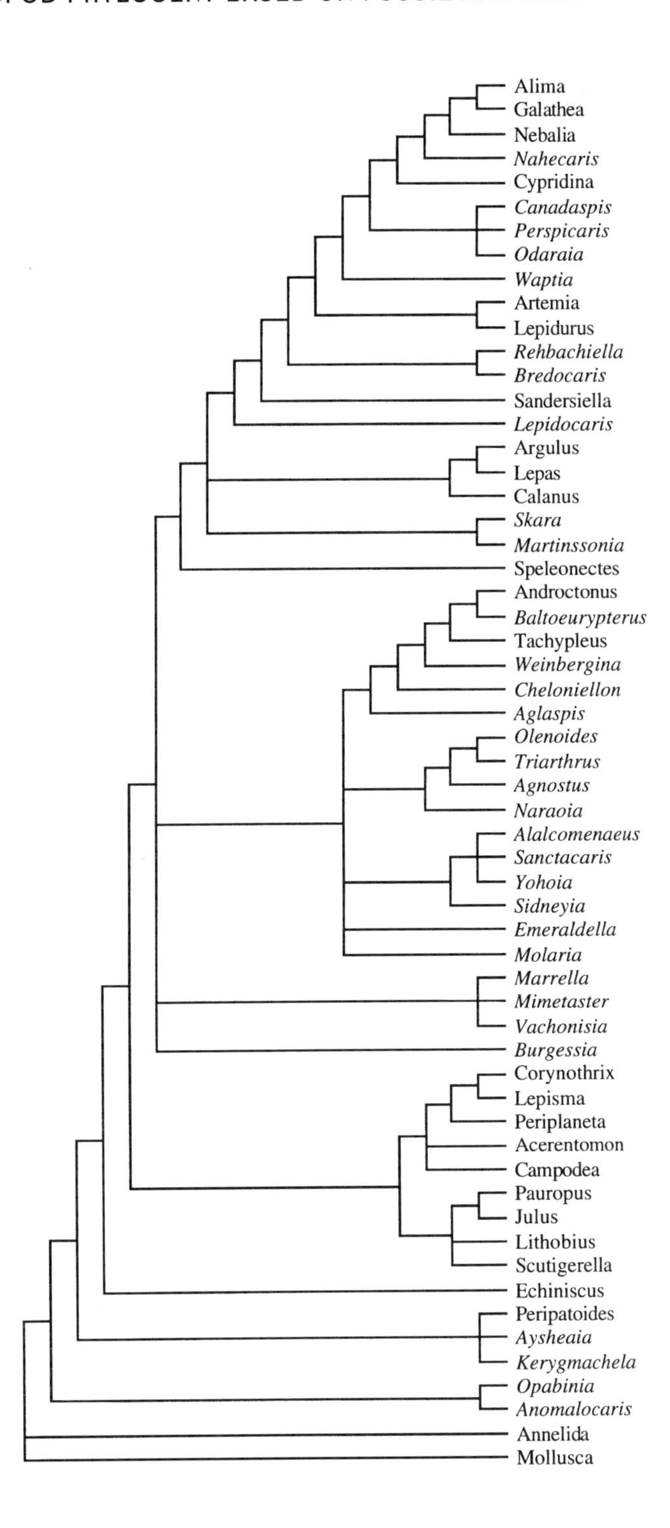


FIGURE 2.3

Primary derivative RCC tree (most fully resolved) of Recent and fossil arthropods (six taxa excluded). Third derivative formed from the forced fusion of four basic trees. All characters weighted equally.

extent. Many of the deepest divisions within the cladogram see the splitting of smaller groups at the base. The lobopods are no more closely related to the rest of the uniramians than any of the other arthropod clades (a position proposed by Sharov [1966], Baccetti [1979], Boudreaux [1979], and Weygoldt [1979]).

The four major arthropod groups (trilobites, chelicerates, crustaceans, uniramians) certainly do not occupy an equivalent level in the hierarchy. The basalmost division within the euarthropods separates a clade of hexapods, myriapods, and *Kalbarria* (Atelocerata or Uniramia *sensu* Harvey and Yen [1989]) from those arthropods with primitively biramous appendages (Schizoramia). This suggests that the atelocerates should be given relatively high taxonomic status. Nonetheless, they appear to be primitive with respect to a large number of characters.

In most fundamentals, the deepest split within the biramous arthropods is between a clade composed of *Branchiocaris*, *Vachonisia*, *Mimetaster*, and *Marrella* and one including all remaining taxa. This second clade broadly parallels Cisne's (1974) grouping of the trilobites, chelicerates, and crustaceans (the TCC lineage; see also Hessler and Newman 1975), although he did not explore relationships at lower levels within this branch. Biramy does not appear to be a primitive feature of arthropods in general (as advocated by Gould 1989; Simonetta and Delle Cave 1975, 1980) but rather a derived feature for a large clade within them. The deepest division within the TCC clade separates the crustaceans and arachnomorphs.

Previous treatments of the cladistic relationships of arthropods (Briggs et al. 1993; Wills et al. 1994) distinguished between crustaceans and "crustaceanomorphs." A number of arthropods (e.g., *Bredocaris*, *Odaraia*, *Skara*, *Waptia*) have been excluded from the Crustacea because they lack the requisite five head appendages or possess some novel features, despite close phenetic similarity with members of the class (Müller 1983; Briggs 1983). However, since these forms are thoroughly interspersed within the crustacean clade, the distinction does not appear to be useful.

The Arachnomorpha (Størmer 1944) accommodates most non-bivalved Cambrian problematica in addition to the trilobites and chelicerates. Despite separating at a low level within the arachnomorph clade, the trilobites appear highly derived with respect to the arthropods as a whole. This is in marked contrast to their frequently assumed position as *de facto* ancestors. The chelicerates emerge as even more highly derived.

The basic topology inevitably echoes elements from the schemes of other workers. The position of the Onychophora is consistent with the interpretations of Sharov (1966), Weygoldt (1979), Baccetti (1979), and Boudreaux (1979). The placement of the atelocerates as a sister group to all other euarthropod taxa was proposed by Schram and Emerson (1991) from a consideration of Arthropod Pattern Theory characters.

The union of the chelicerates and trilobites was also proposed, for example, by Sharov (1966), Paulus (1979), and Weygoldt (1979). However, in all these cases, the scheme also incorporates the hypothesis of a mandibulate clade comprising the myriapods, hexapods, and crustaceans. A mandibulate clade has not emerged from the present analysis. Close affinity to the trilobites of several taxa previously allied within the Trilobitoidea (Størmer 1944) is vindicated, although most of these taxa (with the exception of *Burgessia*) form a less inclusive clade with the chelicerates.

Many elements of the gross topology of the dendrogram presented by Emerson and Schram (1990) are reproduced by the present cladistic analysis. These are: (1) the union of chelicerates and trilobites interspersed with Burgess Shale arachnomorphs; (2) the opposition of this clade to the crustaceans, the whole constituting the schizoramians; (3) the basal status of the remipedes within the crustaceans; (4) the opposition of myriapods and hexapods to the schizoramians (although Schram and Emerson [1991] placed the Euthycarcinoidea in opposition to all the other euarthropods); and (5) the basal location of the onychophorans. This provides strong support for the topology revealed as the two sets of results were arrived at by dissimilar methods. Schram and Emerson (1991) focused on APT characters for their analysis.

*Anomalocaris* and *Opabinia*. *Anomalocaris* (Whittington and Briggs 1985) and *Opabinia* (Walcott 1912; Whittington 1975) constitute a small, extinct clade in opposition to all other ingroup taxa. Cladistically, therefore, this is a group of enormous distinction united by characters including the presence of stalked, compound eyes situated dorsally on the cephalon, flaplike, unsclerotized appendages, and a three-lobed tail fan. Chen et al. (1994) also commented on the shared ventral location of the mouth (although in *Anomalocaris* this is directed ventrally, while in *Opabinia* it faces posteriorly) and on the presence of spine-bearing anterior appendages (a dubious homology). If *Anomalocaris* is replaced by the anomalocaridid Laggania (which lacks a tail fan), the resulting cladograms are identical.

*Opabinia* shows no close affinity with the crustaceans, never emerging with or near the branchiopods (Walcott 1912; Hutchinson 1930; Raymond 1935; see also Fedotov 1925; Richter 1932; Linder 1946; Tiegs and Manton 1958). Neither does it group closely with "trilobitomorph" genera (trilobed forms with jointed appendages and antennae [Størmer 1944]).

The flaplike exopods of *Anomalocaris* and *Opabinia* are acquired independently on the cladogram from the outer rami of euarthropods, despite the fact that originally they had been considered homologous in this analysis and elsewhere (Budd 1993). The supposed mode of feeding in *Opabinia* (bringing food captured with the anterior appendages back under the animal in the vertical plane to the mouth) is strikingly similar to the strategy believed to have been employed by *Anomalocaris*. It seems unlikely that *Kerygmachela* ingested food in this manner, since the anterior appendages diverge anterolaterally from the head (Budd 1993), and the long, distal spines would presumably have restricted their ventral movement.

Lobopodia: *Aysheaia, Peripatoides, Kerygmachela*. These three taxa form a clade in opposition to the euarthropods and the tardigrade *Echiniscus* and are united by the presence of lobopodous inner rami, a wrinkled dorsal cuticle, and mobile terminal appendages. The close association of *Aysheaia* (Whittington 1978) with the modern Onychophora is hardly controversial (Hutchinson 1930; Størmer

1944; Cuénot 1949; Vandel 1949; Dechaseaux 1953; Tiegs and Manton 1958; Moore 1959; Robison 1985). The relatively high degree of cephalization and anterior appendage specialization observed in the modern Onychophora appears to have derived independently from that in euarthropods, despite superficial similarities (Manton 1977).

Hutchinson (1930) proposed a new group (Protonychophora) to accommodate *Aysheaia*, opposing all Recent Onychophora in the Euonychophora. Delle Cave and Simonetta (1975) and Simonetta (1976) speculated that *Aysheaia* might be an intermediate between tardigrades and modern Onychophora. Whittington (1978) regarded it as a representative of an early group of soft-bodied forms from which the tardigrades and onychophorans were both derived. The present analysis supports none of these concepts.

The location of *Kerygmachela* here differs fundamentally from that proposed by Budd (1993) in his original description of the animal. Many of the features presumed to unite the animal with the Schizoramia emerge as convergences, particularly the apparently biramous nature of the appendages. *Kerygmachela* therefore appears to be a relatively highly derived lobopod, evolving many of its unique features in isolation. The present analysis suggests that this fossil taxon is of relatively little import in resolving higher arthropod phylogeny, and its existence does not necessarily imply a diphyletic origin for the group (cf. Budd 1993).

Tardigrada. As observed above, the tardigrade *Echiniscus* stands in opposition to the euarthropods in all fundamentals, a placement proposed by Brusca and Brusca (1990), Nielsen (1995), and Monge-Nájera (1995). In many respects tardigrades are morphologically intermediate between more basal soft-bodied taxa and the euarthropods, having acquired several of the features often cited as diagnostic for this latter group (most importantly a more heavily sclerotized cuticle forming sclerites in some forms [Ruhberg 1985], a reduced number of nephridia, and one or no anatomically pre-oral limb-bearing segment in the larva). Other authors have suggested additional characters that may link the tardigrades with the euarthropods, including the presence of "arthropod" setae, the shift from lobopodal locomotion to true leg-gait movements, rectal pads, peritrophic membrane, the structure of the reproductive system, and the loss of sheets of annelid-like musculature (Marcus 1929; Riggin 1962; Brusca and Brusca 1990; Kinchin 1994). The euarthropods themselves are still supported by an impressive array of characters absent in tardigrades: presence of a labrum, advanced cephalization, trunk appendages that are strongly sclerotized, an intersegmental tendon system, striation of all muscles, sacculate nephridia, ecdysis glands, and specialized ommatidial structures.

Atelocerata: Hexapoda, Myriapoda, Euthycarcinoidea. The Atelocerata together with the euthycarcinoids consistently emerge as a clade, united by the loss of their second antennae, the presence of whole-limb mandibles, tracheae, ectodermal Malpighian tubules, and specialized ommatidial structures. Within this clade the myriapods oppose a clade of hexapods plus *Kalbarria*. Myriapods are united by characters including the presence of a gnathal lobe, repugnatorial glands, organs of

Tömösváry, and medial coalescence of the fourth appendages. *Pauropus* and *Julus* are grouped by their four cephalic appendages, gnathochilarium, diplosegments, collum, and suppression of the appendages of the sixth somite. The clade of hexapods plus *Kalbarria* is united by post-cephalic tagmosis (considered of paramount importance by McNamara and Trewin 1993), the nature of the trunk limb geniculation, and the presence of two primary pigment cells in the ommatidia. The location of *Kalbarria* with respect to the hexapods is variable, and it is therefore not included in figure 2.3. Nonetheless, it always resolves as an atelocerate (Bergström 1980; Schram and Rolfe 1982; McNamara and Trewin 1993) and never with the crustaceans (see Handlirsch 1914; Gall and Grauvogel 1964) or arachnomorphs (see Størmer 1955; Bergström 1971; Starobogatov 1988). This provides another Silurian representative of the clade (the earliest undoubted myriapods occur in the Upper Silurian of the U.K.; see Ross and Briggs 1993).

Marrellomorpha: *Marrella, Mimetaster, Vachonisia,* and *?Branchiocaris. Marrella, Mimetaster, Vachonisia,* and *Branchiocaris* constitute a clade at the base of the schizoramians in the vast majority of the fundamentals or are a sister group to the crustaceans in a small subset. The status of the clade as a group within the schizoramians is supported in all fundamentals. The marrellomorphs (including Branchiocaris) are united by a similar number of cephalic appendages (except in *Vachonisia),* the large number of trunk somites, the regular decrease in the length of the trunk appendages, and the division of trunk endopods into five podomeres. Within this grouping, *Vachonisia* opposes *Marrella* and *Mimetaster,* but all are united by the presence of a doublure and the organization of the filamentous trunk exopods around a longitudinal rhachis. These three taxa were grouped in the same way by Stürmer and Bergström (1976) on the basis of many of the same characters. However, these authors regarded *Marrella, Mimetaster,* and *Vachonisia* as probable basal arachnomorphs rather than as basal to this clade and the crustaceans or to the crustaceans alone.

The grouping of *Branchiocaris* with the marrellomorphs (in over 90% of the fundamentals) is a little equivocal. In previous analyses, it has resolved as a highly derived crustacean. The bivalved carapace, relatively large number of trunk somites, and the homonomous appendages decreasing in length down the trunk served to unite *Branchiocaris* with *Odaraia* (Briggs 1981) in a previous analysis (Briggs et al. 1993). The mobility of *Branchiocaris* within a small percentage of the fundamentals is such that its omission permits much greater resolution of the remaining tree in the RCC (fig. 2.3).

Whether the marrellomorphs (*Vachonisia*, *Mimetaster*, *Marrella*, and possibly *Branchiocaris*) constitute a sister taxon to all the remaining schizoramians or just to the crustaceans, they certainly represent a high-level bodyplan (at least as distinct as, for example, the trilobites, chelicerates, and crustaceans). This group persists to the Devonian.

Arachnomorpha, including Chelicerata and Trilobita. All the major clades emerging from this analysis contain problematic genera, but the majority of them resolve within the arachnomorphs. The post-Cambrian extinction of problematic genera had its greatest effect within this group. With the demise of the trilobites by the end of the Permian (Romano et al. 1993), the chelicerates were markedly isolated from their nearest remaining sister group, the crustaceans. What once lay between the chelicerates and the base of the crustacean clade undoubtedly represents the largest "phylogenetic gap" in the modern fauna.

The arachnomorph clade is supported by an impressive array of characters, including the ventral location of the anus in the penultimate division of the trunk, the development of a styliform terminal projection, the presence of trunk gut diverticulae, a marginal rim on the cephalic shield, the incorporation of four (or more) appendages into the cephalon, and the presence of six podomeres in the inner rami of the trunk appendages. All these characters, however, are subject to reversal higher in the tree. *Burgessia* (Walcott 1931; Hughes 1975) resolves at the base of this clade in most of the fundamentals but is located as parsimoniously elsewhere in a subset of trees. The remaining arachnomorphs are united by the trilobation, overlapping pleurae and markedly concave posterior margins of their tergites, a cephalic doublure, spinose and gnathobasic telopods, and filamentous exopods.

Basal within the Arachnomorpha are the trilobites (including *Naraoia* and *Agnostus*, cf. Ramsköld and Edgecombe 1991; Shergold 1991). The trilobite clade is supported by the rounded terminal division bearing the anus and by the unique tagmosis of the posterior trunk (pygidium). Above *Naraoia* (the basal trilobite) the remaining genera are united by the presence of genal spines, visible ecdysial sutures, a natant labrum, and the nature of the trunk exopods. *Molaria* falls in opposition to the trilobites in most fundamentals (see also Wills et al. 1994). The close affinity previously proposed between *Molaria* and *Emeraldella* (Walcott 1912; Størmer 1944; Simonetta and Delle Cave 1975; Whittington 1981) does not emerge from the majority of fundamentals, nor has this pairing been supported in any previous cladistic analyses (Briggs et al. 1992a,b, 1993). The characters often quoted as uniting *Molaria* and *Emeraldella* usually serve to group a more inclusive clade of arachnomorphs or are plesiomorphic for an even larger group (e.g., the absence of trilobation and the absence of eyes).

The chelicerates form a cohesive group in all the fundamentals, united by the presence of chelicerae, visible and marginal ecdysial sutures, tergites that are straight in lateral aspect, and the absence of inner rami on the trunk appendages. They are represented by the modern scorpion *Androctonus*, the eurypterid *Baltoeurypterus*, and the xiphosurans *Tachypleus* and *Weinbergina* in a pectinate series. Both the merostomes and xiphosurans are paraphyletic.

Two problematic fossils, the Devonian *Cheloniellon* (Stürmer and Bergström 1978) and the Cambrian *Aglaspis* (Briggs et al. 1979), are closely associated with this clade. The former is grouped by the incorporation of six appendages into the cephalon, the detachment of the labrum, and the presence of gnathobases on the trunk appendages. The node below *Aglaspis* is supported by a trilobate and tuberculate dorsal cuticle, a cardiac lobe and spinose trunk appendages. *Cheloniellon* and

*Aglaspis* are excluded from the chelicerates principally by the absence of chelicerae (Raasch 1939 v. Briggs et al. 1979, for discussion of *Aglaspis*), despite the fact that they have many characters in common with the merostomes. *Sanctacaris*, considered to be the oldest chelicerate by Briggs and Collins (1988), never groups closely with this clade. Rather it most often forms a tight grouping with *Yohoia* (Whittington 1974) and *Alalcomenaeus* and has a more distant affinity with *Sidneyia*. *Yohoia* and *Sanctacaris* emerged very proximately (though grouped paraphyletically) in several previous cladistic studies (Briggs et al. 1993; Wills et al. 1994). Whatever the precise affinities of *Yohoia*, there is no evidence to support its grouping with the anostracan branchiopods (*contra* Walcott 1912; Raymond 1920; Fedotov 1925; Henriksen 1928) or as allied to the crustaceans more generally (Simonetta 1970).

*Leanchoilia* (Bruton and Whittington 1983) is located within the arachnomorph clade in all topologies and in no case shows more inclusive affinity with the crustaceans (contrary to Walcott 1912, 1931; Fedotov 1925; Simonetta and Delle Cave 1980; Bergström 1992). It is most closely related to *Habelia* (Whittington 1981) in the majority of fundamentals (see also Wills et al. 1994), united by the absence of both a doublure and compound, lateral eyes, and the incorporation of three pairs of appendages into the cephalon. A previous analysis (Briggs et al. 1992a, 1993) paired *Leanchoilia* with *Alalcomenaeus* (Whittington 1981) (a relationship suggested by Bruton and Whittington [1983]), but principally as the result of recognizing several synapomorphies within the "Great Appendage Arthropods." In the present analysis, these arthropods are united in a clade composed of *Sarotrocercus*, *Leanchoilia*, *Habelia*, *Alalcomenaeus*, *Yohoia*, and *Sanctacaris* for which one synapomorphy is the presence of "great appendages," although some taxa have lost them (ACCTRAN optimization). *Habelia* does not appear to show close affinity with *Molaria* in the majority rule consensus (*contra* Whittington 1981).

The closest taxon to the *Habelia*/*Leanchoilia* grouping in most fundamentals is *Sarotrocercus*, united by the absence of cephalic and trunk gnathobases and the development of the terminal trunk division as a styliform extension (although it is modified rather differently in each of these three taxa). *Sarotrocercus* has proved particularly difficult to place cladistically (see Briggs et al. 1993; Wills et al. 1994).

Crustacea. Despite the absence of several "diagnostic" crustacean characters in some taxa within this clade (and uncertainty over others), the Crustacea are supported by the presence of lateral eyes that are free or pedunculate beneath the carapace, the differentiation of antennae and gnathobasic mandibles, the organization of the outer rami of the trunk appendages around a longitudinal rhachis, the presence of mobile telson appendages, and development through a nauplius or egg-nauplius stage.

Resolution within the crustaceans is problematical (see Briggs et al. 1993; Wills et al. 1994). In the present analysis, many of the nodes in the majority rule consensus are not represented in all the fundamentals (fig. 2.1), including that supporting the entire group. This anomaly is caused principally by the mobility of *Branchiocaris*, a taxon that has proved particularly difficult to assign. *Branchiocaris* is interpreted as parsimoniously at various locations within the Crustacea as it is within the marrellomorphs. Rescoring the trunk appendages of *Branchiocaris* in accordance with the interpretation of Schram and Emerson (1991) does nothing to resolve this problem. Pruning *Branchiocaris* from all the fundamentals results in a strict consensus in which the Crustacea are supported with considerable resolution. Pruning the mystacocarid *Derocheilocaris* in addition results in the strict consensus resolution of the crustacean clade (which takes the same form as that in fig. 2.3 [RCC primary derivative]).

The most primitive crustacean (not considering *Branchiocaris*) is *Speleonectes* (see also Schram 1986; Schram and Emerson 1991; Briggs et al. 1993; Wills et al. 1994). More derived forms have a differentiated trunk, and the trunk appendages tend to be pendent beneath the body rather than laterally displaced. The maxillopods *Argulus*, *Lepas*, *Calanus*, and *Derocheilocaris* emerge in a basal clade that also contains the "Orsten" taxa *Skara* and *Martinssonia* (united by the natant condition of the labrum). The ostracod *Cypridina* does not appear to be closely allied to this grouping. Basal within the opposing clade (outer rami of trunk appendages unsegmented, trunk appendages with protopodal endites), the Devonian *Lepidocaris* and Recent *Sandersiella* (Brachypoda; Schram 1986) are adjacent but not united. More derived taxa possess lateral eyes. The remainder of the tree is a pectinate arrangement of small clades and single taxa. The "Orsten" taxa *Rehbachiella* and *Bredocaris* come next, followed by taxa with predominantly uniramous antennae (endopods only) and trunk appendages organized around a flattened lamellate or lobate region. Basalmost are the branchiopods *Artemia* and *Lepidurus*, followed by the problematic Burgess Shale form *Waptia* (not forming a clade with the branchiopods [Briggs 1983], but close to them). The following clade contains phyllocarids, eumalacostracans, the ostracod, and the Burgess Shale form *Odaraia*. A clade of *Canadaspis*, *Odaraia*, and *Perspicaris* opposes the remainder, united by the similar form of the carapace and the vestigial first antenna. The eumalacostracans and the phyllocarids *Nahecaris* and *Nebalia* are supported by the presence of an articulating rostral spine, biramous antennule, and the ventral location of the anus.

The crustaceans contain a more even interspersion of fossil and Recent taxa than the arachnomorphs or atelocerates. Post-Cambrian extinction does not appear to have completely removed any large clades or contiguous portions of the tree. The large number of extant crustaceans means that more exhaustive and detailed comparisons will be necessary to investigate fully the relationships within this group. This is probably not true for the arachnomorphs.

## 2. Fossil Taxa Excluded, with all Characters Weighted Equally

The analysis excluding all fossil taxa yielded 54 trees of 211 steps (CI = 0.55; RI = 0.72). The greatest mobility of taxa is within the crustaceans. The gross topology of a majority rule consensus (fig. 2.4) is similar to that resulting from the analysis of all taxa together, with all the fossils subsequently pruned. The lobopod *Peripatoides* re-

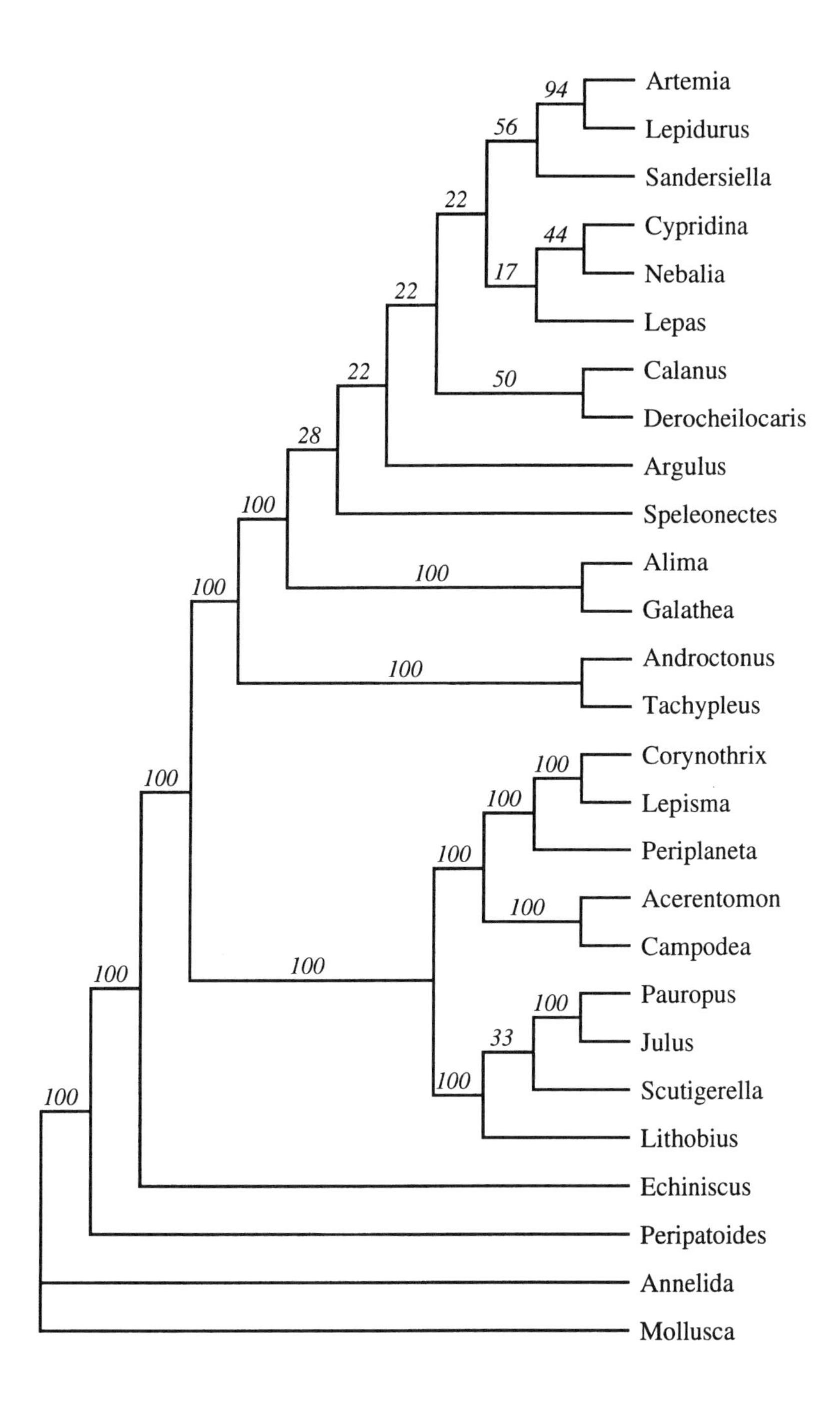


FIGURE 2.4

50% majority rule consensus tree, plus all compatible groupings. All characters weighted equally. Recent taxa only (54 fundamentals). Figures indicate the percentage of fundamentals in which a grouping occurs.

solves basally, followed by the tardigrade *Echiniscus*. The euarthropods resolve into atelocerates and schizoramians. The topology of the atelocerate clade is modified a little, with *Kalbarria* being pruned, and *Scutigerella* and *Lithobius* falling as successive sister taxa to the clade of *Julus* and *Pauropus*. Within the Schizoramia, the arachnomorphs (represented only by the Recent chelicerates *Androctonus* and *Tachypleus*) still oppose the crustaceans. The Eumalacostraca fall basally within the crustaceans, but resolution within the rest of the group is rather poor.

## 3. Living Taxa Excluded, with all Characters Weighted Equally

The analysis excluding all Recent taxa yielded 150 trees of 294 steps (CI = 0.36; RI = 0.61). The topology of the majority rule consensus (fig. 2.5) shows many similarities to that for a pruned analysis of all taxa, but with a number of important differences. The clade of *Anomalocaris* and *Opabinia* falls basally, followed by a clade of *Aysheaia* plus *Kerygmachela*. In the absence of the atelocerates, the basalmost euarthropod is *Burgessia*. The remainder divide into two large clades, one composed of crustaceans, the other of arachnomorphs and marrellomorphs.

The topology of the lower section of the crustacean clade is identical to that in the full analysis with the Recent taxa subsequently pruned. A small clade of *Skara* plus *Martinssonia* lies basal to *Lepidocaris* and to a small clade of *Rehbachiella* and *Bredocaris*. The remainder of the group is composed of bivalved taxa in a pectinate sequence.

A clade of *Marrella*, *Mimetaster*, and *Vachonisia* falls basal to a grouping of the remaining arachnomorphs, the position proposed for them by Stürmer and Bergström (1976). *Aglaspis* and *Cheloniellon* group next in the majority of fundamentals, but they do not form a clade with the chelicerates *Weinbergina* and *Baltoeurypterus*, which plot next in opposition to *Kalbarria*. The placement of *Kalbarria* with the chelicerates in the absence of the atelocerates merits comment. Arachnomorph affinities for the euthycarcinoids were proposed (indirectly) by Starobogatov (1988) and Schram (1971), but these arthropods assume this position here only in the absence of Recent taxa. Even so, contrary to Starobogatov, the closest relatives do not appear to be the aglaspids. The remaining arachnomorph taxa fall into two pectinate clades. One is composed of the trilobites plus *Molaria* in the same order as that resulting from the original analysis, with *Emeraldella* basal to them. The other contains the taxa grouped in opposition to the "chelicerate" clade in the full analysis, albeit in a different order.

## 4. All Taxa, with all Characters Successively Weighted According to the Maximum Value of the Rescaled Consistency Index

The topology stabilized after two successive weightings, nine trees resulting. A 50% majority rule consensus plus compatible groupings (fig. 2.6) differs little from a

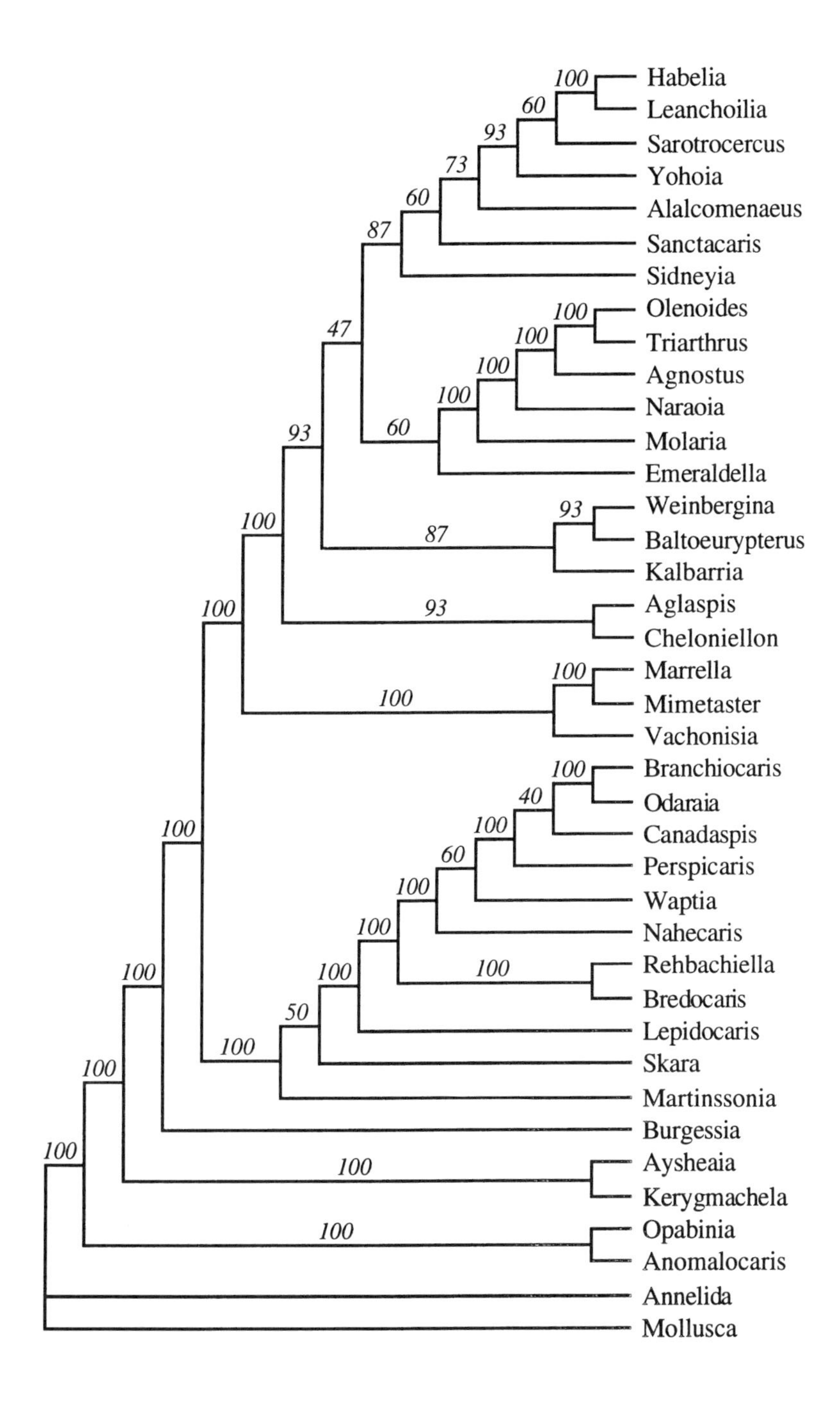


FIGURE 2.5

50% majority rule consensus trees, plus all compatible groupings. All characters weighted equally. Fossil taxa only (150 fundamentals). Figures indicate the percentage of fundamentals in which a grouping occurs.

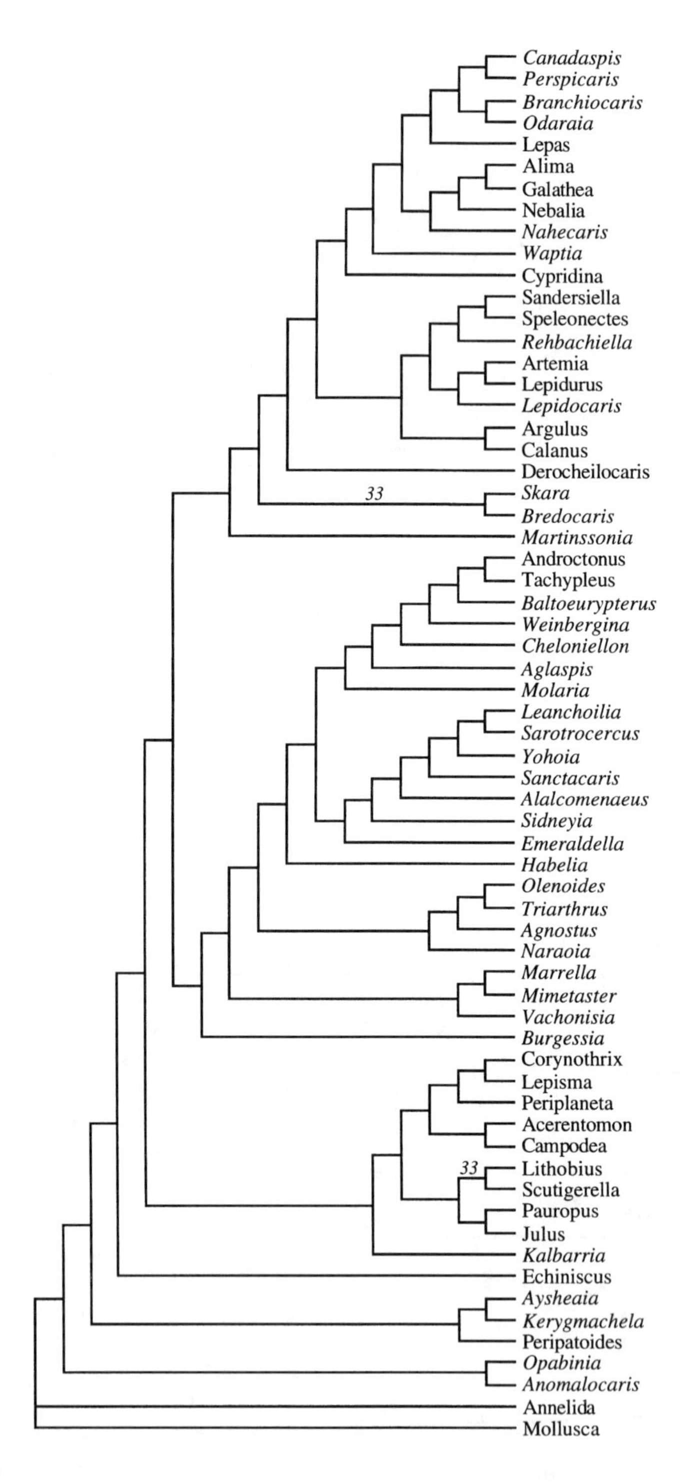


FIGURE 2.6

Successive approximations weighting of data for Recent and fossil arthropods. Characters reweighted by the maximum value of the rescaled consistency index. Two iterations required to converge on nine trees. 50% majority rule consensus trees, plus all compatible groupings. All branches in 100% of fundamentals except where otherwise indicated.

strict consensus. Gross topology is similar to that resulting from the flat-weighted runs. The atelocerates, arachnomorphs (except *Burgessia*), and crustaceans all form clades. Detailed relationships are identical up to and including the positioning of *Echiniscus* in opposition to the euarthropods.

The topology within the atelocerates is identical, except that *Kalbarria* is pruned from its location within the hexapods and now falls basally. A small clade of *Marrella*, *Mimetaster*, and *Vachonisia* is dissociated from *Branchiocaris* (now within the crustaceans) and no longer falls at the base of the schizoramians, but rather at the base of all the arachnomorphs. *Burgessia* resolves in opposition to the marrellomorphs and arachnomorphs.

Much of the topology of the arachnomorph clade remains the same after weighting. The trilobites still form a relatively basal clade, although *Molaria* no longer groups with them. Above this, *Habelia* falls in opposition to all the remaining arachnomorphs, which form two opposing, pectinate clades. One of these contains the chelicerates, plus *Cheloniellon*, *Aglaspis*, and *Molaria*. The positions of *Tachypleus* and *Baltoeurypterus* are reversed compared to that in the flat-weighted run. The other pectinate clade is composed of *Emeraldella*, *Sidneyia*, *Alalcomenaeus*, *Sanctacaris*, *Yohoia*, *Sarotrocercus*, and *Leanchoilia* (compared with a clade excluding *Emeraldella* but containing *Habelia* in the flat-weighted run).

The topology of the crustacean clade is markedly different here from that in the flat-weighted run. *Martinssonia* falls at the base of the group, rather than *Speleonectes*. This position for *Martinssonia* is consistent with that proposed by Walossek and Szaniawski (1991), who regarded it as a probable stem-lineage form. This is followed by a small clade of two "Orsten" forms, *Skara* and *Bredocaris*, falling before the mystacocarid, *Derocheilocaris*. The maxillopods do not group. The remainder of the crustaceans form two opposing clades, one containing predominantly bivalved forms and the other univalved or carapace-less taxa. In the former, the eumalacostracans plus the phyllocarids *Nebalia* and *Nahecaris* separate from the Burgess phyllocarids *Canadaspis* and *Perspicaris* in opposition to *Branchiocaris* and *Odaraia*. This relocation of *Branchiocaris* from a clade at the base of the schizoramians to a highly derived position within the crustaceans (as in the cladogram of Wills et al. [1994]) is one of the more marked differences in topology between the weighted and unweighted runs.

The approach adds no new information to the analysis but provides an objective criterion by which some aspects of the data become self-reinforcing and can be amplified relative to others. Where weighting results in a topology identical to that obtained without it, such internal consistency reinforces the result. Similarly, where weighting yields only a subset of the topologies observed from the original analysis, it provides an ancillary criterion for selecting from equally parsimonious interpretations of the unweighted data (Farris 1969). However, where weighting produces a new topology, alternative aspects in the structure of the data have been emphasized, overriding those previously determining the outcome of the analysis. In this case both the successive approximations weighting and at least some of the compatibility-based weightings retain the gross relationships of major clades, even though relationships approaching the terminals are less stable.

### Phenetic Clustering

Several major clades are also identified by phenetic clustering (fig. 2.7). These are the Atelocerata (containing a distinct Hexapoda and Myriapoda), Marrellomorpha, Crustacea, and Arachnomorpha (containing a distinct Trilobita minus *Naraoia* and Chelicerata). However, the higher-level organization of these groups does not correspond to that derived from the cladograms. The Schizoramia do not form a cluster because the Marrellomorpha, Crustacea, and Arachnomorpha dissociate. The group closest to the atelocerates is the marrellomorphs, and both in turn cluster with the crustaceans. The lobopods, *Echiniscus*, *Anomalocaris*, and *Opabinia*, form a cluster ("protarthropods") with no cladistic equivalent. This grouping opposes all the arthropods except the arachnomorphs, which cluster with the rest of the tree at the lowest level of similarity.

The behavior of several taxa deserves comment. *Kalbarria* clusters not with the hexapods, but in opposition to all the other atelocerates (the position occupied in the successively weighted cladogram). The eumalacostracans fall basally within the crustaceans, reflecting their phenetic distinctness. The maxillopods (including *Cypridina*, cf. cladograms) cluster tightly, in opposition to *Speleonectes* and *Sandersiella*. Phyllocarid taxa (plus *Odaraia* and *Waptia*) cluster, while "Orsten" forms (Müller 1983) also associate, but not exclusively. Next to the Eumalacostraca the most dissimilar crustacean group is the branchiopods (see the cladistic position proposed by Fryer 1992). Within the arachnomorphs, *Aglaspis* and *Cheloniellon* cluster alongside the chelicerates (as in the cladograms), while *Sanctacaris* associates with *Alalcomenaeus* and *Yohoia*. *Burgessia* falls deep within the arachnomorph cluster, rather than at the base of the group.

The close phenetic association of the crustaceans and atelocerates (only the fossil marrellomorphs intercede) parallels their grouping as mandibulates in many classifications and phylogenies based primarily on the extant fauna. Only the parsimony networking and rooting serves to separate them. The "protarthropods" are phenetically rather similar, despite their pectinate arrangement at the base of the cladogram. Nonetheless, *Kerygmachela* groups more closely with the other lobopods than with *Anomalocaris* and *Opabinia*.

## Conclusions

1. The role of fossils in cladistics has excited a good deal of discussion. This investigation clearly demonstrates that the inclusion of fossils is necessary to fully appreciate the complexity of early branching events (see Gauthier et al. 1988). Coding of character combinations that are present only in living taxa yields trees that differ

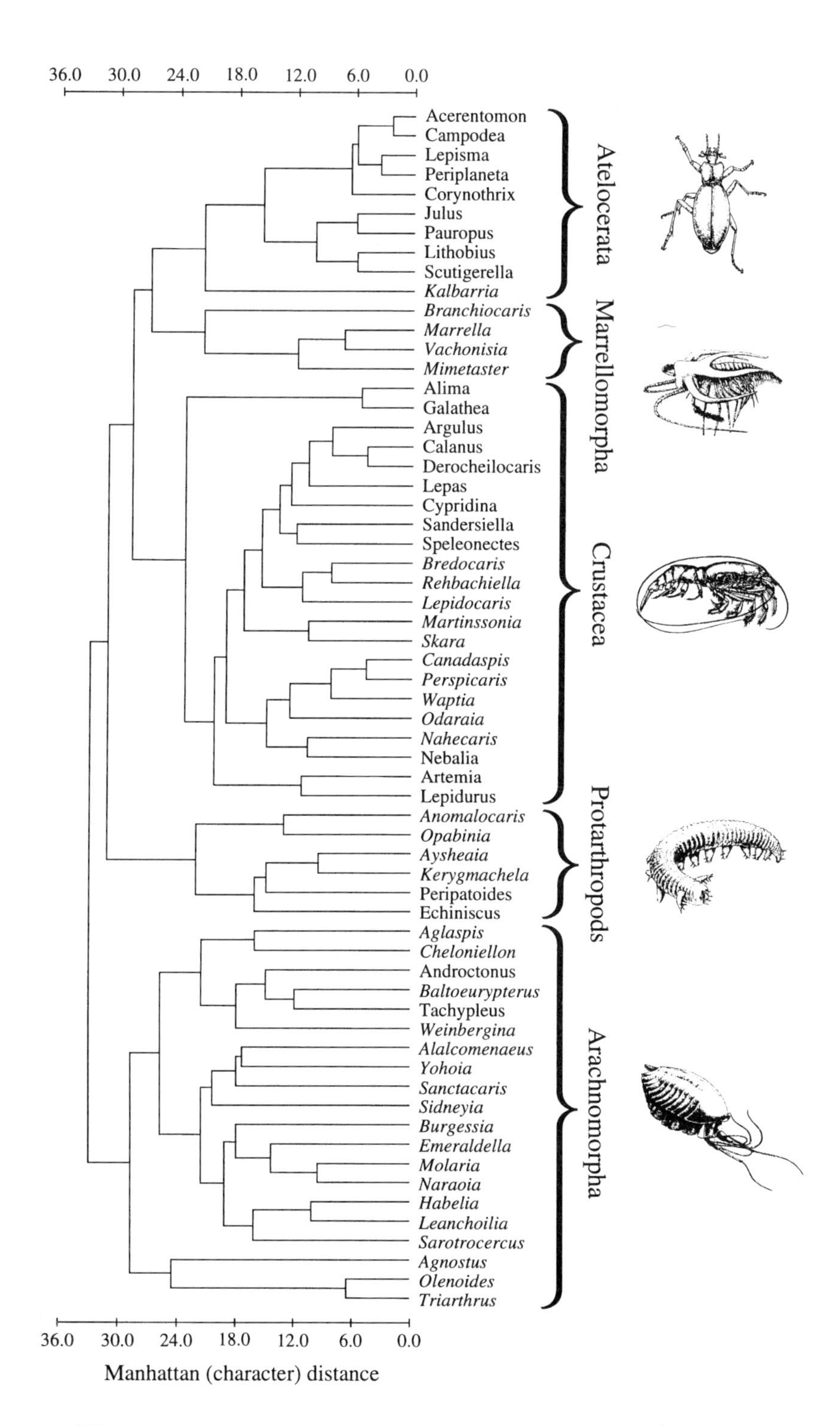


FIGURE 2.7

UPGMA clustering of Recent and fossil arthropods, based on a simple Manhattan distance matrix. $r^2 = 0.568$.

significantly from those that additionally treat the fossils (compare fig. 2.4 with figs. 2.1–2.3). Conversely, analyses based on character combinations solely present in fossils may also be misleading. *Kalbarria*, for example, falls within the Arachnomorpha when only fossil taxa are considered (fig. 2.5) whereas with the inclusion of living forms, it is clearly allied to the Atelocerata. Likewise analysis of the fossils alone groups the marrellomorphs with the arachnomorphs whereas they form a separate clade when the complete data set is considered.

2. Paleontological data have only recently been factored directly into analyses of arthropod relationships (e.g., Briggs and Fortey 1989; Bergström 1992; Wills et al. 1994; Robison and Wiley 1995). The present study represents the most extensive attempt yet to combine data for Recent and fossil taxa. The emerging pattern differs significantly from at least some studies based on molecular data (Abele et al. 1992; Wheeler at al. 1993; Boore et al. 1995; Friedrich and Tautz 1995). However, at present only a small selection of taxa has been sampled for molecular data, and the results are not stable. It will be interesting to see to what extent a larger molecular data set supports or contradicts the results of paleontological/morphological analyses in the future.

3. Despite the large number of equally parsimonious trees resulting from the primary analysis of the data used here, the deeper branching structure is relatively stable and consistent with that revealed by previous, more constrained analyses (Briggs et al. 1992a, 1993; Wills et al. 1994; see also Schram and Emerson 1991). The lack of resolution is largely a function of the mobility of a small number of genera, and a reduced cladistic consensus approach (Wilkinson 1994) reveals the common cladistic structure across the group of fundamentals.

4. The euarthropods consistently form a clade, with several small groups of "protarthropodan" grade falling in a pectinate series below them. The basalmost branch comprises the anomalocaridids and *Opabinia*, a group in which sclerotized, jointed, frontal appendages were derived independently. The next group in sequence contains the Recent and fossil Onychophora (*Peripatoides* and *Aysheaia*), plus the Greenland lobopod *Kerygmachela*. Immediately below the euarthropods are the tardigrades, which have acquired a number of important euarthropod characters, with the notable exception of articulating trunk appendages and a well-differentiated cephalon.

5. Among euarthropod groups, the pattern (Hexapoda + Myriapoda + Euthycarcinoidea) v. ((Marrellomorpha) v. (Crustacea v. Arachnomorpha)) is supported. The modern Chelicerata and fossil Trilobita both emerge as clades, although at different levels in the hierarchy. Recent crustaceans are interspersed with a number of fossils previously excluded from the group, while the Atelocerata also include the euthycarcinoids.

6. The largest number of Cambrian problematica are accommodated within the arachnomorph clade, which also includes the chelicerates and trilobites as separate groups. A small number of problematica are apparently closely allied to the chelicerates. The problematic marrellomorphs and *Branchiocaris* form a clade basal to

all the remaining Schizoramia. This clade represents an extinct bodyplan of a high order.

7. The results of the present analyses are consistent with arthropod monophyly. Several characters of the euarthropods are shared with the tardigrades, although the euarthropods themselves are distinguished by many additional synapomorphies. A case for polyphyly would require the demonstration of convincing homologies between various arthropod groups and specific groups of annelids, mollusks, or other protostomes. Such evidence is unlikely to come from morphological comparison (Iwanoff 1928; Sharov 1966; Weygoldt 1979), but the possibility that it might emerge with sufficient molecular data has yet to be eliminated.

8. Phenetic clustering reveals many of the same major groupings (e.g., Atelocerata, Arachnomorpha, Crustacea) as the cladograms, but the hierarchical pattern of similarity differs markedly. The "protarthropods" constitute a phenon (*sensu* Sneath and Sokal 1973), although they do not emerge as a clade in the cladistic analyses. The crustaceans and uniramians show phenetic similarity even though the cladistic analyses revealed no evidence of phylogenetic relatedness between them.

## ACKNOWLEDGMENTS

M. A. W.'s research was funded by a Smithsonian Postdoctoral Fellowship, and by Leverhulme Trust Grant F/182/AK awarded to Prof. M. J. Benton.

## REFERENCES

Abele, L. G., W. Kim, and B. E. Felgenhauer. 1989. Molecular evidence for the inclusion of the phylum Pentastomida in the Crustacea. *Molecular Biology and Evolution* 6:685–691.

Abele, L. G., T. Spears, W. Kim, and M. Applegate. 1992. Phylogeny of selected maxillopodan and other crustacean taxa based on 18S ribosomal nucleotide sequences: A preliminary analysis. *Zoologica Scripta* 73:373–382.

Anderson, D. T. 1973. *Embryology and Phylogeny in Annelids and Arthropods*. Oxford: Pergamon Press.

Anderson, S. O. and T. Weis-Fogh. 1964. Resilin: A rubberlike protein in arthropod cuticle. *Advances in Insect Physiology* 2:1–66.

Averof, M. and M. Akam. 1995. Insect-crustacean relationships: Insights from comparative developmental and molecular studies. *Philosophical Transactions of the Royal Society of London*, series B, 347:293–303.

Ax, P. 1987. *The Phylogenetic System*. Chichester: Wiley.

Baccetti, B. 1979. Ultrastructure of sperm and its bearing on arthropod phylogeny. In A. P. Gupta, ed., *Arthropod Phylogeny*, pp. 609–644. New York: Van Nostrand Reinhold.

Baccetti, B. and F. Rosati. 1971. Electron miscroscopy on tardigrades, part 3: The integument. *Journal of Ultrastructural Research* 34:214–243.

Ballard, J. W. O., G. J. Olsen, D. P. Faith, W. A. Odgers, D. M. Rowell, and P. W. Atkinson. 1992. Evidence from 12S ribosomal RNA that onychophorans are modified arthropods. *Science* 258:1345–1348.

Bergström, J. 1971. *Paleomerus*: Merostome or merostomoid? *Lethaia* 4:393–401.

Bergström, J. 1973. Organization, life, and systematics of trilobites. *Fossils and Strata* 4:1–69.

Bergström, J. 1979. Morphology of fossil arthropods as a guide to phylogenetic relationships. In A. P. Gupta, ed., *Arthropod Phylogeny*, pp. 3–56. New York: Van Nostrand Reinhold.

Bergström, J. 1980. Morphology and systematics of early arthropods. *Abhandlungen des naturwissenschaftlichen Vereins in Hamburg (NF)* 23:7–42.

Bergström, J. 1992. The oldest arthropods and the origin of the Crustacea. *Acta Zoologica* 73:287–291.

Bergström, J., D. E. G. Briggs, E. Dahl, W. D. I. Rolfe, and W. Stürmer. 1987. *Nahecaris stuertzi*, a phyllocarid crustacean from the Lower Devonian Hunsrück Slate. *Paläontologische Zeitschrift* 61:273–298.

Birenheide, R. 1971. Beobachtungen am "Scheinstern" *Mimetaster* aus dem Hunsrück-Schiefer. *Senckenbergiana Lethaea* 52:77–91.

Blower, J. G., ed. 1974. *Myriapoda*. London: Academic Press.

Boore, J. L., T. M. Collins, D. Stanton, L. L. Daehler, and W. M. Brown. 1995. Deducing the pattern of arthropod phylogeny from mitochondrial DNA rearrangements. *Nature* 376:163–165.

Boudreaux, H. B. 1979. *Arthropod Phylogeny, with Special Reference to Insects*. New York: Wiley.

Bowman, T. E. 1971. The case of the nonubiquitous telson and the fraudulent furca. *Crustaceana* 21:165–175.

Boxshall, G. A. 1983. A comparative functional analysis of the major maxillopodan groups. In F. R. Schram, ed., *Crustacean Issues*. Vol. 1, *Crustacean Phylogeny*, pp. 121–143. Rotterdam: Balkema.

Briggs, D. E. G. 1976. The arthropod *Branchiocaris* n. gen., Middle Cambrian, Burgess Shale, *British Columbia. Bulletin of the Geological Survey of Canada* 264:1–29.

Briggs, D. E. G. 1978. The morphology, mode of life, and affinities of *Canadaspis perfecta* (Crustacea: Phyllocarida), Middle Cambrian, Burgess Shale, British Columbia. *Philosophical Transactions of the Royal Society of London*, series B, 281:439–487.

Briggs, D. E. G. 1981. The arthropod *Odaraia alata* Walcott, Middle Cambrian, Burgess Shale, British Columbia. *Philosophical Transactions of the Royal Society of London*, series B, 291: 541–585.

Briggs, D. E. G. 1983. Affinities and early evolution of the Crustacea: The evidence of the Cambrian fossils. In F. R. Schram, ed., *Crustacean Issues*. Vol. 1, *Crustacean Phylogeny*, pp. 1–22. Rotterdam: Balkema.

Briggs, D. E. G. 1994. Giant predators from the Cambrian of China. *Science* 264:1283–1284.

Briggs, D. E. G., D. L. Bruton, and H. B. Whittington. 1979. Appendages of the arthropod *Aglaspis spinifer* (Upper Cambrian, Wisconsin) and their significance. *Palaeontology* 22:167–180.

Briggs, D. E. G. and H. B. Whittington. 1985. Modes of life of arthropods from the Burgess Shale, British Columbia. *Transactions of the Royal Society of Edinburgh, Earth Sciences* 76:149–160.

Briggs, D. E. G. and D. Collins. 1988. A Middle Cambrian chelicerate from Mount Stephen, British Columbia. *Palaeontology* 31:71–73.

Briggs, D. E. G. and R. A. Fortey. 1989. The early radiation and relationships of the major arthropod groups. *Science* 246:241–243.

Briggs, D. E. G., R. A. Fortey, and M. A. Wills. 1992a. Morphological disparity in the Cambrian. *Science* 256:1670–1673.

Briggs, D. E. G., R. A. Fortey, and M. A. Wills. 1992b. Cambrian and Recent morphological disparity. Response to Foote and Gould, and Lee. *Science* 258:1817–1818.

Briggs, D. E. G., R. A. Fortey, and M. A. Wills. 1993. How big was the Cambrian explosion? A taxonomic and morphologic comparison of Cambrian and Recent arthropods. In D. R. Lees and D. Edwards, eds., *Evolutionary Patterns and Processes,* pp. 33–44. London: Linnean Society Symposium Series, Academic Press.

Broili, F. 1933. Ein zweites Exemplar von *Cheloniellon. Sitzungsberichte Bayerischen Akademie der Wissenschaften* 1933:11–32.

Brusca, R. C. and G. J. Brusca. 1990. *Invertebrates.* Sunderland, Mass.: Sinauer.

Bruton, D. L. and H. B. Whittington. 1983. *Emeraldella* and *Leanchoilia*: Two arthropods from the Burgess Shale, British Columbia. *Philosophical Transactions of the Royal Society of London,* series B, 300:553–585.

Budd, G. 1993. A Cambrian gilled lobopod from Greenland. *Nature* 364:709–711.

Callahan, P. S. 1979. Evolution of antennae, their sensilla, and the mechanism of scent detection in Arthropoda. In A. P. Gupta, ed., *Arthropod Phylogeny,* pp. 259–298. New York: Van Nostrand Reinhold.

Camatini, M., ed. 1980. *Myriapod Biology.* New York: Academic Press.

Carpenter, J. M. 1988. Choosing among multiple equally parsimonious cladograms. *Cladistics* 4:291–296.

Chen, J.-Y, L. Ramsköld, and G.-Q. Zhou. 1994. Evidence for monophyly and arthropod affinity of Cambrian giant predators. *Science* 264:1305–1308.

Cisne, J. L. 1974. Trilobites and the origin of arthropods. *Science* 186:13–18.

Cisne, J. L. 1982. Origin of the Crustacea. In L. G. Abele, ed., *The Biology of the Crustacea.* Vol. 1, *Systematics, the Fossil Record, and Biogeography,* pp. 65–92. New York: Academic Press.

Clarke, K. U. 1979. Visceral anatomy and arthropod phylogeny. In A. P. Gupta, ed., *Arthropod Phylogeny,* pp. 467–549. New York: Van Nostrand Reinhold.

Collins, D. 1996. The "Evolution" of *Anomalocaris* and its classification in the arthropod class Dinocarida (nov.) and order Radiodonta (nov.). *Journal of Paleontology* 70:280–293.

Crepet, W. L. and K. C. Nixon. 1989. Earliest megafossil evidence of Fagaceae: Phylogenetic and biogeographic implications. *American Journal of Botany* 76:842–845.

Cuénot, L. 1949. Les Onychophores. In P. P. Grassé, ed., *Traité de Zoologie.* Vol. 6, *Onychophores — Tardigrades — Arthropodes — Trilobitomorphes — Chélicérates,* pp. 3–37. Paris: Masson.

Cutler, B. 1980. Arthropod cuticle features and arthropod monophyly. *Experientia* 36:953.

Cuvier, G. 1817. *Le règne animal distribué d'après son organisation, pour servir de base à l'histoire naturelle des animaux et d'introduction à l'anatomie comparée.* Vol. 2. Paris: Deterville.

Dahl, E. 1983. Phylogenetic systematics and the Crustacea Malacostraca: A problem of prerequisites. *Abhandlungen des naturwissenschaftlichen Vereins in Hamburg (NF)* 26:355–371.

Dahl, E. 1991. Crustacea, Phyllopoda, and Malacostraca: A reappraisal of cephalic dorsal shield and fold systems. *Philosophical Transactions of the Royal Society of London*, series B, 344:1–26.

Dechaseaux, C. 1953. Onychophores. In J. Piveteau, ed., *Traité de Paléontologie*. Vol. 3, *Les Formes Ultimés d'Invertébrés, Morphologie et Évolution: Onychophores, Arthropodes, Échinodermes, Stomocordés,* pp. 4–7. Paris: Masson.

Delle Cave, L. and A. M. Simonetta. 1975. Notes on the morphology and taxonomic position of *Aysheaia* (Onychophora?) and of *Skania* (undetermined phylum). *Monitore Zoologico Italiano*, n.s., 9:67–81.

Donoghue, M. J., J. A. Doyle, J. Gauthier, A. G. Kluge, and T. Rowe. 1989. The importance of fossils in phylogeny reconstruction. *Annual Reviews of Ecology and Sytematics* 29:431–460.

Doyle, J. A. and M. J. Donoghue. 1987. The importance of fossils in elucidating seed plant phylogeny and macroevolution. *Reviews in Palaeobotany and Palynology* 52:321–431.

Dzik, J. and G. Krumbiegel. 1989. The oldest "onychophoran" *Xenusion*: A link connecting phyla? *Lethaia* 22:169–181.

Eldredge, N. 1974. Revision of the Suborder Synziphosurina (Chelicerata, Merostomata), with remarks on merostome phylogeny. *American Museum Novitates* 2543:1–41.

Emerson, M. J. and F. R. Schram. 1990. The origin of crustacean biramous appendages and the evolution of Arthropoda. *Science* 250:667–669.

Fahrenbach, W. H. 1971. The morphology of the *Limulus* visual system, part 4: The lateral optic nerve. *Zeitschrift für zoologische Systematik und Evolutionsforschung* 87:278–290.

Farris, J. S. 1969. A successive approximations approach to character weighting. *Systematic Zoology* 18:374–385.

Farris, J. S. 1989. The retention index and the rescaled consistency index. *Cladistics* 5:417–419.

Fedotov, D. 1925. On the relations between the Crustacea, Trilobita, Merostomata, and Arachnida. *Izvestiya Akademii Nauk SSSR* 1924:383–408.

Field, K. G., G. J. Olsen, D. J. Lane, S. J. Giovannoni, M. T. Ghiselin, E. C. Raff, N. R. Pace, and R. A. Raff. 1988. Molecular phylogeny of the animal kingdom. *Science* 239:748–753.

Fitch, W. M. and E. Margoliash. 1967. Construction of phylogenetic trees. *Science* 155:279–284.

Foote, M., and S. J. Gould. 1992. Cambrian and Recent morphological disparity. *Science* 258:1816.

Fortey, R. A. 1990. Ontogeny, hypostome attachment, and trilobite classification. *Palaeontology* 33:529–576.

Fretter, V. and M. A. Graham. 1962. *British Prosobranch Molluscs*. London: Ray Society.

Friedrich, M. and D. Tautz. 1995. Ribosomal DNA phylogeny of the major extant arthropod classes and the evolution of myriapods. *Nature* 376:165–167.

Fryer, G. 1992. The origin of the Crustacea. *Acta Zoologica* 73:273–286.

Gall, J.-C. and L. Grauvogel. 1964. Un arthropode peu connu le genre *Euthycarcinus* Handlirsch. *Annales de Paléontologie, Invertébrés* 50:3–18.

Gauthier, J., A. G. Kluge, and T. Rowe. 1988. Amniote phylogeny and the importance of fossils. *Cladistics* 4:105–209.

Ghiselin, M. T. 1988. The origin of molluscs in the light of molecular evidence. *Oxford Surveys of Evolutionary Biology* 5:66–95.

Gould, S. J. 1989. *Wonderful Life: The Burgess Shale and the Nature of History.* New York: Norton.

Gould, S. J. 1991. The disparity of the Burgess Shale arthropod fauna and the limits of cladistic analysis: Why we must strive to quantify morphospace. *Paleobiology* 17:411–423.

Gould, S. J. 1993. How to analyze Burgess Shale disparity: A reply to Ridley. *Paleobiology* 19:522–523.

Gupta, A. P., ed. 1979. *Arthropod Phylogeny.* New York: Van Nostrand Reinhold.

Haeckel, E. 1866. *Generelle Morphologie der Organismen.* Berlin: Reimer.

Handlirsch, A. 1914. Eine interessante Crustaceenform aus der Trias der Vogesen. *Verhandlungen der Zoologischen-Botanischen Gesellschaft in Wien* 64:1–18.

Harrington, H. J. 1959. General description of Trilobita. In R. C. Moore, ed., *Treatise on Invertebrate Paleontology, Part O: Arthropoda* , 1:O38–O117. Boulder, Lawrence: Geological Society of America, University of Kansas Press.

Harvey, M. S. and A. L. Yen. 1989. *Worms to Wasps.* Oxford: Oxford University Press.

Hennig, W. 1966. *Phylogenetic Systematics.* Urbana: University of Illinois Press.

Henriksen, K. L. 1928. Critical notes on some Cambrian arthropods described by Charles D. Walcott. *Videnskabelige Meddeleser fra Dansk naturhistorisk Forening* 86:1–20.

Hessler, R. R. 1969. Cephalocarida. In R. C. Moore, ed., *Treatise on Invertebrate Paleontology, Part R: Arthropoda,* 4 (1): R120–R128. Boulder, Lawrence: Geological Society of America, University of Kansas Press.

Hessler, R. R. and W. A. Newman. 1975. A trilobitomorph origin for the Crustacea. *Fossils and Strata* 4:437–459.

Hessler, R. R. and H. L. Sanders. 1973. Two new species of *Sandersiella,* including one from the deep sea. *Crustaceana* 13:181–196.

Heyler, D. 1981. Un très riche gisement fossilifère dans la Carbonifère supérieure de Montceau-les Mines. *Compte Rendu Hebdomadaire des Séances de l'Academie des Sciences,* series D, 292:169–171.

Hillis, D. M. 1987. Molecular versus morphological approaches to systematics. *Annual Reviews of Ecology and Systematics* 18:23–42.

Hoffman, R. L. 1969. Myriapoda, exclusive of Hexapoda. In R. C. Moore, ed., *Treatise on Invertebrate Paleontology, Part R: Arthropoda,* 4 (1): R572–R606. Boulder, Lawrence: Geological Society of America, University of Kansas Press.

Hou, X.-G. and J. Bergström. 1995. Cambrian lobopodians: Ancestors of extant onychophorans? Zoological *Journal of the Linnean Society* 114:3–19.

Hughes, C. P. 1975. Redescription of *Burgessia bella* from the Middle Cambrian Burgess Shale, British Columbia. *Fossils and Strata* 4:415–464.

Hutchinson, G. E. 1930. Restudy of some Burgess Shale fossils. *Proceedings of the United States National Museum* 78:1–11.

Iwanoff, P. P. 1928. Die Entwicklung der Larvalsegmente bei den Anneliden. *Zeitschrift für Morphologie und Ökologie der Tiere* 10:62–161.

Jones, C. J., J. Nolte, and J. E. Brown. 1971. The anatomy of the median ocellus of *Limulus. Zeitschrift für Zellforschung und Mikroskopische Anatomie* 118:297–309.

Kaestner, A. 1970. *Invertebrate Zoology.* Vol. 2. New York: Wiley.

Kinchin, I. M. 1994. *The Biology of Tardigrades.* London: Portland Press.

Kluge, A. 1989. A concern for evidence and a phylogenetic hypothesis of relationships among *Epicrates* Boidae, Serpentes. *Systematic Zoology* 38:7–25.

Kozloff, E. N. 1990. *Invertebrates*. Philadelphia: Saunders.

Kristensen, R. M. 1978. On the structure of *Batillipes nøerrevangi* Kristensen, 1978, part 2: The muscle attachments and true cross-striated muscles. *Zoologischer Anzeiger* 200: 173–184.

Kunze, P. 1979. Apposition and superposition eyes. In H. Autrum, ed., *Comparative Physiology and Evolution of Vision in Invertebrates, A: Invertebrate Photoreceptors*. Vol. 7/6A of *Handbook of Sensory Physiology*. Berlin, New York: Springer-Verlag.

Lake, J. A. 1987. A rate independent technique for analysis of nucleic acid sequences: Evolutionary parsimony. *Molecular Biology and Evolution* 4:167–191.

Lake, J. A. 1990. Origin of the Metazoa. *Proceedings of the National Academy of Science, USA* 87:763–766.

Lauterbach, K.-E. 1983. Zum Problem der Monophylie der Crustacea. *Abhandlungen des naturwissenschaftlichen Vereins in Hamburg (NF)* 26:293–320.

Lee, M. S. Y. 1992. Cambrian and Recent morphological disparity. *Science* 258:1816–1817.

Lehmann, W. M. 1955. *Vachonisia rogeri*: Ein Branchiopod aus dem unterdevonischen Hunsrückschiefer. *Paläontologische Zeitschrift* 29:126–130.

Lemche, H. 1959a. Molluscan phylogeny in the light of *Neopilina*. *Proceedings of the 15th International Congress on Zoology*, 380–381.

Lemche, H. 1959b. Protostomian relationships in the light of *Neopilina*. *Proceedings of the 15th International Congress on Zoology*, 381–389.

Lewis, J. G. E. 1981. *The Biology of Centipedes*. New York: Cambridge University Press.

Linder, F. 1946. Affinities within the Branchiopoda, with notes on some dubious fossils. *Arkiv Zoologie* 34:1–28.

Marcus, E. 1929. Tardigrada. In H. G. Bronn, ed., *Klassen und Ordnungen des Tierreichs*, 5:1–608. Leipzig: Akademische Verlagsgesellschaft.

Manton, S. M. 1977. *The Arthropoda: Habits, Functional Morphology, and Evolution*. Oxford: Clarendon Press.

McLaughlin, P. A. 1980. *Comparative Morphology of Recent Crustacea*. San Francisco: Freeman.

McNamara, K. J. and N. H. Trewin. 1993. A euthycarcinoid arthropod from the Silurian of Western Australia. *Palaeontology* 36:319–335.

McShea, D. W. 1993. Arguments, tests, and the Burgess Shale: A commentary on the debate. *Paleobiology* 19:399–403.

Mettam, C. 1967. Segmental musculature and parapodial movement of *Nereis diversicolor* and *Nephthys hombergi*. *Journal of Zoology* 153:245–275.

Millecchia, R, J. Bradbury, and M. Mauro. 1966. Simple photoreceptors of *Limulus polyphemus*. *Science* 154:1199–1201.

Miyamoto, M. M. 1985. Consensus cladograms and general classifications. *Cladistics* 1:186–189.

Monge-Nájera, J. 1995. Phylogeny, biogeography, and reproductive trends in the Onychophora. *Zoological Journal of the Linnean Society* 114:21–60.

Moore, R. C. 1959. Protarthropoda. In R. C. Moore, ed., *Treatise on Invertebrate Paleontology, Part O: Arthropoda*, 1:O16–O21. Boulder, Lawrence: Geological Society of America, University of Kansas Press.

Moore, R. C. and L. McCormack. 1969. General features of Crustacea. In R. C. Moore, ed., *Treatise on Invertebrate Paleontology, Part R: Arthropoda*, 4 (1): R57–R120. Boulder, Lawrence: Geological Society of America, University of Kansas Press.

Müller, K. J. 1983. Crustacea with preserved soft parts from the Upper Cambrian of Sweden. *Lethaia* 16:93–109.

Müller, K. J. and D. Walossek. 1985. Skaracarida: A new order of Crustacea from the Upper Cambrian of Västergötland. *Fossils and Strata* 17:1–65.

Naef, A. 1913. Studien zur generellen Morphologie der Mollusken, part 2: Das Cölomsystem in seinen topographischen Beziehungen. *Ergebnisse Zoologica* 3:329–462.

Naef, A. 1924. Studien zur generellen Morphologie der Mollusken, part 3: Die typischen Beziehungen der Weichtierklassen untereinander und das Verhältnis ihrer Urformen zu anderen Cölomaten. *Ergebnisse Zoologica* 6:28–124.

Nelson, G. J. 1978. Ontogeny, phylogeny, paleontology, and the biogenetic law. *Systematic Zoology* 27:324–345.

Nielsen, C. 1995. *Animal evolution: Interrelationships of the Living Phyla.* Oxford: Oxford University Press.

Nixon, K. C. and Q. D. Wheeler. 1992. Extinction and the origin of species. In M. J. Novacek and Q. D. Wheeler, eds., *Extinction and Phylogeny,* pp. 119–143. New York: Columbia University Press.

Patterson, C. 1981. Significance of fossils in determining evolutionary relationships. *Annual Reviews of Ecology and Systematics* 12:195–223.

Paulus, H. F. 1979. Eye structure and the monophyly of the Arthropoda. In A. P. Gupta, ed., *Arthropod Phylogeny*, pp. 299–383. New York: Van Nostrand Reinhold.

Pelseneer, P. 1906. Mollusca. In E. R. Lankester, ed., *A Treatise on Zoology,* 5:1–354. London: Adam and Charles Black.

Platnick N. I., C. E. Griswald, and J. A. Coddington. 1991. On missing entries in cladistic analysis. *Cladistics* 7:337–343.

Raasch, G. O. 1939. Cambrian Merostomata. *Special Papers of the Geological Society of America* 19:1–146.

Raff, R. A., K. G. Field, G. J. Olsen, S. J. Giovannoni, D. L. Lane, M. T. Ghiselin, N. R. Pace, and E. C. Raff. 1989. Metazoan phylogeny based on analysis of 18S ribosomal RNA. In B. Fernholm, K. Bremer, and H. Jörnvall, eds., *The Hierarchy of Life,* pp. 247–260. Amsterdam: Elsevier.

Ramsköld, L. 1992a. The second leg row of *Hallucigenia* discovered. *Lethaia* 25:221–224.

Ramsköld, L. 1992b. Homologies in Cambrian Onychophora. *Lethaia* 25:443–460.

Ramsköld, L. and G. D. Edgecombe. 1991. Trilobite monophyly revisited. *Historical Biology* 4:267–283.

Ramsköld, L. and X.-G. Hou. 1991. New Early Cambrian animal and onychophoran affinities of enigmatic metazoans. *Nature* 351:225–228.

Raymond, P. E. 1920. The appendages, anatomy, and relationships of trilobites. *Memoirs of the Connecticut Academy of Arts and Sciences* 7:1–169.

Raymond, P. E. 1935. *Leanchoilia* and other mid-Cambrian Arthropoda. *Bulletin of the Museum of Comparative Zoology, Harvard* 76:205–230.

Richter, R. 1932. Crustacea. In *Handwörterbuch der Naturwissenschaften.* 2d ed. Jena: Fischer.

Ridley, M. 1993. Analysis of the Burgess Shale. *Paleobiology* 19:519–522.

Riek, E. F. 1964. Merostomoidea Arthropoda, Trilobitomorpha from the Australian Middle Triassic. *Records of the Australian Museum* 26:327–332.

Riek, E. F. 1968. Re-examination of two arthropod species from the Triassic of Brookvale, New South Wales. *Records of the Australian Museum* 27:313–332.

Riggin, G. T. 1962. Tardigrada from southwest Virginia: With the addition of a new marine species from Florida. *Virginia Agriculture Experiment Station Technical Bulletin* 152:1–145.

Robison, R. A. 1985. Affinities of *Aysheaia* (Onychophora), with description of a new Cambrian species. *Journal of Paleontology* 59:226–235.

Robison, R. A. 1990. Earliest-known uniramous arthropod. *Nature* 343:163–164.

Robison. R. A. and E. O. Wiley. 1995. A new arthropod, *Meristosoma*: More fallout from the Cambrian explosion. *Journal of Paleontology* 69:447–459.

Rolfe, W. D. I. 1969. Phyllocarida. In R. C. Moore, ed., *Treatise on Invertebrate Paleontology, Part R: Arthropoda,* 4:R296–R331. Boulder, Lawrence: Geological Society of America, University of Kansas Press.

Rolfe, W. D. I. 1985. Les Euthycarcinoides des Montceau et Mazon Creek. *Bulletin de la Société d'Histoire Naturelle d'Autun* 115:71–72.

Rolfe, W. D. I., F. R. Schram, G. Pacaud, D. Sotty, and S. Secretan. 1982. A remarkable Stephanian biota from Montceau-les-Mines, France. *Journal of Paleontology* 56:426–428.

Romano, M., W. T. Chang, W. T. Dean, G. D. Edgecombe, R. A. Fortey, D. J. Holloway, P. D. Lane, A. W. Owen, R. M. Owens, A. R. Palmer, A. W. A. Rushton, J. H. Shergold, D. J. Siveter, and M. A. Whyte. 1993. Arthropoda: Trilobita. In M. J. Benton, ed., *The Fossil Record 2*, pp. 279–296. London: Chapman and Hall.

Ross, A. J. and D. E. G. Briggs. 1993. Arthropoda: Euthycarcinoidea and Myriapoda. In M. J. Benton, ed., *The Fossil Record* 2, pp. 357–361. London: Chapman and Hall.

Ruhberg, H. 1985. Die Peripatopsidae Onychophora: Systematik, Ökologie, Chorologie, und phylogenetische Aspekte. *Zoologica* 137:1–184.

Schaefer, C. W. and R. A. B. Leschen. 1993. *Functional Morphology of Insect Feeding.* Lanham, Maryland: Entomological Society of America.

Scheuring, L. 1913. Die Augen der Arachnoideen, part 1. *Zoologische Jahrbücher, Abteilung für Anatomie* 33:335–636.

Schram, F. R. 1971. A strange arthropod from the Mazon Creek of Illinois and the Trans-Permo-Triassic Merostomoidea (Trilobitoidea). *Fieldiana, Geology* 20:85–102.

Schram, F. R. 1978. Arthropods: A convergent phenomenon. *Fieldiana, Geology* 39:61–108.

Schram, F. R. 1982. The fossil record and evolution of the Crustacea. In L. G. Abele, ed., *Biology of Crustacea.* Vol. 1, *Systematics, the Fossil Record, and Biogeography,* pp. 93–147. New York: Academic Press.

Schram, F. R. 1986. *Crustacea.* New York: Oxford University Press.

Schram, F. R. and M. J. Emerson. 1991. Arthropod Pattern Theory: A new approach to arthropod phylogeny. *Memoirs of the Queensland Museum* 31:1–18.

Schram, F. R. and W. D. I. Rolfe. 1982. New euthycarcinoid arthropods from the Upper Pennsylvanian of France and Illinois. *Journal of Paleontology* 56:1434–1450.

Secretan, S. 1980. Comparaison entre des Crustacés à céphalon isolé, à propos d'un beau matérial de Syncarides du Paléozoique, implications phylogéniques. *Geobios* 13:411–433.

Sharov, A. G. 1966. *Basic Arthropodan Stock, with Special Reference to Insects.* Oxford: Pergamon Press.

Shergold, J. H. 1991. Protaspid and early meraspid growth stages of the eodiscoid trilobite *Pagetia ocellata* Jell, and their implications for classification. *Alcheringa* 15:65–86.

Shultz, J. W. 1990. Evolutionary morphology and phylogeny of Arachnida. *Cladistics* 6:1–38.

Simonetta, A. M. 1970. Studies of non-trilobite arthropods of the Burgess Shale Middle Cambrian. *Palaeontographica Italica* 66:35–45.

Simonetta, A. M. 1976. Remarks on the origin of the Arthropoda. *Atti della Società Toscana di Scienze Naturali,* memoir series B, 82:112–134.

Simonetta, A. M. and L. Delle Cave. 1975. The Cambrian non-trilobite arthropods from the Burgess Shale of British Columbia: A study of their comparative morphology, taxonomy, and evolutionary significance. *Palaeontographia Italica* 69:1–37.

Simonetta, A. M. and L. Delle Cave. 1980. The phylogeny of the Palaeozoic arthropods. *Bolletino di Zoologica* 47:1–19.

Sneath, P. H. A. and R. R. Sokal. 1973. *Numerical Taxonomy: The Principles and Practice of Numerical Classification.* San Francisco: Freeman.

Snodgrass, R. E. 1938. The evolution of the Annelida, Onychophora, and Arthropoda. *Smithsonian Miscellaneous Collections* 97:1–59.

Snodgrass, R. E. 1950. Comparative studies of the jaws of mandibulate arthropods. *Smithsonian Miscellaneous Collections* 116:1–85.

Snodgrass, R. E. 1952. *A Textbook of Arthropod Anatomy.* Ithaca: Cornell University Press.

Snodgrass, R. E. 1958. Evolution of arthropod mechanisms. *Smithsonian Miscellaneous Collections* 138:1–77.

Starobogatov, Y. I. 1988. O sisteme evikartsinid Arthropoda Trilobitoidees. *Byulletin Geologii* 3:65–74.

Størmer, L. 1944. On the relationships and phylogeny of fossil and Recent Arachnomorpha: A comparative study on Arachnida, Xiphosura, Eurypterida, Trilobita, and other fossil Arthropoda. *Skrifter Utgitt av det Norske Videnskaps-Academi i Oslo, I. Matematisk-Naturvidenskapelig. Klasse* 5:1–158.

Størmer, L. 1955. Chelicerata, Merostomata. In R. C. Moore, ed., *Treatise on Invertebrate Paleontology, Part P: Arthropoda,* 2:P1–P41. Boulder, Lawrence: Geological Society of America, University of Kansas Press.

Stürmer, W. and J. Bergström. 1976. The arthropods *Mimetaster* and *Vachonisia* from the Devonian Hunsrück Shale. *Paläontologische Zeitschrift* 50:78–111.

Stürmer, W. and J. Bergström. 1978. The arthropod *Cheloniellon* from the Devonian Hunsrück Shale. *Paläontologische Zeitschrift* 52:57–81.

Swofford, D. L. 1990. PAUP: Phylogenetic Analysis Using Parsimony, version 3.0: Manual and computer program distributed by the Illinois Natural History Survey.

Tasch, P. 1969. Branchiopoda. In R. C. Moore, ed., *Treatise on Invertebrate Paleontology, Part R: Arthropoda,* 4 (1): R128–R191. Boulder, Lawrence: Geological Society of America, University of Kansas Press.

Tiegs, O. W. 1947. The development and affinities of the Pauropoda, based on a study of *Pauropus sylvaticus. Quarterly Journal of the Microscopical Society* 88:165–336.

Tiegs, O. W. and S. M. Manton. 1958. The evolution of the Arthropoda. *Biological Reviews of the Cambridge Philosophical Society* 33:255–337.

Turbeville, J. M., D. M. Pfeifer, K. G. Field, and R. A. Raff. 1991. The phylogenetic status of arthropods, as inferred from 18S RNA sequences. *Journal of Molecular Evolution* 8:669–686.

Vandel, A. 1949. Embranchement des arthropodes: Généralités, composition de l'embranchement. In P.-P. Grassé, ed., *Traité de Zoologie,* 6:79–158. Paris: Masson.

Wägele, J. W. 1993. Rejection of the "Uniramia" hypothesis and implications of the Mandibulata concept. *Zoologische Jahrbücher, Abteilung für Systematik, Ökologie, und Geographie der Tiere* 120:253–228.

Walcott, C. D. 1908. Mount Stephen rocks and fossils. *Canadian Alpine Journal* 1:232–248.

Walcott, C. D. 1912. Middle Cambrian Branchiopoda, Malacostraca, Trilobita, and Merostomata: Cambrian geology and paleontology, part 2. *Smithsonian Miscellaneous Collections* 57:109–144.

Walcott, C. D. 1931. Addenda to description of Burgess Shale fossils. *Smithsonian Miscellaneous Collections* 85:1–46.

Walossek, D. and H. Szaniawski. 1991. *Cambrocaris baltica* n. gen. n. sp.: A possible stem-lineage crustacean from the Upper Cambrian of Poland. *Lethaia* 24:363–378.

Weygoldt, P. 1979. Significance of later embryonic stages and head development in arthropod phylogeny. In A. P. Gupta, ed., *Arthropod Phylogeny,* pp. 107–135. New York: Van Nostrand Reinhold.

Weygoldt, P., and H. F. Paulus. 1979. Untersuchungen zur Morphologie, Taxonomie, und Phylogenie der Chelicerata, part 2: Cladogramme und die Entfaltung der Chelicerata. *Zeitschrift für zoologische Systematik und Evolutionsforschung* 17:117–200.

Wheeler, W. C., P. Cartwright, and C. Y. Hayashi. 1993. Arthropod phylogeny: A combined approach. *Cladistics* 9:1–39.

Whittington, H. B. 1971. Redescription of *Marrella splendens* Trilobitoidea from the Burgess Shale, Middle Cambrian, British Columbia. *Geological Survey of Canada Bulletin* 209:1–24.

Whittington, H. B. 1974. *Yohoia* Walcott and *Plenocaris* n. gen.: Arthropods from the Burgess Shale, Middle Cambrian, British Columbia. *Geological Survey of Canada Bulletin* 231:1–21.

Whittington, H. B. 1975. The enigmatic animal *Opabinia regalis*, Middle Cambrian, Burgess Shale, British Columbia. *Philosophical Transactions of the Royal Society of London,* series B, 271:1–43.

Whittington, H. B. 1978. The lobopod animal *Aysheaia pedunculat*a Walcott, Middle Cambrian, Burgess Shale, British Columbia. *Philosophical Transactions of the Royal Society of London,* series B, 284:165–197.

Whittington, H. B. 1981. Rare arthropods from the Burgess Shale. *Philosophical Transactions of the Royal Society of London,* series B, 292:329–357.

Whittington, H. B. 1985. *The Burgess Shale.* New Haven: Yale University Press.

Whittington, H. B. 1988. Hypostomes and ventral cephalic sutures in Cambrian trilobites. *Palaeontology* 31:577–610.

Whittington, H. B. 1992. *Trilobites.* Woodbridge: Boydell Press.

Whittington, H. B. and D. E. G. Briggs. 1985. The largest Cambrian animal, *Anomalocaris*, Burgess Shale, British Columbia. *Philosophical Transactions of the Royal Society of London,* series B, 309:569–609.

Wilkinson, M. 1992. Consensus, Compatibility, and Missing Data in Phylogenetic Inference. Ph.D. diss., University of Bristol.

Wilkinson, M. 1994. Common cladistic information and its consensus representation: Reduced Adams and reduced cladistic consensus trees and profiles. *Systematic Biology* 43:343–368.

Wills, M. A., D. E. G. Briggs, and R. A. Fortey. 1994. Disparity as an evolutionary index: A comparison of Cambrian and Recent arthropods. *Paleobiology* 20:93–130.

Wills, M. A., D. E. G. Briggs, R. A. Fortey, and M. Wilkinson. 1995. The significance of fossils in understanding arthropod evolution. *Verhandlungen der deutschen zoologischen Gesellschaft* 88:203–215.

Wills, M. A. and J. J. Sepkoski, Jr. 1993. Problematica. In M. J. Benton, ed., *The Fossil Record 2,* pp. 543–554. London: Chapman and Hall.

Wilson, G. D. F. 1992. Computerized analysis of crustacean relationships. *Acta Zoologica* 73:383–389.

## APPENDIX 2.1

**Apomorphies (using accelerated and delayed character transformation) for the cladogram in figure 2.1. Numbers 1–61 refer to nodes of that cladogram.**

## ACCTRAN

1. Lateral eyes present. Single pair of post-acronal somites incorporated into the cephalon. Ventrolateral arthropod or lobopod appendages present. Cuticle of alpha chitin and protein. Resilin present. Gonads generally elongate and dorsally situated. Reduced coelom. No cilia except in photoreceptors and/or sperm. Superficial blastoderm formation.
2. Compound eyes present. First cephalic appendages of *Anomalocaris* type. Outer ramus of second cephalic appendage present. Fifteen somites in the cephalon and trunk. Outer ramus of eighth appendage present. Terminal division with characteristic three-lobed tail fan. Gut diverticulae present.
3. Inner ramus of second cephalic appendage present. Inner ramus of eighth appendage present.
4. Wrinkling of the dorsal cuticle. Mobile appendages present on terminal division.
5. Lateral eyes absent.

EUARTHROPODA and TARDIGRADA 6. Cuticle sclerotized and forming sclerites. Outer ramus of eighth trunk appendages segmented. Suppression of all circular body wall musculature. Number of nephridia reduced to four or less. No more than one, anatomically pre-oral limb-bearing segment in the larva if not the adult.

EUARTHROPODA 7. Labrum present. Lateral eyes absent. Five post-acronal somites incorporated into the cephalon. Trunk appendages strongly sclerotized and rigid or composed of a number of articulating podomeres. Inner ramus of eighth appendages not clawed. Similar intersegmental tendon system present. All muscle striated. Nephridia with sacculi. Cephalic (at least anterior) ecdysis glands present. Specialized ommatidial structures.

ATELOCERATA 8. Inner ramus of appendages of second post-acronal somite absent. Appendages of third somite modified as whole-limb jaws (mandibles). Appendages of fifth somite fused to form a labium. Nineteen somites in the cephalon and trunk. Tracheae present. Pretarsal segment of inner ramus of trunk appendages with only a single muscle. Malpighian tubules present and ectodermal.

MYRIAPODA 9. Organs of Tömösváry present. Gnathal lobe of appendage of third somite present. Appendages of fourth somite medially coalesced. Appendages of sixth appendages modified as poison fangs. Myriapod repugnatorial glands present.

10. Four post-acronal somites incorporated into the cephalon. Appendages of fourth somite fused to form a gnathochilarium. More pairs of segments than trunk divisions (duplosegments). Six podomeres in inner ramus of eighth trunk appendage. Suppression of appendages of sixth somite. Collum present.

11. No characters with this optimization.
HEXAPODA 12. Differentiation of an abdomen. Major trunk limb geniculation as joint. Two primary pigment cells in ommatidia.
13. No characters with this optimization.
14. Lateral eyes present and compound.
15. Three median ocelli present.
16. Mobile appendages on terminal trunk division.
SCHIZORAMIA 17. Cephalic shield present and entire. Outer ramus of trunk appendage present. Anterior to posterior ordering of fate map tissues as midgut-mesoderm-stomodaeum.
MARRELLOMORPHA 18. Two post-acronal somites incorporated into the cephalon. Twenty-five or more somites in the cephalon and trunk. Relative length of trunk limbs diminishing anterior to posterior. Inner rami of trunk appendages composed of five podomeres.
19. Doublure present. Outer rami of trunk appendages filamentous and organized around a longitudinal rhachis.
20. Prominent, thick, and armlike lateral extensions from the carapace.
ARACHNOMORPHA and CRUSTACEA 21. Orientation of anterior esophagus posterior to anterior. Outer ramus of second cephalic appendage present. Gnathobases present on the appendages of the third post-acronal somite.
ARACHNOMORPHA 22. Marginal rim present on carapace. Four post-acronal somites incorporated into the cephalon. Six podomeres in the inner ramus of the eighth appendage. Outer ramus of the eighth appendage segmented. Anus located ventrally in the penultimate division of the trunk. Terminal division of the trunk styliform. Gut diverticulae present. Variable and high number of ommatidial cells. Embryos with four gangliar post-oral segments.
23. Post-cephalic articulation with overlapping pleurae. Dorsal cuticle trilobed. Doublure present. Posterior trunk tergites strongly curved or semicircular in dorsal aspect. Inner ramus of eighth trunk appendages spinose. Filaments present on the outer rami of the trunk appendages. Gnathobases present on trunk appendages.
24. No characters with this optimization.
TRILOBITA 25. Pygidium present. Posterior trunk tergites straight in dorsal aspect. Nine cephalic and trunk somites. Anus ventral in the terminal division of the body. Terminal division of the body rounded.
26. Ecdysial sutures visible and marginal. Genal spines present. Labrum natant. Outer ramus of trunk appendages organized around a longitudinal rhachis.
27. Ecdysial sutures visible and dorsal. Eye ridges present. Articulating half-rings present. Posterior trunk tergites strongly curved or semicircular in dorsal aspect. Four median eyes present. Lateral eyes present and compound. Differentiation of trunk tagmata.
28. Marginal rim of carapace absent. Outer rami of second cephalic appendages absent. Fifteen cephalic and trunk somites. Gut diverticulae absent.
29. Lateral eyes present and compound. Four podomeres in inner rami of trunk appendages.
30. Cuticle tuberculate. Cuticle trilobed. Cardiac lobe present. Inner rami of trunk appendages spinose.
31. Labrum natant. Six post-acronal somites incorporated into the cephalon. Gnathobases absent from the trunk appendages.
CHELICERATA 32. Ecdysial sutures visible and marginal. Posterior trunk tergites straight or slightly curved in dorsal aspect. Appendages of first post-acronal somite chelicerae. Fourteen cephalic and trunk somites. Inner rami of trunk appendages absent.

33. Cuticle smooth. Post-cephalic articulation without overlapping pleurae. Marginal rim present on cephalic shield. Two median eyes present. Appendages of second post-acronal somite pedipalps. Outer rami of trunk appendages lacking filaments.
34. Dorsal cuticle not trilobed. Eighteen cephalic and trunk somites. Trunk tagmatized. Outer rami of trunk appendages absent. Endodermal Malpighian tubules present.
35. Lateral eyes free or pedunculate beneath the carapace. Outer ramus of second cephalic appendage present. Eight podomeres in inner ramus of trunk appendages. Anus ventral in the terminal division of the trunk. Terminal trunk division paddle-shaped or platelike.
36. Anterior esophagus directed anteriorly or ventrally. Appendages of first post-acronal somite "great appendages." Trunk appendages pendent beneath the body. Border of terminal trunk division flanked with fine spines.
37. Two post-acronal somites incorporated into the cephalon. Gnathobases absent from appendages of cephalon. Gnathobases absent from appendages of trunk. Terminal division of trunk styliform.
38. Doublure absent. Lateral eyes absent. Compound eyes absent. Three post-acronal somites incorporated into the cephalon.
39. Margins of posterior trunk tergites straight or slightly curved in dorsal aspect.
40. Seventeen cephalic and trunk somites. Post-cephalic tagmatization. Outer rami of trunk appendages lacking filaments.
CRUSTACEA 41. Lateral eyes free or pedunculate beneath the carapace. Appendage of second post-acronal somite differentiated from those that follow as an antenna. Appendage of third post-acronal somite differentiated from those that follow as a gnathobasic mandible. Outer rami of trunk appendages organized around a simple, longitudinal rhachis. Mobile appendages present on the terminal trunk division. Nauplius or egg nauplius present in development.
42. Differentiation of trunk somites into tagmata. Trunk appendages pendent or located ventrally beneath the body. One podomere in inner rami of trunk appendages.
43. Appendages of second post-acronal somite undifferentiated from those that follow. Outer rami of trunk appendages absent.
44. Inner rami of trunk appendages absent. "Orsten"-type appendages.
45. Labrum natant. Three median eyes. Fifteen cephalic and trunk somites.
46. Appendages of second post-acronal somite modified as antennae. Outer rami of trunk appendages present.
47. Outer rami of appendages of second post-acronal somite absent. Mobile appendages absent from telson.
48. Twenty-two somites in cephalon and trunk. Outer rami of trunk appendages unsegmented. Protopodal endites present on trunk appendages.
49. Compound eyes present. Twenty-four somites in cephalon and trunk. Carapace adductor muscles present. Gut diverticulae present.
50. Lateral eyes present.
51. Four podomeres in inner rami of trunk appendages.
52. Outer rami of appendages of second post-acronal somite absent. Outer rami of trunk appendages organized around a flattened lamellate or lobate region.
53. Three median ocelli. Lateral eyes on dorsal cuticle of cephalon. First appendages vestigial antennae. Trunk appendages bearing epipodites.
54. Cephalic shield bivalved. Labrum natant. Twenty somites in cephalon and trunk. Five podomeres in inner rami of trunk appendages. Gut diverticulae absent.

55. Carapace all-enveloping.
56. Posterior margin of carapace with a sharp, V-shaped invagination. First appendages vestigial antennules.
57. Mobile appendages absent from terminal trunk division.
58. Outer rami of second cephalic appendages present. Gnathobases absent from trunk appendages.
59. Anterior margin of carapace produced into an articulating rostral spine. First appendages biramous. Anus ventral in the terminal division of the trunk.
60. Post-cephalic articulation with overlapping pleurae. Trunk appendages bearing epipodites.
61. Cephalic shield entire. Carapace not all-enveloping. Cephalic doublure present. Well-developed epistome present. Three median eyes present. Outer rami of appendages of second post-acronal somite developed as a scaphocerite. Nineteen cephalic and trunk somites. Outer rami of trunk appendages absent. Mobile appendages absent from the terminal trunk division. Uropods forming a tail fan with the telson. Carapace adductor muscles absent.

## DELTRAN

1. Single pair of appendages incorporated into the cephalon. Ventrolateral arthropod or lobopod appendages present.
2. Lateral eyes present. Compound eyes present. Outer rami of second appendages present. Outer ramus of eighth appendage present. Terminal division with characteristic three-lobed tail fan.
3. Inner ramus of second cephalic appendage present. Inner ramus of eighth appendage present. Cuticle of alpha chitin and protein. Resilin present. Gonads generally elongate and dorsally situated. Superficial blastoderm formation.
LOBOPODIA 4. Wrinkling of the dorsal cuticle. Mobile appendages present on terminal division.
5. No characters with this optimization.
EUARTHROPODA and TARDIGRADA 6. Cuticle sclerotized and forming sclerites. Suppression of all circular body wall musculature. Number of nephridia reduced to four or less. No more than one anatomically pre-oral limb-bearing segment in the larva if not the adult.
EUARTHROPODA 7. Labrum present. Five post-acronal somites incorporated into the cephalon. Trunk appendages strongly sclerotized and rigid or composed of a number of articulating podomeres. Inner ramus of eighth appendage without terminal claws. Similar intersegmental tendon system present. All muscle striated. Nephridia with sacculi. Cephalic (at least anterior) ecdysis glands present. Reduced coelom. No cilia except in photoreceptors and/or sperm.
ATELOCERATA 8. Inner ramus of appendages of second somite absent. Appendages of third somite modified as whole-limb jaws (mandibles). Appendages of fifth somite fused to form a labium. Nineteen somites in the cephalon and trunk. Tracheae present. Pretarsal segment of inner ramus of trunk appendages with a single muscle. Malpighian tubules ectodermal. Specialized ommatidial structures (two corneagen cells [usually producing a cone with four parts], and a retinula with eight cells [one or two of which may be modified as short basal cells]).
MYRIAPODA 9. Organs of Tömösváry present. Gnathal lobe present on appendages of third somite. Myriapod repugnatorial glands.

10. Four post-acronal somites incorporated into the cephalon. Appendages of fourth somite fused to form a gnathochilarium. Trunk segmentation distinct, with more pairs of appendages than somites (duplosegments). Suppression of the appendages of the sixth somite. Collum.
11. Appendages of fourth post-acronal somite medially coalesced. Appendages of sixth somite modified as poison fangs.
HEXAPODA 12. Trunk somites differentiated into two or more tagmata.
13. Major trunk limb geniculation as joint.
14. Lateral eyes present and compound.
15. Three median ocelli. Major trunk limb geniculation as joint. Two primary pigment cells in ommatidia.
16. Mobile appendages present on terminal trunk division.
SCHIZORAMIA 17. Cephalic shield present and entire. Outer ramus of trunk appendage present.
MARRELLOMORPHA 18. Two somites incorporated into the cephalon. Twenty-five or more somites in the cephalon and trunk. Relative length of trunk limbs diminishing anterior to posterior. Outer rami of trunk appendages segmented.
19. Inner rami of trunk appendages composed of five podomeres. Outer rami of trunk appendages filamentous and organized around a longitudinal rhachis.
20. Prominent, thick, and armlike lateral extensions from the carapace.
ARACHNOMORPHA and CRUSTACEA 21. Orientation of anterior esophagus posterior to anterior. Outer ramus of second cephalic appendage present. Anterior to posterior ordering of fate map tissues as midgut-mesoderm-stomodaeum.
ARACHNOMORPHA 22. Four post-acronal somites incorporated into the cephalon. Six podomeres in the inner ramus of the eighth appendage. Anus located ventrally in the penultimate division of the trunk. Terminal division of the trunk styliform.
23. Post-cephalic articulation with overlapping pleurae. Doublure present. Gnathobases present on the appendages of the third post-acronal somite. Inner ramus of eighth trunk appendages spinose. Filaments present on the outer rami of trunk appendages. Gnathobases present on trunk appendages. Embryo with four gangliar post-oral segments.
24. Dorsal cuticle trilobed. Marginal rim present on carapace. Gut diverticulae present.
TRILOBITA 25. Pygidium present. Anus ventral in the terminal division of the body. Terminal division of the body rounded.
26. Genal spines present. Outer ramus of trunk appendages organized around a longitudinal rhachis.
27. Ecdysial sutures visible and dorsal. Eye ridges present. Articulating half-rings present. Posterior trunk tergites strongly curved or semicircular in dorsal aspect. Four median eyes present. Lateral eyes present and compound. Differentiation of trunk tagmata.
28. Posterior trunk tergites strongly curved or semicircular in dorsal aspect.
29. Lateral eyes present and compound. Four podomeres in inner rami of trunk appendages. Fifteen cephalic and trunk somites.
30. Cuticle tuberculate. Cuticle trilobed. Outer rami of second cephalic appendages absent. Inner rami of trunk appendages not spinose. Four podomeres in inner rami of trunk appendages.
31. Labrum natant. Six post-acronal somites incorporated into the cephalon. Gnathobases absent from the trunk appendages.

CHELICERATA 32. Ecdysial sutures visible and marginal. Cardiac lobe present. Posterior trunk tergites straight or slightly curved in dorsal aspect. Appendages of first, post-acronal somite chelicerae.
33. Cuticle smooth. Post-cephalic articulation without overlapping pleurae. Two median eyes present.
34. Dorsal cuticle not trilobed. Eighteen cephalic and trunk somites. Trunk tagmatized. Outer rami of trunk appendages absent. Endodermal Malpighian tubules present.
35. Lateral eyes free or pedunculate beneath the carapace. Eight podomeres in inner ramus of trunk appendages. Anus ventral in the terminal division of the trunk.
36. Trunk appendages pendent beneath the body.
37. Dorsal cuticle trilobed. Gnathobases absent from appendages of cephalon. Gnathobases absent from appendages of trunk.
38. Doublure absent. Lateral eyes absent. Compound eyes absent. Three post-acronal somites incorporated into the cephalon. Border of terminal trunk division flanked with fine spines.
39. Margins of posterior trunk tergites straight or slightly curved in dorsal aspect. Appendages of first post-acronal somite "great appendages." Terminal trunk division paddle-shaped or platelike. Border of terminal trunk division flanked with fine spines.
40. Dorsal cuticle trilobed. Seventeen cephalic and trunk somites. Post-cephalic tagmatization. Outer rami of trunk appendages lacking filaments.
CRUSTACEA 41. Appendage of third post-acronal somite differentiated from those that follow as a gnathobasic mandible. Outer rami of trunk appendages organized around a simple, longitudinal rhachis. Mobile appendages present on the terminal trunk division.
42. Lateral eyes free or pedunculate beneath the carapace. Differentiation of trunk somites into tagmata. Trunk appendages pendent or located ventrally beneath the body. One podomere in inner rami of trunk appendages. Specialized ommatidial structures. Nauplius or egg nauplius present in development.
43. No characters with this optimization.
44. Inner rami of trunk appendages absent. Outer rami of trunk appendages absent. "Orsten"-type appendages.
45. Labrum natant.
46. Two median eyes. Appendages of second post-acronal somite modified as antennae. Outer rami of trunk appendages segmented.
47. Outer rami of appendages of second post-acronal somite absent. Mobile appendages absent from telson.
48. Appendages of second post-acronal somite modified as antennae. Protopodal endites present on trunk appendages.
49. Compound eyes present. Twenty-four somites in cephalon and trunk. Carapace adductor muscles present.
50. Lateral eyes present.
51. Four podomeres in inner rami of trunk appendages.
52. Outer rami of appendages of second post-acronal somite absent. Outer rami of trunk appendages organized around a flattened lamellate or lobate region.
53. Lateral eyes on dorsal cuticle of cephalon. First appendages vestigial antennae. Trunk appendages bearing epipodites. Gut diverticulae present.
54. Cephalic shield bivalved. Labrum natant. Twenty somites in cephalon and trunk. Five podomeres in inner rami of trunk appendages.

55. Carapace all-enveloping. Five podomeres in inner rami of trunk appendages.
56. Posterior margin of carapace with a sharp, V-shaped invagination. First appendages vestigial antennules.
57. Mobile appendages absent from terminal trunk division.
58. Outer rami of second cephalic appendages present. Gnathobases absent from trunk appendages.
59. Anterior margin of carapace produced into an articulating rostral spine. First appendages biramous.
60. Post-cephalic articulation with overlapping pleurae. Trunk appendages bearing epipodites.
61. Cephalic shield entire. Carapace not all-enveloping. Cephalic doublure present. Well-developed epistome present. Three median eyes present. Outer rami of appendages of second post-acronal somite developed as a scaphocerite. Nineteen cephalic and trunk somites. Outer rami of trunk appendages absent. Mobile appendages absent from the terminal trunk division. Uropods forming a tail fan with the telson. Carapace adductor muscles absent. Anus ventral in the terminal division of the trunk.

# CHAPTER 3

## Cambrian Lobopodians: Morphology and Phylogeny

*Lars Ramsköld and Chen Junyuan*

## Abstract

An analysis is presented of the morphology and phylogeny of Cambrian lobopodians known from soft-body preservation. New morphological information is given for Early Cambrian lobopodians from Chengjiang. *Onychodictyon* has a pair of anterior appendages and a sclerotized head shield but no antennae or sclerotized jaws. *Cardiodictyon* possesses paired, not multiple, claws. *Luolishania* lacks differentiated anterior appendages; the head and legs are described. The recently suggested presence of bivalved head shields in *Hallucigenia* and *Cardiodictyon* is refuted. The anterior end in *Xenusion* is reinterpreted. The alleged alga *Acinocricus stichus* is reinterpreted as a lobopodian with close affinity to the form popularly known as Collins's monster. New characters supporting Ramsköld's anteroposterior orientation of *Microdictyon sinicum* and *Hallucigenia sparsa* are detailed. A cladistic analysis based on nineteen characters indicates that Cambrian lobopodians form two clades that form an unresolved trichotomy with extant onychophorans. These three clades are here formalized into Alphonychophora *taxon novum*, Betonychophora *taxon novum*, and a redefined Euonychophora Hutchinson, 1930. For the names a phylogenetic, stem-based definition is used, and the taxa lack a given rank in the Linnean hierarchy. The three clades share four synapomorphies and are together taken to constitute the Onychophora. No tentative synapomorphies between tardigrades and Cambrian lobopodians are found.

Knowledge of Cambrian lobopodians has increased dramatically in the last few years. Some fossil lobopodians were already described early this century, but the group as such was unknown before 1991 (Ramsköld and Hou 1991). The Early and Middle Cambrian lobopodians known from soft-body preservation now number fifteen species (eleven named) of twelve genera (nine named). Several additional species, and at least two more genera (see below) are recognized only from their mineralized sclerites. Cambrian lobopodians have proven to constitute an important and diverse group during and shortly after the explosive radiation of Metazoa in the Early Cambrian.

Most of the Cambrian lobopodians have been discovered in either the Canadian Burgess Shale fauna or the Chinese Chengjiang fauna, the two most diverse Cambrian faunas with soft-body preservation. Two lobopodians have been described from the Middle Cambrian Burgess Shale: *Aysheaia pedunculata* Walcott, 1911 (see Whittington 1978; fig. 3.1A herein), and *Hallucigenia sparsa* (Walcott, 1911) (see Conway Morris 1977; Ramsköld 1992a,b; figs. 3.1B, 3.2A herein). Two further lobopodians not belonging to any genera thus far established have been discovered in the Burgess Shale and are awaiting description (Collins 1986, and pers. comm.). The richest lobopodian assemblage is known from the Early Cambrian Chengjiang fauna, containing six lobopodians: *Luolishania longicruris* Hou and Chen, 1989 (figs. 3.4, 3.5B herein), *Microdictyon sinicum* Chen, Hou, and Lu, 1989 (see Chen et al. 1995a; fig. 3.8 herein), *Cardiodictyon catenulum* Hou, Ramsköld, and Bergström, 1991 (figs. 3.9B, C, 3.10 herein), *Onychodictyon ferox* Hou, Ramsköld, and Bergström, 1991 (see Ramsköld and Hou 1991; figs. 3.5A, C, 3.6, 3.7 herein), *Paucipodia inermis* Chen, Zhou, and Ramsköld, 1995b (fig. 3.9D herein), and *Hallucigenia fortis* Hou and Bergström, 1995 (figs 3.2B–D, 3.3 herein).

Outside these faunas, the Middle Cambrian Wheeler Formation in Utah has yielded a second species of *Aysheaia*, *A. prolata* Robison, 1985. The proposed chlorophyte *Acinocricus stichus* Conway Morris and Robison, 1988, from the slightly older Spence Shale is reinterpreted here as another lobopodian. The Early Cambrian *Xenusion auerswalde* Pompeckj, 1927 (see Dzik and Krumbiegel 1989; fig. 3.9A herein) is known from Lower Cambrian erratics of the Kalmarsund Sandstone in Sweden. In addition, a form close to *Xenusion* occurs in the Early Cambrian Sirius Passet Formation of Greenland (Budd, pers. comm.). Specimens from the Cambrian "Orsten" faunas of Sweden that are possibly lobopodian are currently under study by Müller, Walossek, and Ramsköld.

In addition to *Microdictyon sinicum*, which has soft-body preservation, there are several *Microdictyon* species known only from isolated trunk plates. Six of the named species are considered here as valid (see Chen et al. 1995a), and there are several additional unnamed ones. Most *Microdictyon* species are of Early Cambrian age, but a few are Middle Cambrian. The genus is cosmopolitan. Another genus, *Quadratapora* Hao and Shu, 1987, is known only from isolated trunk plates found in Lower Cambrian strata in China and Siberia (Bengtson et al. 1986; Hao and Shu 1987). *Fusuconcharium* Hao and Shu, 1987, was also described based on isolated plates from the Lower Cambrian of China. Finally, a form possibly related to *Microdictyon* has been found in the Lower Cambrian of Greenland (Bengtson 1991).

After the Middle Cambrian, lobopodians disappear from the fossil record for a long time. A tentatively identified Ordovician specimen (Rhebergen 1990) turned out to be a pelmatozoan column (Rhebergen and Donovan 1994). The only fossil lobopodians between the Cambrian and the Recent come from the Upper Carboniferous, where *Ilyodes* Scudder, 1890, and *Helenodora* Thompson and Jones, 1980, are found in Illinois and possibly also in France (Rolfe et al. 1982). The morphology of *Helenodora* indicates that lobopodians had made the transition to land before the Upper Carboniferous.

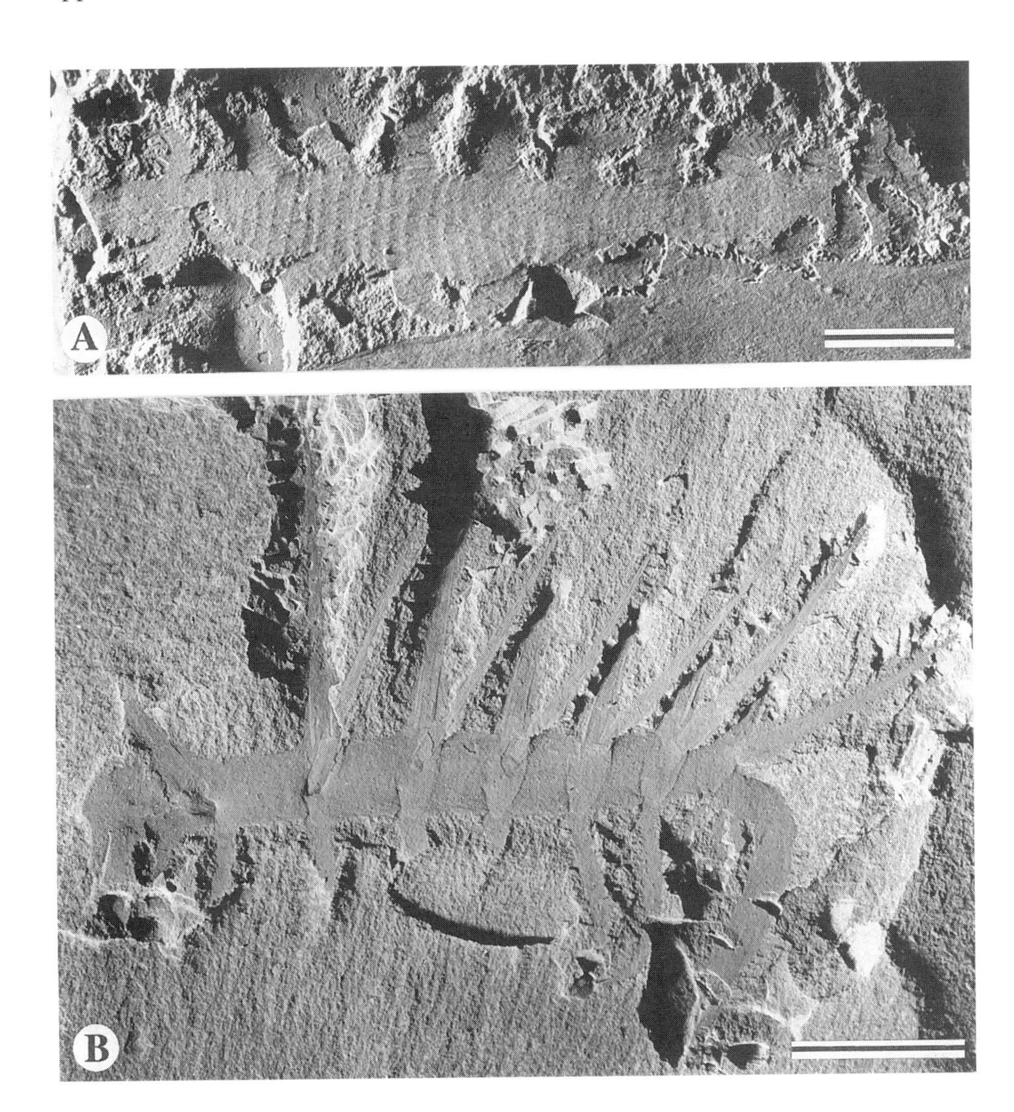


FIGURE 3.1

Lobopodians from the Middle Cambrian "Phyllopod Bed" (Stephen Formation), Burgess Quarry, British Columbia, Canada. Specimens coated with ammonium chloride. *(A)* Holotype (USNM 57655) of *Aysheaia pedunculata* Walcott, 1911, anterior to left. Low light from W to emphasize surface sculpture. Scale bar 5 mm. *(B)* Holotype (USNM 83935) of *Hallucigenia sparsa* (Walcott, 1911). For camera lucida drawing see figure 3.2A. Anterior to right, light from WSW. Scale bar 5 mm.

In total, there are over twenty fossil lobopodian species known, nearly all from the Early and Middle Cambrian. The forms display a wide range of morphologies and sizes (from 20–200 mm in adults). The affinities of the group itself have repeatedly been proposed to lie with the Onychophora, the extant velvet worms. This paper aims at a phylogenetic analysis of both the relationships within the group and between the Cambrian lobopodians and the extant Onychophora, with a discussion of the similarities to Tardigrada.

Terminology employed herein follows Chen et al. (1995a). Figured specimens are housed in the collection of the Early Life Research Centre (ELRC) at the Nanjing Institute of Geology and Palaeontology (NIGPAS), People's Republic of China; the National Museum of Natural History (USNM), Washington, D.C.; and the Museum für Naturkunde der Humboldt Universität, Berlin. Unless noted otherwise, all photographs are of dry, uncoated specimens. The localities yielding the figured Chengjiang fossils are given as code names, explained in Chen et al. (1995a).

## Previous Views on the Phylogenetic Position of Lobopodians

The two earliest discovered Cambrian animals now recognized as lobopodians, *Aysheaia pedunculata* and *Canadia* (now *Hallucigenia*) *sparsa,* were first described as polychaetes (Walcott 1911). Similarities between *Aysheaia* and Recent onychophorans were pointed out to Walcott immediately after publication of the description (Resser in Walcott 1931). Hutchinson (1930, 1969) assigned *Aysheaia* to a separate, extinct order of the Onychophora. This view was subsequently adopted in major textbooks. Delle Cave and Simonetta (1975) and Simonetta (1976) pointed out some similarities between *Aysheaia* and the Tardigrada, suggesting that *Aysheaia* was intermediate between Tardigrada and Onychophora. In his detailed revision of *Aysheaia,* Whittington (1978) supported this general view but refrained from taking a more precise phylogenetic standpoint. Robison (1985), however, questioned the synapomorphic nature of the characters forming the basis for the phylogenetic hypothesis. Later, *Xenusion* was also compared with the tardigrades, and it was suggested as ancestral both to Recent Onychophora and to Tardigrada (Dzik and Krumbiegel 1989).

*Aysheaia* and *Xenusion* were recently united with *Hallucigenia* and three forms from the Chengjiang fauna into a single group of Cambrian lobopodians (Ramsköld and Hou 1991), and a widened onychophoran concept was suggested to accommodate both fossil and Recent taxa. Ramsköld (1992b) later surveyed the then known Cambrian lobopodians in detail for homologous features and described twelve such characters. He did not, however, state the group to be monophyletic, but cautioned that Recent Onychophora could make the group paraphyletic. The reinterpretation of *Hallucigenia* as an onychophoran, and with that the general affinity of the Cambrian taxa with the Recent Onychophora, appears to be accepted (see references in Chen et al. [1995a:19]).

The detailed phylogenetic position of the fossil forms relative to extant Onychophora and more distantly related taxa has, however, remained unclear. Recently, two studies have been published that set the Cambrian lobopodians and extant Onychophora in a phylogenetic context (Hou and Bergström 1995; Monge-Nájera 1995). Their results are not upheld herein, largely because the character support for the two different phylogenies includes much morphological data shown below to be erroneous. From a cladistic perspective, both studies rely on inadequate methodology, and neither study includes a published data matrix. In Monge-Nájera's (1995) study the cladograms were stated to be constructed by hand. Unfortunately, manual construction of cladograms often produces results not supported by base data. One such problem encountered in the cladogram with the fossil taxa (Monge-Nájera 1995, fig. 2) is that the only clade (branch 7, uniting *Cardiodictyon* + *Microdictyon*) recognized within the grade group of Cambrian lobopodians lacks character support (as it is defined by "Synapomorphies yet undetermined"). Another problem, the lack of support for the crown group (Monge-Nájera 1995, fig. 2), is detailed below. The phylogeny of Hou and Bergström (1995) derived from intuitive ordering of taxa, on the basis of explicitly stated ad hoc assumptions about character evolution (ibid.: 9), and no cladistic analysis was made. The authors presented a phylogeny (Hou and Bergström 1995, fig. 7) where *Helenodora* and extant onychophorans are ingroup taxa among Cambrian marine lobopodians, yet they assigned all the Cambrian lobopodians to the Class Xenusia Dzik and Krumbiegel, 1989, the latter thus being explicitly paraphyletic. We emphasize here that rigorous cladistic procedures are essential in phylogenetic analyses, and we reject intuitive procedures as well as the validity of paraphyletic taxa.

## New Morphological Data and their Interpretation

Much new morphological information on lobopodians, mainly on the forms from the Chengjiang fauna, has come to light since the last comprehensive study of Cambrian lobopodians was published (Ramsköld 1992b). Some of this new information has been published elsewhere (Chen et al. 1995a,b; Hou and Bergström 1995) or is being prepared by the authors. Further new data, both on Chengjiang lobopodians and on others, are presented below together with a reinterpretation of several features. This information is used in the subsequent phylogenetic analysis.

### *Hallucigenia fortis* (figs. 3.2B–D, 3.3A–E)

The presence of *Hallucigenia* in the Chengjiang fauna was first reported by Chen (1991). The species was recently briefly described (Hou and Bergström 1995) from an incomplete specimen and a fragment and was named *H. fortis*. Our collection of fifteen well preserved specimens of *H. fortis* shows that the accompanying reconstruction (Hou and Bergström 1995, fig. 5C) contains several errors, and a published popular reconstruction (Gore 1993; Nash 1995) based on Hou and Bergström's data differs further from the actual morphology. Our material forms the basis for a comprehensive treatment of the species (Chen et al. in prep.), and only some important data are presented here in anticipation of that publication.

Some amendments and additions to the brief description by Hou and Bergström (1995) are necessary. The head is preserved in twelve of our fifteen specimens (e.g., fig. 3.3B, C). There is no indication in any of these of the presence of bivalved sclerites covering the head as proposed by Hou and Bergström (1995). We accordingly regard their interpretation as erroneous. Their reconstruction (Hou and Bergström 1995, fig. 5C) lacks the constriction separating the head from the trunk, and the shape of the head is incorrect. Hou and Bergström (1995) identified the two pairs of appendages anterior to the first plate pair, but their incomplete specimen led them to state that only the first pair was very long. Our material (e.g., fig. 3.3C) shows that both pairs of anterior appendages are exceedingly long and slender. Also the succeeding leg pairs are much longer than shown by Hou and Bergström (1995). The plates in their reconstruction are too small and too prominent, and the observed change in direction between the four anterior and three posterior spine pairs (fig. 3.3A) is lacking. The last plate is erroneously set right above the penultimate leg pair instead of more posteriorly, and the trunk portion behind this plate is drawn twice its actual length (cf. fig. 3.3A), creating a tail that does not exist. The comparison made by Hou and Bergström (1995) with *H. sparsa* (see figs. 3.1B, 3.2A) is confused, because their anteroposterior orientation of the latter species is regarded here as back to front (see "Anteroposterior orientation" below).

## *Luolishania* (figs. 3.4A–C, 3.5B)

*Luolishania* was described from a single specimen (Hou and Chen 1989). We have collected a second, less well preserved specimen, which confirms most of the original observations. In addition, the holotype has been further prepared, with new details revealed (figs. 3.4A, 3.5B).

There is a fairly well defined head in both specimens. The head is rounded, slightly longer than it is wide, and defined posteriorly by a slight constriction of the body. In the holotype there is a large, rounded structure that occupies most of the anterior part of the head. Anteriorly within this area is a transversely arranged pair of tubercles. These structures are only seen in the holotype, and they may simply be preservational artifacts. Although there is some possibility that they are ventral structures (if so, they are pits), their preservation indicates that they are more likely to be dorsal structures. They are therefore probably the base of some paired structures protruding from the body. Such structures can be either legs, eyes, or antennae.

Their anterior position favors one of the latter interpretations (as does the dorsal position, if correct), which if proven true, would be the only known instance of either eyes or antennae in Cambrian lobopodians. Eyes have not yet been identified in any described Cambrian lobopodians, but they may be present in an undescribed form from the Burgess Shale (Collins, pers. comm.). The structures in *Luolishania* are defined by their relief rather than by staining, which agrees better with antennae than with eyes. Antennae are not known in any Cambrian lobopodian, and previously suggested antennae in *Microdictyon* and *Onychodictyon* (Ramsköld and Hou 1991) turned out to be other structures (Ramsköld 1992b). We are reluctant to interpret the structures in *Luolishania,* if they are indeed real, as antennae.

The structure in the holotype originally tentatively identified as an anterior appendage has been prepared, and in addition the corresponding opposing appendage has been exposed (fig. 3.5B: Lg1R and Lg1L). The size of these legs is similar to that of the other legs. They are based just behind the rounded head, and there is no indication of more anteriorly set appendages; indeed, there is hardly space for such because they would have to be based on the head itself. It is therefore most likely that *Luolishania* lacked anterior appendages.

The number of legs is sixteen, one more than originally stated, as was also recognized by Hou and Bergström (1995). The leg identified as lobopod 2 on the right side by Hou and Chen (1989, fig. 2) is actually Lg3R, which is confirmed after the previously unexposed legs Lg1R, Lg2R, Lg4R, and Lg5R have now been excavated (figs. 3.4A, 3.5B).

A new reconstruction of *Luolishania longicruri*s was recently published (Hou and Bergström 1995). In it, the number of leg pairs is correctly amended to sixteen, but the legs are shown as becoming shorter anteriorly. Our preparation of the holotype has shown that the anterior legs are at least as long as those centrally in the trunk.

## *Onychodictyon* (figs. 3.5A, C, 3.6A–F, 3.7)

Ten new specimens (ELRC 33001–33010), collected by Chen and coworkers between 1990 and 1994, have provided much new information and prompted a reinterpretation of several characters.

Originally, eleven leg pairs were described (Ramsköld and Hou 1991; Hou et al. 1991) in *Onychodictyon.* An additional anterior leg revealed during subsequent preparation of the holotype was first regarded as the previously missing leg Lg1R (R1 of Ramsköld 1992b). A newly discovered, more completely preserved specimen (figs. 3.5C, 3.6A, B, E), the first to show the head region, shows the presence of a previously unknown leg pair anterior to the pair beneath the first plate pair. The leg figured by Ramsköld (1992b, fig. 4) belongs to this pair. The legs of this anteriormost pair are built similar to other legs, but appear to be based more laterally on the trunk. Both this pair and Lg1 (the one beneath P1) are markedly smaller than other

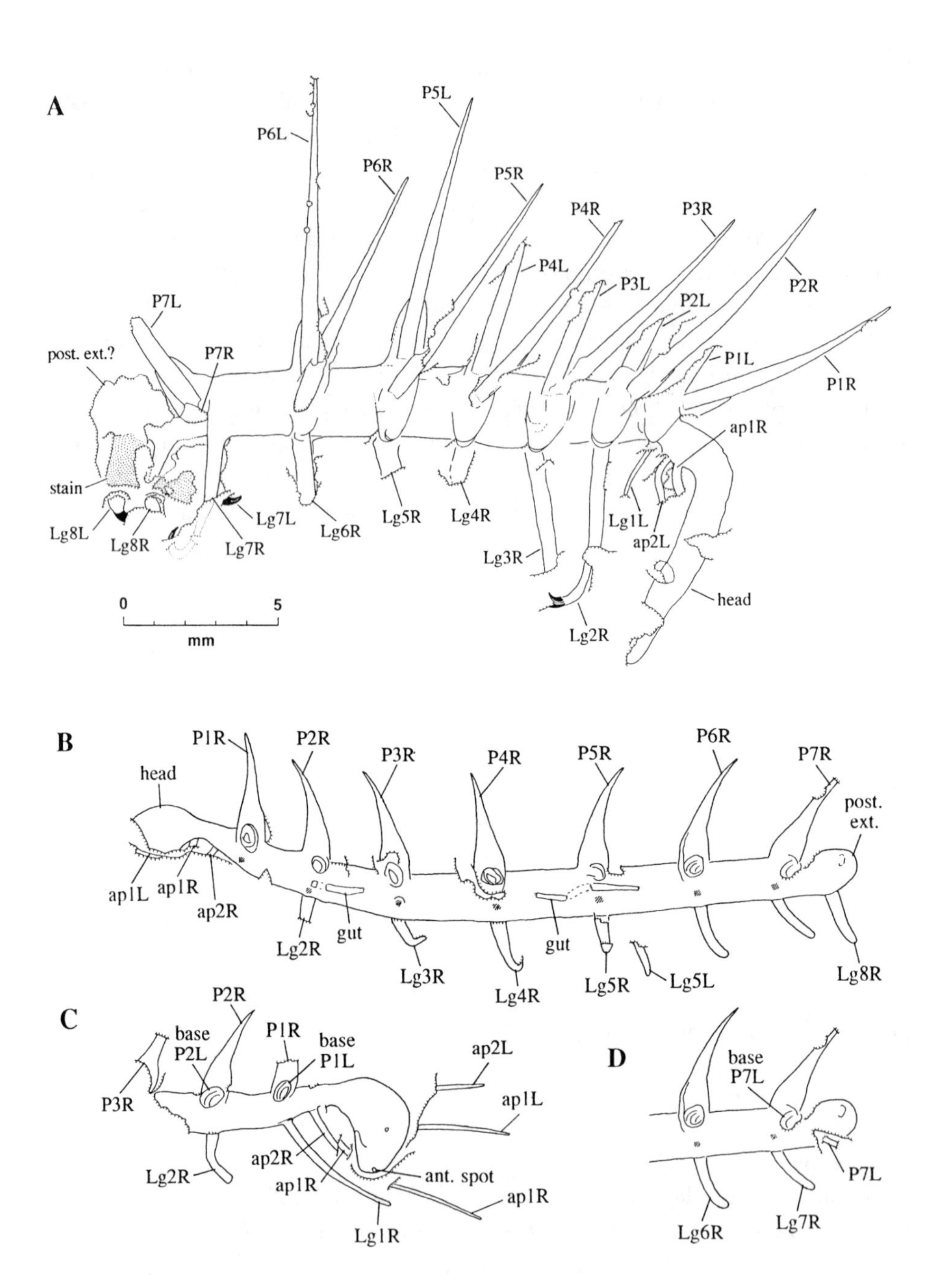


FIGURE 3.2

*(A)* Camera lucida drawing of holotype (USNM 83935) of *Hallucigenia sparsa* (Walcott, 1911). Stippled areas are parts of a patch of stain earlier considered to be the head. The posterior extension is preserved with ragged margins due to the split of the rock but may be essentially complete. Scale applies to *A–D*. *(B–D) Hallucigenia fortis* Hou and Bergström, 1995, from Chengjiang. Camera lucida drawings of ELRC 32008, from locality MN5. *(B)* Part ELRC 32008a (see fig. 3.3A). *(C)* Counterpart ELRC 32008b (see fig. 3.3B). *(D)* Posterior part before further preparation exposed Lg8R through removing the remains of P7L; note orientation of P7L relative to P7R.

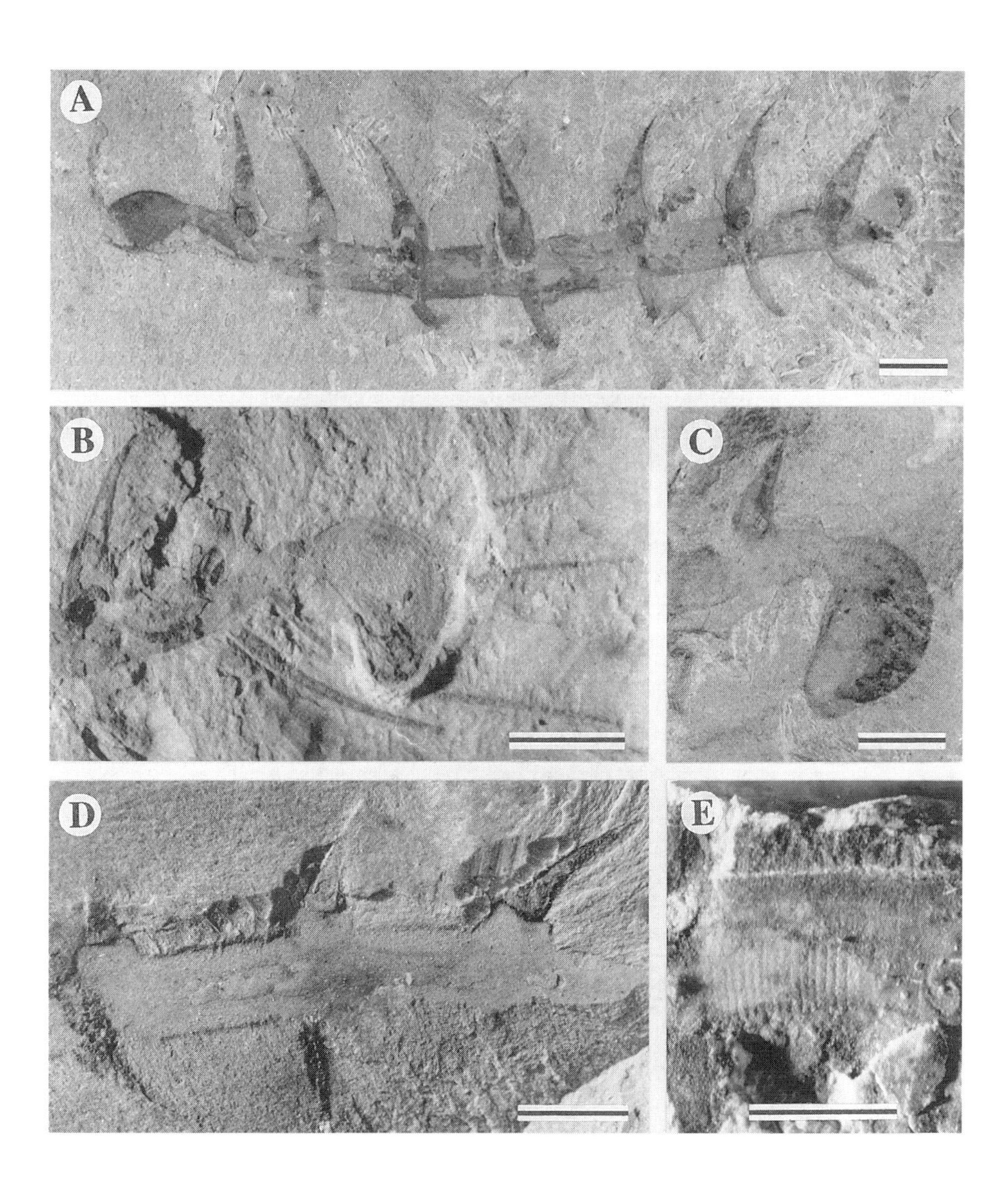


FIGURE 3.3

*Hallucigenia fortis* Hou and Bergström, 1995, from Chengjiang. All scale bars 2 mm. *(A, B)* Nearly complete individual ELRC 32008, from MN5 (for camera lucida drawings see fig. 3.2B–D). *(A)* Part ELRC 32008a. The legs are shortened due to oblique preservation. High light. *(B)* Counterpart ELRC 32008b, showing full length of anterior appendages. Light from W. *(C)* Head of ELRC 32005a from locality ME6. High light. *(D)* Central part of body (between P4 and P6) of ELRC 32015a from locality MW2. Anterior to left. Note spiral-shaped sediment infill in part of gut, and how spines protrude from dorsal bulges of trunk. Light from NW. *(E)* ELRC 32004a from locality MQ1. Anterior to left. Annulation between legs 4 and 5 (Lg4R at lower right, Lg5R at lower right; bases of Lg4L and Lg5L seen as raised areas near lower margin of trunk). Light from W.

leg pairs. Their length is about 65% that of Lg3 and subsequent legs, and the claws are of the same smaller relative size (see Ramsköld 1992b, fig. 5). Lg2 (the leg pair beneath P2) carries claws of the same (large) size as subsequent legs, but the leg appears to be shorter than Lg3 and subsequent legs both in a new specimen (fig. 3.7) and in the holotype (Ramsköld and Hou 1991, fig. 2).

The anterior appendages are set at a greater distance from Lg1 than Lg1 is set from Lg2 (as implied by leg and plate positions; fig. 3.5C). This anterior position of the anterior appendages is also present in *Aysheaia* (fig. 3.1A). This was interpreted by Whittington (1978) as indicating a leg-free somite between the anterior appendages and Lg1. Ramsköld (1992b) made a comparison with *Xenusion* and suggested that the presence in both taxa of more numerous annuli anteriorly was perhaps due to a need for increased flexibility in this area (in order to enable extensive head movements). Of the small number of reasonably complete *Onychodictyon* specimens known, at least three (the holotype; ELRC 33002, fig. 3.5A; ELRC 33004, fig. 3.7) are strongly bent in this area (but not elsewhere), which supports this functional interpretation. Whether or not this functional requirement was met

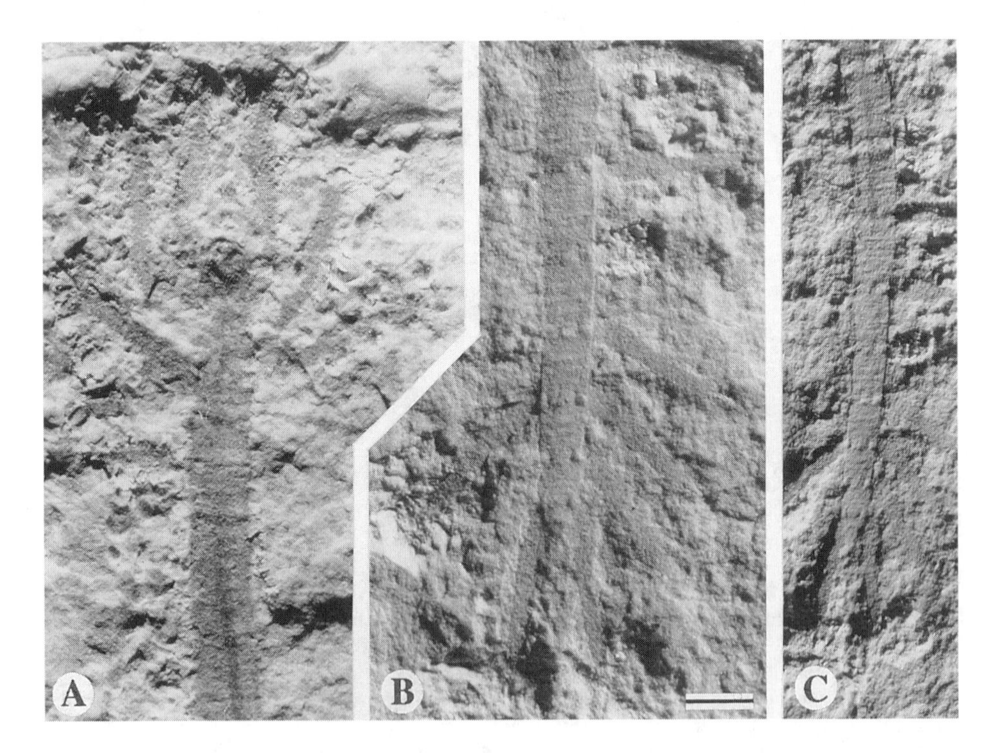


FIGURE 3.4

*Luolishania longicruris* Hou and Chen, 1989, from Chengjiang, holotype NIGPAS 108741, from locality MQ1, after preparation by L. R. revealing anterior legs (for camera lucida drawing see fig. 3.5B). *(A)* Anterior portion of part NIGPAS 108741A, ventral view. Light from N. *(B)* Posterior part of counterpart NIGPAS 108741B, dorsal view. Light from NNE. Scale bar 1 mm, applies to *A* and *B*. *(C)* Posterior part of counterpart NIGPAS 108741B, dorsal view. Low light from NW, enhancing surface relief.

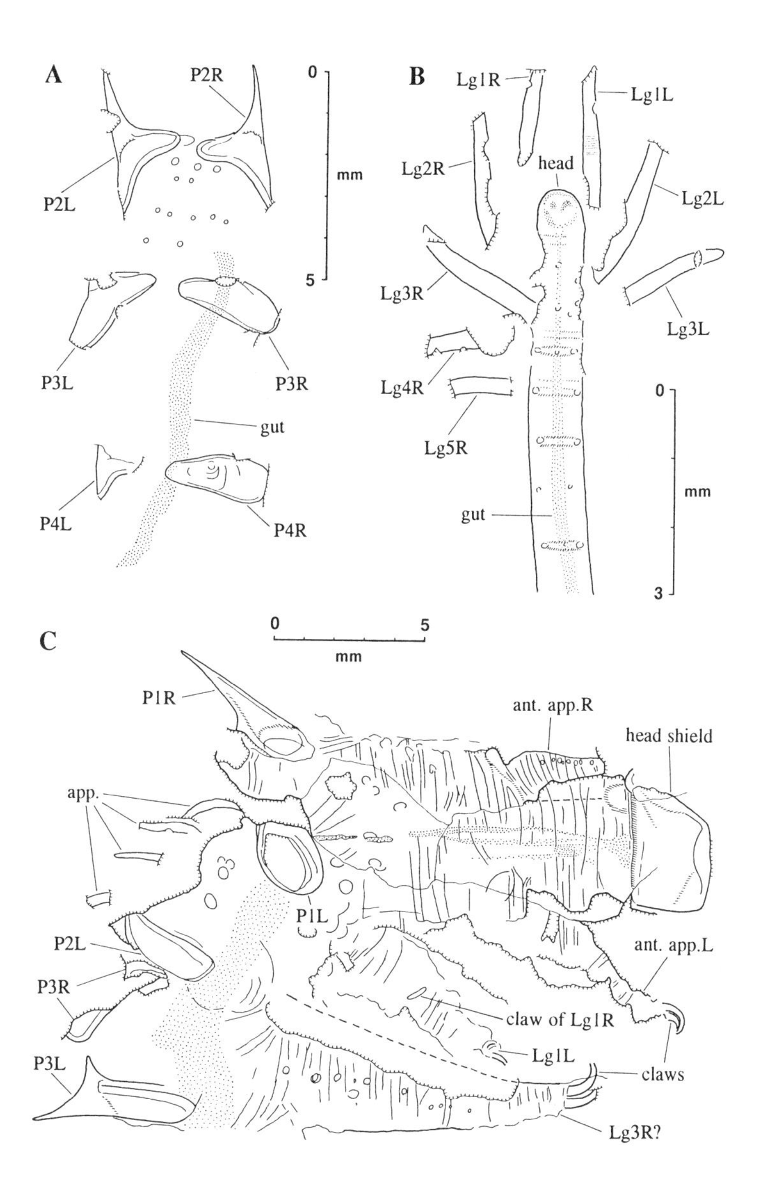


FIGURE 3.5

*(A, C) Onychodictyon ferox* Hou, Ramsköld, and Bergström, 1991, from Chengjiang. *(A)* Camera lucida drawing of anterior part of trunk of dorsoventrally compacted, nearly complete individual, ELRC 33002, from locality MN5, dorsal view (see fig. 3.6D). P2 is exposed in nearly posterior view. The head is turned down into the sediment and is not exposed. *(C)* Camera lucida drawing of anterior part of obliquely compacted, nearly complete individual, ELRC 33001a, from locality MQ1, ventrolateral view (see fig. 3.6B). Some dorsal appendicules (app.) are indicated. The margins of the right anterior appendage (ant. app.R) and the right third (?) leg (Lg3R?) are added from the counterpart as dashed lines. The position in the counterpart of one claw of Lg1R is shown, and the tips of the claws in Lg3R? and ant. app.L have been added from the counterpart (see fig. 3.6A, E). *(B) Luolishania longicruris* Hou and Chen, 1989, from Chengjiang. Camera lucida drawing of anterior part of holotype NIGPAS 108741A, from locality MQ1, after preparation by L. R. (see fig. 3.4A). Annulation of legs is indicated only in Lg1L, but is present along the entire length of all legs.

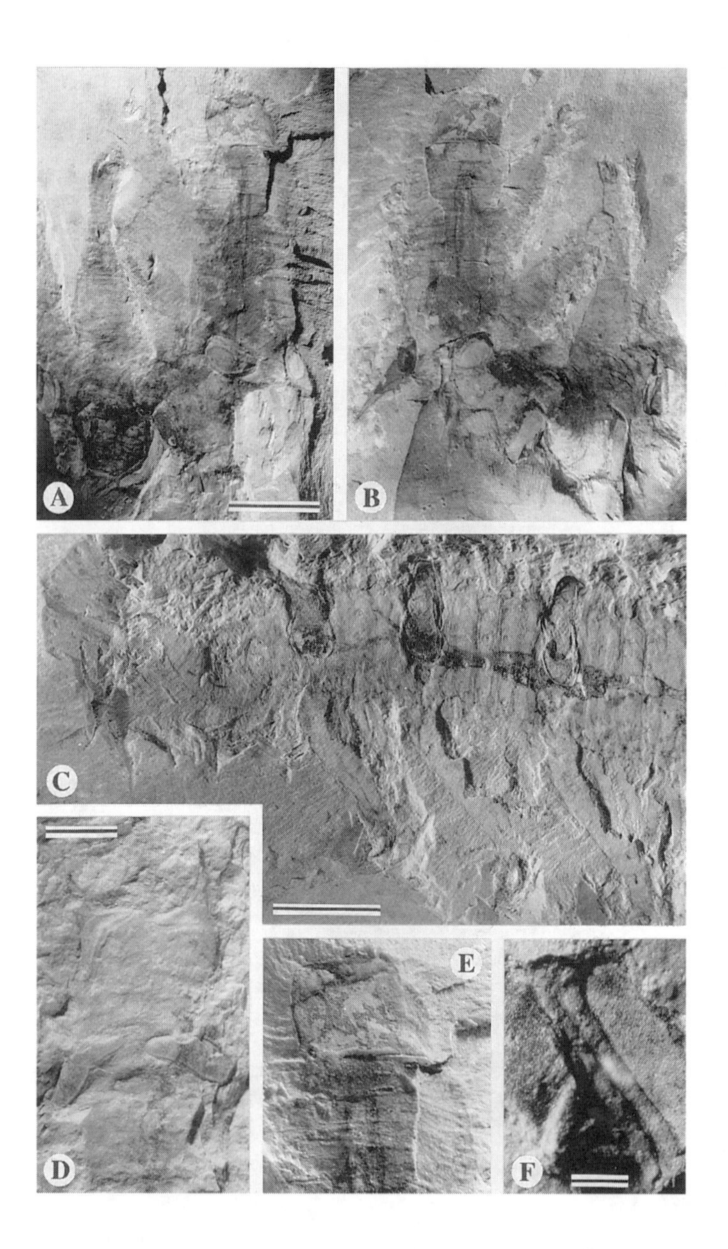

FIGURE 3.6

*Onychodictyon ferox* Hou, Ramsköld, and Bergström, 1991, from Chengjiang. *(A, B, E)* Anterior part of obliquely compacted, nearly complete individual, ELRC 33001, from locality MQ1 (for camera lucida see fig. 3.5C). *(A)* ELRC 33001b, dorsolateral view, light from NNW. Scale bar 5 mm, applies also to *B*. *(B)* ELRC 33001a, ventrolateral view, light from NE. *(E)* detail of *A*, light from NNE. *(C)* Posterior part of nearly complete specimen ELRC 33004a, from locality MN5 (for camera lucida and scale see fig. 3.7), light from W. *(D)* Anterior part of trunk of dorsoventrally compacted, nearly complete individual ELRC 33002, from locality MN5 (for camera lucida see fig. 3.5A), light from NNW. Scale bar 5 mm. *(F)* Plate pair in ELRC 33009. Surface relief preserved as tiny knobs on internal mold of plate (to left; convex toward observer) and as similarly sized pits on external mold (to right; concave toward observer). Light from NW. Scale 0.5 mm.

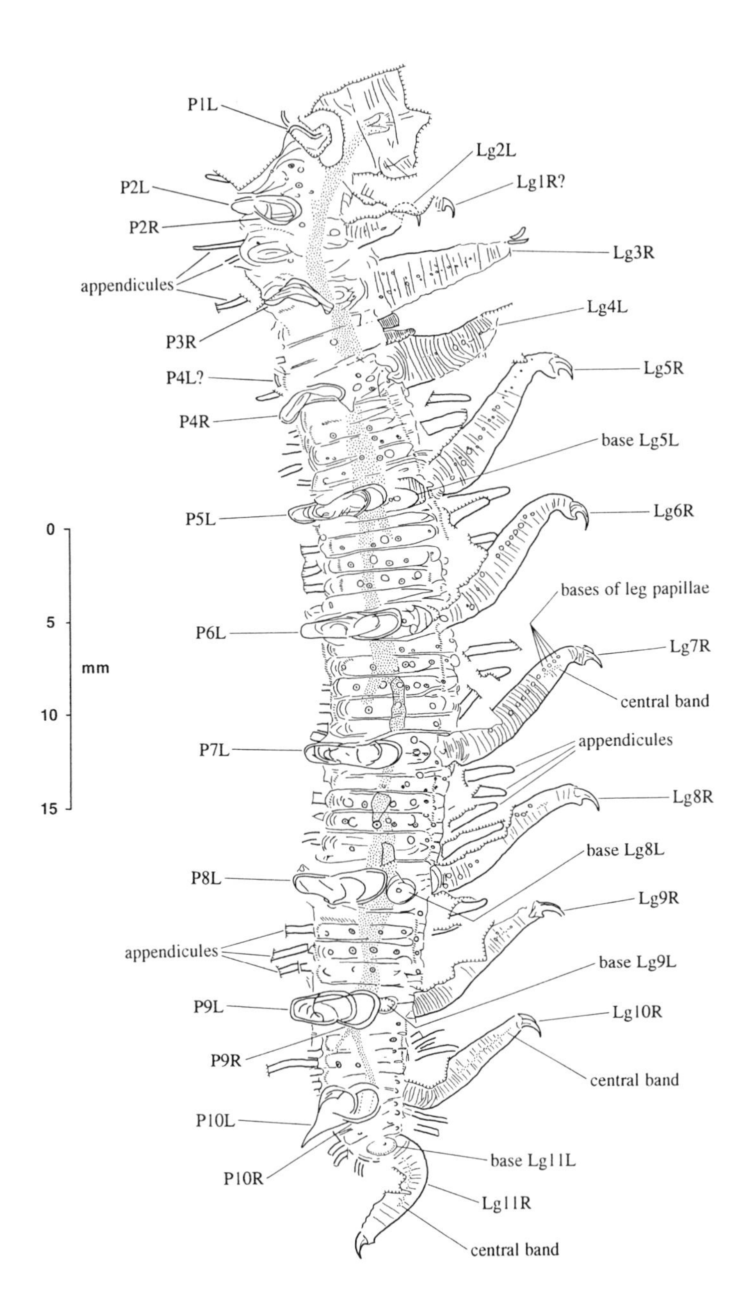


FIGURE 3.7

*Onychodictyon ferox* Hou, Ramsköld, and Bergström, 1991, from Chengjiang. Camera lucida drawing of ELRC 33004a, from locality MN5 (see fig. 3.6C). Head and anterior appendages not preserved. Plates of right side convex toward observer, those of left side concave. Dashed outline of claw of Lg2L added from counterpart. Attachment sites of appendicules indicated, raised ones by open circles and depressed ones by circles with a central dot. Annulation of ventral appendicules always present but indicated only between Lg3 and Lg4. Ventral appendicules between Lg7 and Lg8 preserved in full length. Bases of legs of left side preserved as depressions, some are indicated (Lg5L, Lg8L, Lg9L, Lg11L). Note reversed claw direction in Lg11R.

by complete loss of the legs in one segment or by lengthening of the trunk in this area is uncertain. However, lengthened segments are evidently present in the head region of some taxa *(Hallucigenia sparsa, Microdictyon sinicum)* and differentiated segmental lengths are present in all forms, whereas appendage-free segments in the trunk are not known, circumstances leading us to favor the latter hypothesis (segment lengthening) over an appendage-free segment in *Aysheaia* and/or *Onychodictyon.*

Another important discovery is that of a sclerotized head shield in *Onychodictyon.* This head shield is folded and deformed in the holotype, and only its anteroventral part was partly exposed during the late stages of preparation. It was first tentatively identified as a jaw structure (Ramsköld 1992b, fig. 4), an interpretation accepted by Hou and Bergström (1995), but a new specimen (ELRC 33001, figs. 3.5C, 3.6A, B, E) reveals the structure as a shield covering the head.

The detailed extent and structure of the head shield is only partly known. The portion exposed in the holotype (Ramsköld 1992b, fig. 4) corresponds to the anterior part of the shield in ELRC 33001. This portion appears to be a curved-under (or curved-over) part of the shield (fig. 3.6E). The lateral margins of the shield coincide with the trunk diameter, and the shield clearly covered the entire width of the head. How far it extended down the sides of the head is unknown. The oblique split through the matrix in front of the head shield would almost certainly have revealed any structures protruding there, such as antennae or oral papillae, but no structures have been found to protrude outside the shield. The possibility that the shield is ventral cannot be entirely excluded but appears unlikely on preservational and functional grounds.

The structures tentatively identified as antennae (Ramsköld and Hou 1991) in the holotype are not seen in any other specimen. Following the discovery of complete specimens showing the head, it is clear that in the holotype the body is bent nearly vertically in front of the first plate pair, and that the structures are not based terminally. They may be extended body papillae or preservational artifacts, or perhaps fortuitously associated structures such as algae; the latter are common and were responsible for the erroneous suggestion of antennae in *Microdictyon* (see correction in Ramsköld 1992b). The recent reconstruction of *Onychodictyon* by Hou and Bergström (1995) requires correction in including antennae, in lacking a head shield and an anterior appendage pair, in having anterior legs of the same size as the other legs, and in the direction of the plate spines (see below). In addition, in the reconstruction the head and trunk region in front of P1 is only half its actual length, and there are only three annuli per trunk segment instead of five (see fig. 3.7).

In one dorsoventrally compacted specimen (ELRC 33002, figs. 3.5A, 3.6D) the anterior end is flexed ventrally, so that the plate pair P2 is exposed in a nearly posterior view. The specimen shows that the plate spines were directed dorsally, not obliquely dorsolaterally as previously thought (see Ramsköld 1992b, fig. 3B). ELRC 33001 (fig. 3.5C) also shows vertically directed spines in plates P1R and P3L. The surface structure of the plates is preserved in ELRC 33009 (fig. 3.6F). The entire surface, both in internal and external molds, carries a finely reticulate pattern of granulation (and corresponding pitting of the interior surface).

There are five annuli between each of plate- and leg-pairs 4 to 10, and possibly only three annuli between Lg10 and Lg11. The number anterior to P4 is uncertain, but annuli appear to be about equally numerous in this area as well. Each annulus carries a single row of at least ten widely spaced appendicules around its circumference. In each annulus, the appendicules are set symmetrically relative to the dorsal midline, whereas the positions in successive annuli appears to alternate.

On its outer side, each leg carries along its length a row of at least ten, perhaps up to twenty, small appendicules. When the claws are directed posteriorly, this row is in transverse orientation. There is some evidence also of a similar row on the inner side of the leg, exactly opposite the outer row. The apparently irregular distribution of appendicules in the anterior appendage of Ramsköld (1992b, fig. 4) must be a preservational artifact since in ELRC 33001 the anterior appendages carry appendicules arranged in rows (fig. 3.6E).

The new specimens show that all claws are directed posteriorly, except in Lg11 where the direction is reversed. In ELRC 33004 (figs. 3.6C, 3.7), Lg11 is held in a different way from the other legs, being bent posteriorly and dorsally instead of reaching obliquely forward. The claws point anteriorly. This posture further underlines the adaptation of the posterior leg pair in several Cambrian lobopodians to an "anchoring" function, as suggested by Chen et al. (1995a).

The claws are terminal. The legs are slightly expanded at the base of the claws (fig. 3.7), but there are no "walking pads," in contrast to extant onychophorans, and during locomotion the claws would have rested against the substrate or reached into it. This supports the often repeated idea that lobopodians in general, including *Onychodictyon*, did not spend most of their time on the muddy bottoms where they were later embedded, but that they were climbers.

## *Microdictyon* (fig. 3.8A–E)

A monographic treatment of the Chengjiang material (Chen et al. 1995a) includes all presently known data. The orientation of *Microdictyon* has been established by Ramsköld (1992a,b; see also Chen et al. 1995a), and the recent suggestion of a reversed orientation (Hou and Bergström 1995) is rejected under "Anteroposterior Orientation" below.

*Microdictyon* is unusual among Cambrian lobopodians in lacking anterior appendages. A sister species relationship with *Hallucigenia* proposed below may offer an explanation of this absence. *Microdictyon sinicum* has nine plate pairs, each with a leg pair beneath, plus a posterior leg pair. *Hallucigenia sparsa* and *H. fortis* both have seven plate pairs with corresponding legs, plus two pairs of anterior appendages and a posterior leg pair. All three forms thus have a total of ten leg pairs, the only difference being the absence of plates above the anterior two in *Hallucigenia*. A

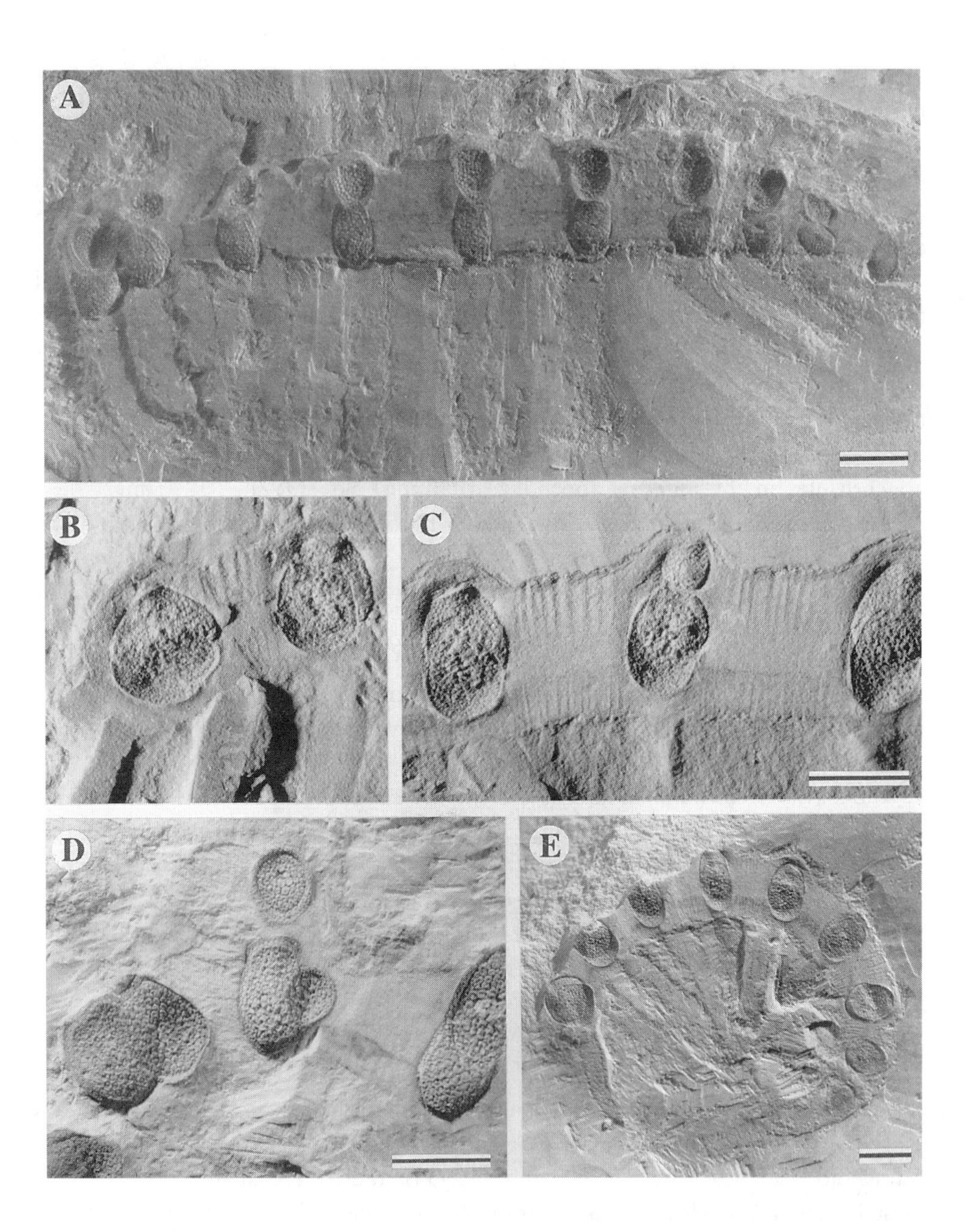


FIGURE 3.8

*Microdictyon sinicum* Chen, Hou, and Lu, 1989, from Chengjiang (for additional views and camera lucida drawings of specimens of *A–E* see Chen et al. 1995a). All scale bars 2 mm, bar in *C* applies also to *B*. (*A*) Nearly complete, extended individual ELRC 30012, from locality MN5. Head curved down into sediment. Light from ENE. (*B, C*) ELRC 30027, from locality MQ1. Anterior to right. Light from W. (*B*) Detail showing posterior trunk extension. (C) Detail showing teratological P5R developed as two small, separate plates. (*D*) ELRC 30042, from locality MQA. Anterior to right. Detail showing teratological P7L developed as two small, separate plates. Light from N. (*E*) Curled-up individual ELRC 30009, from locality MQ1. Light from NNE.

nondevelopment in *Hallucigenia* of the anteriormost plate pairs present in *Microdictyon* would thus yield the configuration found in *H. sparsa* and *H. fortis*. Considering the added evidence of other characters (see under "Anteroposterior Orientation"), we accordingly here set forth the hypothesis, admittedly a very tentative one, of homology of all ten leg pairs in *Microdictyon* and *Hallucigenia*.

Two specimens of *Microdictyon sinicum* show a teratological development of plates, in that instead of one of the plates of a pair, two small plates are present (fig. 3.8C, D). In both cases, these plates are rounded and have a normal hexagonal pattern of ridges and depressions, with the largest structures centrally in the plate. In one specimen (fig. 3.8D) the two plates are of equal size, whereas in the other (fig. 3.8C) there is a smaller dorsal and a larger ventral plate.

## *Xenusion* (fig. 3.9A)

The holotype of *Xenusion auerswalde* was for a long time the only known specimen (fig. 3.9A). It has been described in detail by Pompeckj (1927) and Jaeger and Martinsson (1967). Pompeckj (1927:290, 312, pl. 5, fig. 3) observed that the cuticle is ruptured sagittally in the anterior half of the specimen. This rupture curves adaxially posteriorly and crosses the sixth hump (as numbered by Pompeckj) on the right side before continuing exsagittally backward at least as far as the tenth hump.

The discovery of a second specimen of *Xenusion* (Dzik and Krumbiegel 1989) added much new information on the morphology of this animal but also created some confusion. For the following reasons, we regard the interpretation of the shape of the anterior end of the animal as erroneous, and maintain that the specimen should be interpreted as preserved from the ventral side rather than the dorsal. Note that to view the photographs of the second *Xenusion* specimen (Dzik and Krumbiegel 1989, fig. 1), the reader should turn the illustrations by 180 degrees because the main light in both photographs is from the south, and the relief appears inverted unless this is done.

As recognized by Dzik and Krumbiegel (1989), both the holotype and the new specimen were preserved as empty cuticles ruptured along the ventral midline, thus apparently representing cuticles shed after molting. In the second specimen the dorsal midline is preserved for much of the length of the specimen. It is defined by longitudinal ridges halfway between the "humps" of the left and right sides. These sagittal ridges are strongly impressed as far anteriorly as beside hump 3 (as labeled by Dzik and Krumbiegel). A row of eleven humps is present to the right side of the sagittal line. The corresponding humps on the left side are present posteriorly, but toward the anterior the split of the rock and erosion has removed a progressively wider part of the fossil's left side. Most of hump 6 remains, whereas only the most adaxial portion of hump 5 is preserved. Immediately anterior to hump 5, the margin of the specimen is seen as a straight line. This straight line runs along the side of the median, dorsal ridge. The marginal line is broken due to a stretch of matrix between

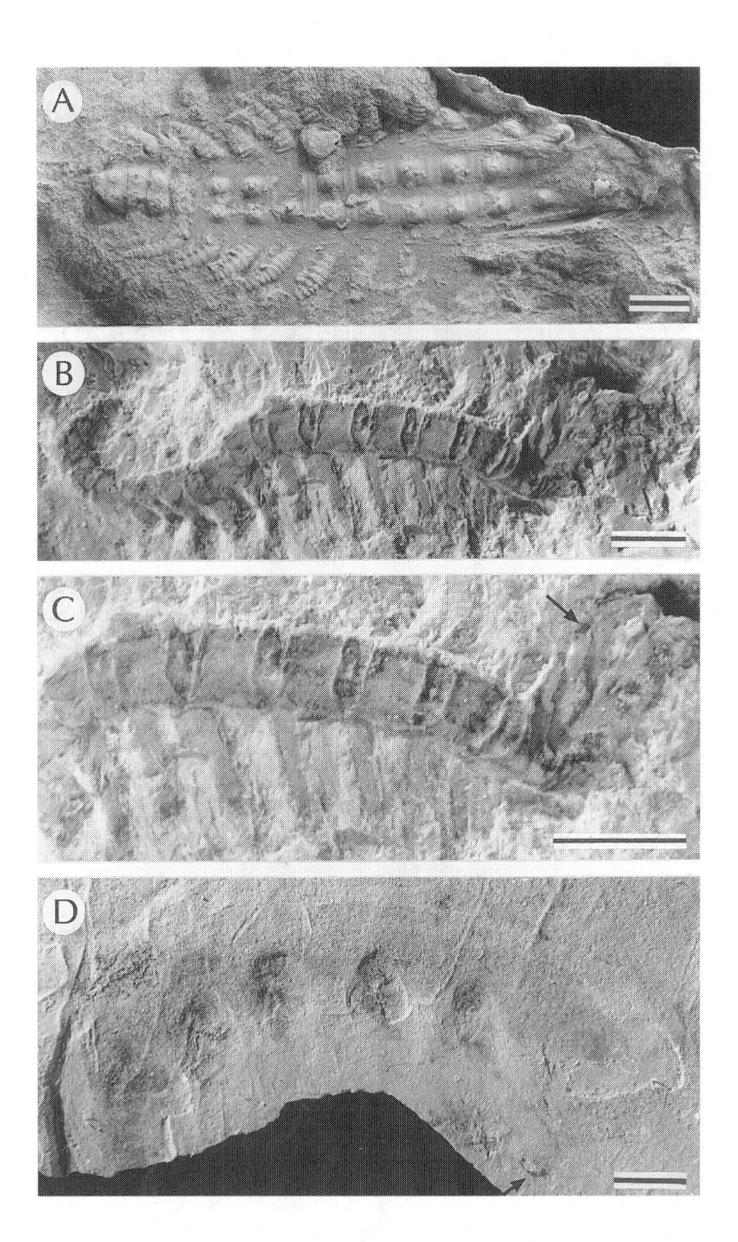


FIGURE 3.9

*(A)* Holotype of *Xenusion auerswalde* Pompeckj, 1927, from a Lower Cambrian Kalmarsund Sandstone erratic found in Germany, now housed in the Museum für Naturkunde der Humboldt Universität, Berlin. Latex cast whitened with ammonium chloride. Light from NNE. Scale bar 10 mm. Photo by U. Samuelsson. *(B, C) Cardiodictyon catenulum* Hou, Ramsköld, and Bergström, 1991, from Chengjiang. Large, twisted, nearly complete specimen (ELRC 31001), from locality MQ1 (for camera lucida drawing see fig. 3.10C). Scale bar 2 mm. *(B)* Light from NNW. *(C)* Detail of *B*, with higher light to emphasize gut and claws. Paired claws in Lg4L arrowed (cf. fig. 3.10C). Scale bar 2 mm. *(D) Paucipodia inermis* Chen, Zhou, and Ramsköld, 1995b, from Chengjiang. Newly identified specimen ELRC 34006, from locality NW5, preserved on an *Eldonia*. Posterior trunk portion with sixth leg pair lost. Claw of Lg1R arrowed. Bases of left side legs seen as rounded depressions; Lg2L prepared a short distance from its base. Light from NW. Scale bar 2 mm.

humps 3 and 2, and the line then reappears beside hump 2, continuing to the front of the animal. In the interpretation of Dzik and Krumbiegel, this line changes from a median to a marginal position between humps 3 and 2. To explain this, they must invoke a strong longitudinal rotation in this area. However, we can see no evidence for rotation, and this line is regarded here as running sagittally along the specimen, also anteriorly. It appears likely that the line is that of a sagittal rupture of the anterior portion of the cuticle, similar to that in the holotype, and that only the right half of the specimen is preserved anterior to hump 4.

All humps on the specimen's right side are exsagittally aligned along the sagittal fold and are set centrally between the fold and the lateral trunk margin. Dzik and Krumbiegel's interpretation of strong rotation anteriorly necessitates that hump 1 is set laterally rather than dorsally, in fact much more so than in their reconstruction.

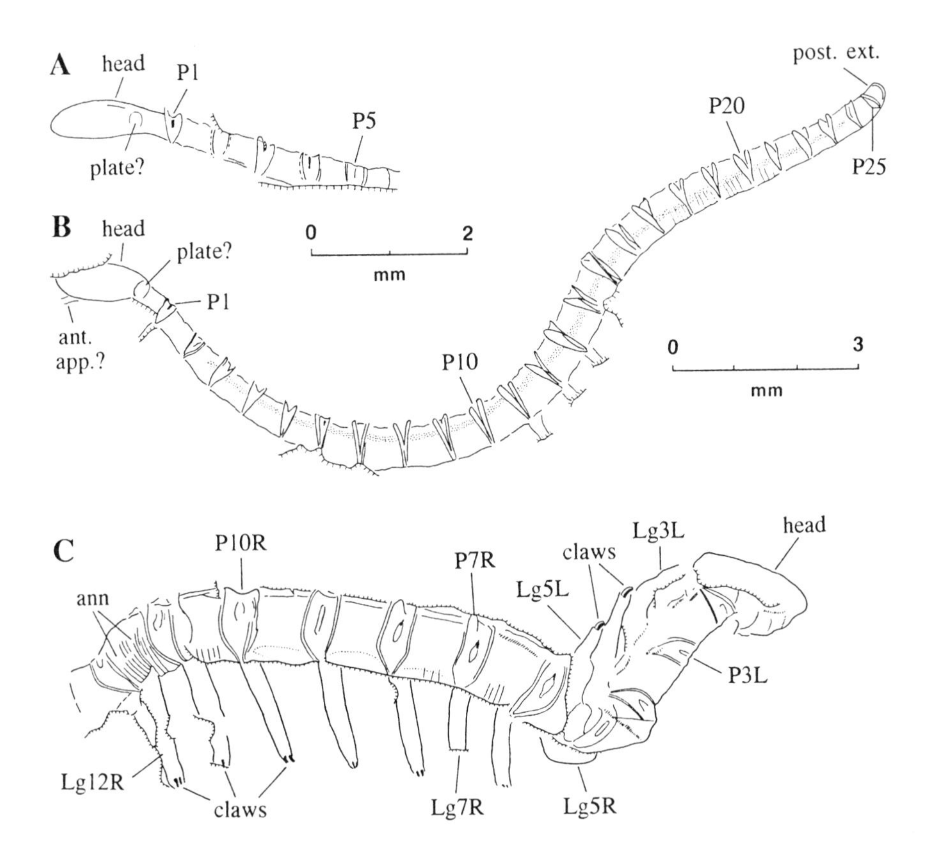


FIGURE 3.10

Camera lucida drawings of *Cardiodictyon catenulum* Hou, Ramsköld, and Bergström, 1991, from Chengjiang. *(A)* Anterior part of incomplete, small specimen (ELRC 31016b), from locality MN5. *(B)* Nearly complete, small specimen (ELRC 31015a), from locality MN4. Posterior trunk extension (post. ext.) and possible anterior appendage (ant. app.?) indicated. *(C)* Anterior portion of specimen ELRC 31001 (see fig. 3.9B–C). Note the annulation (ann), best seen between P11 and P12. The 3 mm scale applies to *B* and *C*.

In our interpretation the hump is set in the same near dorsal position as all other humps and is aligned with these. This agrees with our interpretation that the preserved anterior portion of the second *Xenusion* specimen is only half of the complete width. The anterior end of *Xenusion* was therefore not a very narrow, proboscis-like extension, but was about as wide as the trunk more posteriorly. The anterior termination is unknown.

The dorsoventral orientation of the specimen is of no consequence for the above reinterpretation. However, the excavated hump 1 is directed down into the matrix and has a spine extending further down from its lowest part. Dzik and Krumbiegel acknowledge that the hump is seen from its interior, with which we agree; in this area, the cuticle is therefore unambiguously seen from the ventral side. The other humps are flattened and give no clue as to the side from which they are seen, and neither do the annulations or legs.

However, the median, dorsal ridge indicates a particular dorsoventral orientation. A reasonably stiff exuvium of a cylindrical animal that is spread out flat will form longitudinal folds in areas of weakness. The band-shaped areas with humps will best resist such folding, and folds should form on each side of these bands. Indeed, in the specimen there are longitudinal folds medially and also lateral to the hump-bearing bands. Notably, the median folds should be directed inward, that is, ventrally. The folds in the specimen are convex toward the observer, which is the expected relief if the specimen is seen from the ventral side. As stated by Dzik and Krumbiegel, the specimen was apparently exposed to sea water for a considerable time before becoming completely embedded. Just as in the holotype, in the second specimen the cuticle apparently ruptured anteriorly along the lines of weakness caused by the median dorsal folds, so that the line seen from hump 4 and anteriorly is this rupture, and the cuticle on the other side is lost.

We conclude that the most natural interpretation is that the second *Xenusion* specimen is exposed from the ventral side, that it is not rotated anteriorly but spread out flat, that the humps are aligned exsagittally in two straight rows that are not deflected laterally toward the front, and that there is no evidence for a long, narrow proboscis, but that the anterior portion of the cuticle has ruptured sagittally and that it, as far as is preserved, was of a width approximately similar to that of the trunk more posteriorly.

### *Cardiodictyon* (figs. 3.9B, C, 3.10A–C)

The head in *Cardiodictyon catenulum* was originally unknown (Hou et al. 1991). A specimen with a nearly completely exposed head was illustrated by Ramsköld (1992b); this specimen and a further one were described by Hou and Bergström (1995). The latter authors identify a pair of sclerites covering most of the head, leaving only the anterior tip free. Although the presence of a large (but unpaired) sclerite covering the head in *Onychodictyon* could be taken to support this interpretation, we have seen no evidence for such sclerites in the *Cardiodictyon* material we have studied, which includes several new specimens (e.g., fig. 3.10A, B) and one of the two specimens of Hou and Bergström (1995) where such sclerites are said to be present. We also reject the suggested presence of such sclerites in *Hallucigenia fortis* (see above).

The number of plate pairs in ELRC 31015 is twenty-five (fig. 3.10B). Another complete specimen shows twenty-four plate pairs, and in a complete but twisted specimen (NIGPAS 115274; Ramsköld 1992b, fig. 2) there are most probably twenty-three plate pairs. This may indicate that there is some variation in the number of segments in *Cardiodictyon catenulum,* a variation unknown from other Cambrian lobopodians but with parallels in other arthropods with a high number of segments.

The shape of the plates changes during ontogeny, from V-shaped structures in small adults (fig. 3.10A, B) to ventrally pointed, shield-shaped in large adults (fig. 3.10C). The plates of each pair appear to meet at the dorsal midline, but we have not yet been able to work out their detailed shape. Just behind the head there may be a rounded plate (fig. 3.10A, B) lacking the dorsal spines of other plates. This structure is much weaker than the normal plates, which are always preserved in strong relief. The interpretation as a plate is therefore tentative.

A well-preserved specimen (ELRC 31001; figs. 3.9B, C, 10C) shows annulation of both trunk and legs. There appear to be at least twelve annuli between two plate pairs. The legs are finely and densely annulated. There are unquestionably two claws terminally on each leg (fig. 3.10C). Each claw is curved and pointed, but no further details are seen in the available material. Annulation of the trunk and the lobopods in *Cardiodictyon* was also recently reported by Hou and Bergström (1995). The same authors also reported four or possibly five claws per leg, a report we regard as based on a misinterpretation of more poorly preserved material. Hou and Bergström (1995) also indicated the presence of two pairs of legs anterior to the leg pair beneath the most anterior pair of plates. In our specimens there is only an inconclusive fragment in one specimen of a possible, single, anterior appendage (fig. 3.10B), so we consider the report to be in need of confirmation. If the feature is real, which may well be the case, *Cardiodictyon* shares the configuration of anterior appendages seen in the two *Hallucigenia* species.

## *Acinocricus* (not figured)

The early Middle Cambrian form *Acinocricus stichus* Conway Morris and Robison, 1988, occurs in the Spence Shale of Utah. It was first described, under open nomenclature, from fragmentary material as a possible medusoid (Conway Morris and Robison 1982) owing to similarities to *Peytoia,* a form then treated as a medusoid but later shown to be the mouth apparatus of an anomalocaridid. Better material led to the naming of *Acinocricus* and its reinterpretation as an unusual chlorophyte (Conway Morris and Robison 1988).

*Acinocricus stichus* consists of an elongated main part bearing spinose whorls at regular intervals. Most specimens consist of detached spine rings. None of these form a complete circle, but only about two-thirds of a circle. A few associations of two to four spine-bearing rings occur. The 97-mm-long holotype is the most complete specimen and includes at least ten spine rings. The holotype is also unique in showing five elongated, spinose structures termed subsidiary branches by Conway Morris and Robison (1988). These authors noted that "in the holotype the whorls at either end point in opposite directions" (ibid.: 13). They also observed that the arrangement may be bilaterally symmetrical, but concluded that "the repetitive nature of the nodal spinose whorls, possibly also present in the subsidiary branches, strongly suggests this is some type of alga" (ibid.: 16). We offer here a different interpretation of this material.

In our view, *Acinocricus stichus* shows compelling similarities to several Cambrian lobopodians, in particular to the complete, well-preserved specimen beautifully figured by Collins (1986:39). Collins stated that it has a unique bodyplan, but its affinity to the lobopodians *Aysheaia* and *Xenusion* was recognized by Delle Cave and Simonetta (1991), and it was included among the lobopodians by Chen et al. (1995b). The former authors popularly called the animal "Collins's monster," a term used below in the absence of a formal name. The animal is currently under study by Collins, and its inclusion in a detailed phylogenetic analysis must await its full description. The as yet only figured specimen does, however, allow a comparison with *Acinocricus stichus*.

Collins (1986) stated that the three specimens of the new animal were collected in 1983, all found at locality 9 of Collins et al. (1983). The horizon of this locality is the *Glossopleura* Zone. *Acinocricus stichus* also comes from the *Glossopleura* Zone, and the two forms are approximately coeval.

The figured specimen of Collins's monster is exposed in a nearly lateral view, whereas the holotype of *Acinocricus stichus* shows a similar animal in a roughly dorsal view. The bilateral symmetry of *A. stichus* noted by Conway Morris and Robison (1988) is evident, and a comparison of the row of four segments in another specimen (Conway Morris and Robison 1988, fig. 9.2) indicates that the holotype is rotated to the left (as figured). The spine rings of *A. stichus* correspond to the segmentally arranged spines seen in Collins's monster. The spines are inclined toward the nearest end of the animal, a circumstance paralleled in *Hallucigenia fortis* (see figs. 3.2A, 3.3B).

The "subsidiary branches" in the holotype of *Acinocricus stichus* bear a striking resemblance to the anterior legs in Collins's monster. Only the limbs on the right side are exposed, a circumstance explained by the counterclockwise rotation of the axis of the individual. The spacing of the limbs is similar to that of the adjacent spine sets, but the anterior one or two limbs must be based in front of the first spine set. The presence of two successive similarly spaced tubercles or short spines in front of the first spine set indicates the presence of a few segments anterior to the first one bearing a fully developed set of spines, a feature apparently paralleled in Collins's monster. The limbs thus correspond well to the segmentation expressed by the spine rings. Provided that the specimen exposes a complete individual, the full number of segments is thus twelve or thirteen, the same as that counted by Delle Cave and Simonetta (1991) in Collins's monster. Each limb in *Acinocricus stichus* carries a row of apparently stiff, pointed spines on its distal side, and another row is present, possibly based on the anterior limb side. These limb spines are similar in width, length, and spacing to those in Collins's monster.

The anteroposterior orientation of Collins's monster is unambiguously indicated by limb curvature, the anterior end being to the left in the figured specimen. The six anterior limb pairs bear conspicuous spine-fringes, whereas no spines are seen in the posterior six pairs. The posterior limbs are markedly shorter than the anterior ones. The five spine-bearing limbs in *Acinocricus stichus* agree with the anterior limbs in Collins's monster, and the upper end of the holotype as figured by Conway Morris and Robison (1988, figs. 5.3, 6) is here taken as anterior. The trunk itself is only partially exposed or preserved, but it may have been annulated as is indicated by numerous transverse lines in the "marginal zone" as it is drawn by Conway Morris and Robison (1988, fig. 6). The "central zone" is here interpreted as the gut. The gut is widest centrally in the body and narrows anteriorly. It is clearly seen to extend anteriorly to the two tubercle-bearing segments.

*Acinocricus stichus* differs from other lobopodians in having not one spine pair per segment but a transverse row composed of several spine pairs of alternating lengths. An increase in size and length of the tubercle row on the annules in *Aysheaia* could conceivably result in a similar configuration, provided only one spine set was developed per segment. In Collins's monster a single spine per segment is seen, but additional ones may be present.

We conclude that Collins's monster and *Acinocricus stichus* are two lobopodians, closely similar in both morphology and age. The Canadian material may even represent a new species of *Acinocricus,* but a detailed comparison must await the full description of the material.

## Anteroposterior Orientation

The senior author has, both alone and in joint studies, presented detailed morphological data indicating the anteroposterior orientation of *Hallucigenia* and *Microdictyon* (Ramsköld 1992a,b; Chen et al. 1995a,b). The more recent of these studies were not yet available to Hou and Bergström (1995) when the latter authors presented their suggestion of a return to the old orientation of these forms. However, Hou and Bergström (1995) refrained from explaining why or how they reject the five character sets put forth by Ramsköld (1992b) as evidence for the orientation, although a discussion of current ideas is necessary for the credibility of competing ideas. Hou and Bergström (1995:7) present their reversed orientation of *Microdictyon* on the basis of two characters: (A) "the curling up of the terminal appendages at the end corresponding to the large-sclerite end in *Hallucigenia* [*fortis*]," and (B) the similarity between "the extended, limbless body part in *Microdictyon* and *Hallucigenia [sparsa]*." Character (A) consists of two features, (A1) the curled appendages and (A2) the large sclerite. These three characters will be examined here.

*Character (A1)*

The terminal appendages in *Microdictyon sinicum* are frequently curved, always toward the elongated body end (Ramsköld 1992b, fig. 1; Chen et al. 1995a, figs. 20, 22, 24, 43, 45), with claws curved in the same direction. A similar curvature of the terminal leg pair, opposite to the one of the other legs, is seen in, for example, *Onychodictyon* (fig. 3.7), and Chen et al. (1995a) interpret this morphology as an adaptation for anchoring the posterior body portion while the anterior legs were searching for new holdfasts during locomotion. In forms possessing anterior appendages, these are always preserved outstretched anteriorly (*Hallucigenia fortis*: fig. 3.3B; Hou and Bergström 1995, fig. 1C; *Onychodictyon*: fig. 3.6A, B; *Cardiodictyon:* Hou and Bergström 1995, fig. 1A, B), and the anterior leg pairs in *Microdictyon* are similarly commonly preserved stretched out forward (fig. 3.8A; Chen et al. 1995a, figs. 12, 16, 19, 25, 26, 38). Anterior appendages or anterior leg pairs are not curved posteriorly in any of the eight species of Cambrian lobopodians where they are known. The curvature of the terminal legs in *Microdictyon* is therefore a good argument for their being posterior, not anterior.

*Character (A2)*

The large terminal plate pair in *Microdictyon* (fig. 3.8B) is interpreted by Hou and Bergström (1995) as covering the head and as being homologous with putative head sclerites in *Cardiodictyon* and *Hallucigenia*. Our material, much more extensive than that available to these authors, demonstrates the absence of any type of head sclerites in *Cardiodictyon* (fig. 3.10A, B) and *Hallucigenia* (fig. 3.3B, C). Instead, the comparison of this plate pair in *Microdictyon* with the posteriormost one in *Onychodictyon* made by Ramsköld (1992b) remains valid. The shape of these plates was actually one of the five characters discussed in that study as indicating the reversal of *Microdictyon*.

*Character (B)*

Ramsköld (1992b, character 1) forwarded arguments as to why an extended limbless region indicates homology, but of the anterior body end rather than the posterior. His interpretation (see fig. 3.11 herein) is further supported by the discovery of *Paucipodia,* an additional lobopodian with an elongated head, in the Chengjiang fauna (fig. 3.5D; Chen et al. 1995b) and by comprehensive studies of *Microdictyon* (fig. 3.8; Chen et al. 1995a) and *Hallucigenia fortis* (figs. 3.2B–D, 3.3; Chen et al. in prep.). The arguments put forth by Ramsköld (1992b, character 1) on why the elongated end in *H. sparsa* is the anterior end were not countered by Hou and Bergström (1995) and remain valid.

Hou and Bergström (1995) base their re-reversal of *Hallucigenia sparsa* on a single character, the putative presence of bivalved sclerites covering the head in *Hallucigenia fortis,* and the implicit presence of these also in *H. sparsa.* They mention and

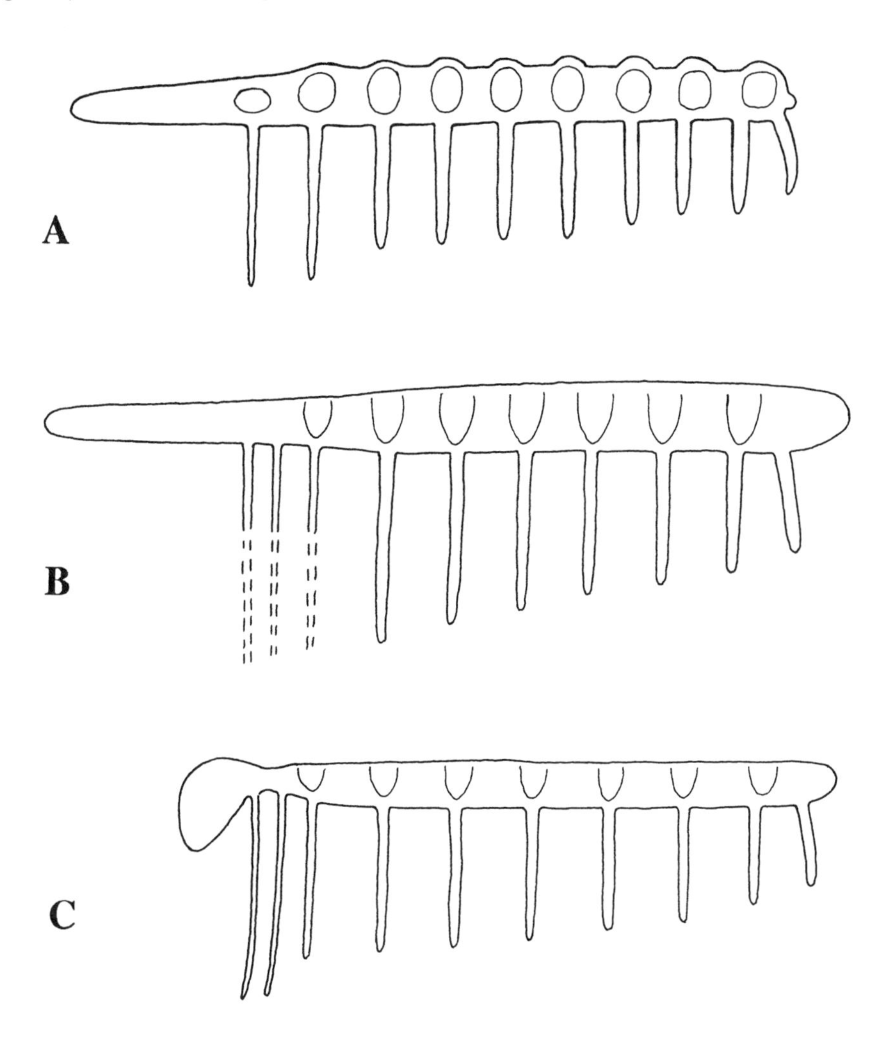


FIGURE 3.11

Stylized representations of alphonychophorans, showing anteroposterior orientation as indicated by anatomical features discussed in the text. Anterior to left. Legs have been drawn vertical to facilitate comparisons, a posture improbable to ever have been held in life. For clarity the spines and claws are omitted, and plates are only schematically indicated. Note anteriorly increasing leg lengths; slender anterior appendages/legs; presence of two leg pairs to posteriormost plate, with the anterior pair set below the anterior edge of the plate; and two pairs of anterior appendages in the two *Hallucigenia* species, set on the most slender trunk portion. *(A) Microdictyon sinicum* Chen, Hou, and Lu, 1989. *(B) Hallucigenia sparsa* (Walcott, 1911). *(C) Hallucigenia fortis* Hou and Bergström, 1995.

reject the possibility that the presence of a long body extension "could be taken as a proboscis very similar to that suggested by Dzik and Krumbiegel (1989) to be present in *Xenusion*" (Hou and Bergström 1995:7). Hou and Bergström (1995) also reject Ramsköld's (1992a) suggestion that the "head" inferred by Conway Morris (1977) is a stain caused by body fluids squeezed out posteriorly. Hou and Bergström (1995) apparently infer the presence of a bivalved head shield in *H. sparsa* in the region described as a stain by Ramsköld. However, there is no evidence of sclerites in this area in any specimen of *H. sparsa*. Instead, the presence in both *Hallucigenia* species of two closely spaced pairs of similar, slender appendages at one body end, distal to the sclerite-bearing portion of the trunk (figs. 3.1B, 3.2A), is a feature unambiguously indicating homology. Hou and Bergström (1995) disregard the known morphology of *H. sparsa* and state in the diagnosis of *Hallucigenia* that "an uncertain number of additional appendages are present in front of the 1st sclerite pair and behind the 7th sclerite pair." The appendages "in front" are present only in *H. fortis* and the ones "behind" only in *H. sparsa*, but Hou and Bergström (1995) overlook that this signals homology and that one of the species must be incorrectly oriented. In the correct orientation, both species have exactly the same number and configuration of legs and anterior appendages, and the only major difference between the two species is the length of the head (fig. 3.11).

The orientation of *Hallucigenia fortis* is established by the presence of the large, bulbous head, a fact recognized both by us and by Hou and Bergström (1995). The task is to find homologies between *H. fortis* on the one hand and *H. sparsa* and *Microdictyon sinicum* on the other in order to orient the latter two. We consider all five characters of Ramsköld (1992b), providing evidence for the anteroposterior orientation of *Hallucigenia* and *Microdictyon,* to be valid. Further support for Ramsköld's interpretation is listed under point (B) above. To Ramsköld's (1992b) characters (numbered 1–5) may now be added some further ones:

6. Leg length. In *Hallucigenia sparsa, H. fortis,* and *Microdictyon sinicum* the leg length increases gradually toward the anterior (fig. 3.11). In *M. sinicum* the anteriormost one or two leg pairs are close to twice as long as the last leg pair. In *Hallucigenia fortis,* the anterior appendages are about twice as long as the posterior leg pairs. In *H. sparsa* the most anterior leg with preserved length is L2, which is more than 50% longer than the posterior leg pairs. In all three forms, the legs increase in length gradually along the trunk toward the anterior.

7. In *Hallucigenia sparsa, H. fortis,* and *Microdictyon sinicum* the anterior legs are more slender than posterior ones. In *Cardiodictyon* the thickness of the anterior appendages (if such are present) and the anteriormost legs is uncertain.

8. On the segment with the most posterior plate, the leg pair is based below the anterior edge of the plate, whereas in all other segments the legs are based centrally below the plates. This character applies to forms with a leg pair behind the last plates, e.g., *Hallucigenia sparsa, H. fortis, Microdictyon sinicum,* and *Onychodictyon.* The forward displacement of the penultimate leg pair is clearly linked to the

presence of the "extra" leg pair posteriorly since in *Cardiodictyon,* which lacks a leg pair behind the last plates, the last legs are based centrally beneath the last plate pair.

In addition, character 5 of Ramsköld (1992b) can be further refined after the discoveries of anterior appendages in *Onychodictyon, Cardiodictyon,* and *Hallucigenia fortis*. The presence of such appendages was predicted by Ramsköld (1992b) on the basis of a morphocline outlined for Cambrian lobopodians. Only *Microdictyon* and *Paucipodia* lack anterior appendages, and parsimony suggests that in both taxa this may be a secondary absence. In the two *Hallucigenia* species, the homology of these appendages is indicated by several features (fig. 3.11): (a) they lack associated plates; (b) there are two pairs; (c) they are longer than any other legs; (d) they are more slender than any other legs; (e) the two pairs are based much closer to each other and to the nearest leg pair beneath plates than are other legs, and (f) they are set on the body part distal to the plate-bearing portion of the trunk that has the smallest diameter. As far as is known, these features may be valid for *Cardiodictyon* as well. The anterior appendage in *Onychodictyon* is described above. Its position agrees with the anterior appendage in *Aysheaia,* and its presence further strengthens the sister taxon relationship proposed here for these forms.

We conclude that the anteroposterior orientation of *Hallucigenia* and *Microdictyon* is the one presented by Ramsköld (1992b).

## Cladistic Analysis

The use of phylogenetic methods has increased greatly in the last two decades, and such methods are now used in all fields in which biological comparisons are made among organisms (Hillis et al. 1994). Proponents of cladistic methodology forcefully argue the merits of the approach, much of which consists of the search for homologies. In that way it concentrates on similarities between taxa and thus avoids a major conceptual pitfall of traditional taxonomy, which emphasizes differences.

It is commonly pointed out that the selection of characters is an important step in the analysis, but only occasionally are the characters analyzed for the likelihood of homology between the taxa studied. The implicit approach is that similar features are assumed to be homologous until proven homoplastic (Hennig 1966). General morphological similarity is not, however, a fully reliable criterion for homology. We emphasize that each character used should be subjected to a homology analysis so as to avoid homoplastic characters. The senior author has previously argued that homology must be proposed on the minimum basis of indications such as ontogenetic development, similarity of structure, and topological relation (Ramsköld and Edgecombe 1991). Knowledge of ontogenetic development greatly aids in evaluating homology, but such information is not available for the fossil lobopodians under study here. In its absence, similarity of structure and topological relation are essential characteristics, and homology should be assumed only when there are two or more mutually supporting indications. In any homology analysis it is inevitable that the suggested homologies rest, with varying support, on a probability basis, and they can be made more or less probable, but only rarely certain.

## Taxa Included

The analysis includes the ten named species of the fifteen Cambrian lobopodians known from soft-body preservation: *Aysheaia pedunculata* Walcott, 1911, *Aysheaia prolata* Robison, 1985, *Cardiodictyon catenulum* Hou, Ramsköld, and Bergström, 1991, *Hallucigenia fortis* Hou and Bergström, 1995, *Hallucigenia sparsa* (Walcott, 1911), *Luolishania longicruris* Hou and Chen, 1989, *Microdictyon sinicum* Chen, Hou, and Lu, 1989, *Onychodictyon ferox* Hou, Ramsköld, and Bergström, 1991, *Paucipodia inermis* Chen, Zhou, and Ramsköld 1995b, and *Xenusion auerswalde* Pompeckj, 1927. Of the remaining two forms (both are new forms from the Burgess Shale, as yet undescribed) only one has been figured (Collins 1986), and too little is known to permit coding more than a few characters. The Carboniferous form *Helenodora inopinata* Thompson and Jones, 1980 (see Rolfe et al. 1982) has also been included. *Ilyodes divisa* Scudder, 1890, which has been suggested as a senior synonym of *H. inopinata* by Rolfe et al. (1982), is too incompletely known for inclusion. Recent Onychophora has also been included, coded from data in Ruhberg (1985).

## Characters

It is common practice to avoid multistate characters by splitting them and defining a separate character of presence/absence for each of the states. Such an approach inevitably introduces weighting of these characters (Carlson 1994). On the other hand, in order to avoid inapplicability of character states to some taxa, fusion of characters into multistate characters has been suggested (Maddison 1993). There is at present no universal solution to this potential problem, and our data include both multistate and presence/absence characters. Contrary to common practice, character state conditions allotted to state 0 should not be interpreted as primitive; states are listed in random order.

1. Head shape: *(0) Short, narrower than trunk behind, directed straight forward; (1) long, diameter half or less of length, tapering or inflated.*

2. Oral papillae: *(0) absent; (1) present.*

3. Plates: *(0) absent; (1) "humps"; (2) present.*

4. Plate spines longitudinally ribbed and strongly sclerotized relative to plates: *(0) absent; (1) present.*

5. Plates differentiated in shape and size, with posterior plate largest, rounded angular: *(0) absent; (1) present.*

6. Number of annuli per segment (state centrally in the trunk coded): *(0) few (up to 6–7 per segment), long (exsag.), continuous around trunk; (1) many (9–25), short (exsag.), weak or absent laterally.*

7. Surface of annuli: *(0) smooth; (1) with tubercles and/or papillae.*

8. Segments progressively shorter posteriorly: *(0) absent; (1) present.*

9. Posterior trunk extension: *(0) absent; (1) present.*

10. Total number of leg pairs: *(0) 6; (1) 10; (2) 11; (3) 12; (4) 16; (5) 23–24; (6) very variable.*

11. Legs of subequal lengths along trunk, with tendency for shortening both toward anterior and posterior: *(0) absent; (1) present.*

12. Legs shorter than half diameter of trunk: *(0) absent; (1) present.*

13. Length cline of legs, with progressively shorter legs posteriorly: *(0) absent; (1) present.*

14. Anterior legs/appendages markedly more slender than other legs, long (> 3 times trunk diameter): *(0) absent; (1) present.*

15. Two pairs of anterior appendages ventrally at base of head: *(0) absent; (1) present.*

16. One pair of anterior appendages, smaller than legs behind, based laterally, set farther from Lg1 than distances between other leg pairs: *(0) absent; (1) present.*

17. Posterior leg arrangement, i.e., number of leg pairs to last leg-bearing trunk section: *(0) one (pair to last plate/segment); (1) two (pairs to last plate/segment, with or without posterior trunk extension).*

18. Leg spines or papillae: *(0) absent; (1) present.*

19. Claws: *(0) paired, talon-shaped; (1) multiple, sickle-shaped.*

*Notes on Characters*

Most characters are extensively discussed in the text or were treated by Ramsköld (1992b). A few additional aspects on character 10, number of leg pairs, may be mentioned here.

The senior author has elsewhere (Ramsköld and Werdelin 1991) cautioned against the "numerical trap," that is, the indiscriminate coding of characters where states are defined as the number of repeated structures. This is particularly true if these structures are segmental. There is often an evolutionary trend toward fixation of segmental numbers (the so-called Rosa's rule; see Ramsköld and Werdelin 1991), such as seen in the head of crustaceans and chelicerates, or the segment number in insects and extant tardigrades. Other groups, for example extant onychophorans, display a wide range of segment numbers, and the character has been regarded as without value in cladistic analyses including Onychophora (Monge-Nájera 1995). Most importantly, early forerunners of living forms show that these numbers were not originally fixed, but variable. For example, the only known Early Paleozoic tardigrade has three pairs of legs (Walossek et al. 1994; Müller et al. 1995), whereas all living forms have four pairs. Fossil euarthropods have anything between one and many segments in the head (see, e.g., character 21 of Wills et al. 1994), and the number in each form cannot by itself be used as a criterion for assignment to a particular group.

In Cambrian lobopodians, the total number of leg pairs ranges from six *(Paucipodia),* through ten *(Microdictyon sinicum, Hallucigenia sparsa* and *H. fortis),* eleven *(Aysheaia pedunculata),* twelve *(Onychodictyon*), sixteen *(Luolishania),* at least twenty *(Xenusion),* to twenty-three to twenty-five *(Cardiodictyon).*

### *Characters Not Used*

We have not used the length of the plate spines in the analysis. Otherwise closely similar plates may have long spines or lack them completely, as is evidenced by *Microdictyon,* which lacks spines or at most has a little central tip on the plate, and the *Microdictyon*-like form figured by Bengtson (1991), which has a long, strong spine. Spines are present both in forms with mineralized and unmineralized plates. Elsewhere in the animal kingdom, dorsally set, paired spines occur in a variety of organisms, ranging from insects to dinosaurs. We regard this character as liable to show considerable homoplastic development in any group under study and thus of doubtful use in the analysis herein.

## Analysis and Results

Usually one or more outgroups are used in a phylogenetic analysis. In the present analysis, it proved impossible to choose and code a relevant outgroup. In an analysis including both fossil and extant lobopodians, the outgroup must, of course, lie outside these taxa. However, there is at present no agreement as to the position of the Onychophora relative to Arthropoda *s.l.,* or as to what group is sister taxon to Onychophora. In a cladistic presentation, Brusca and Brusca (1990, chapter 19, fig. 15) summarized the character evidence supporting the generally held view of onychophorans on the line leading from annelids to arthropods. Similar phylogenies (although tardigrades were not considered) were obtained by Wheeler et al. (1993) from combined morphological and molecular data. A conflicting result was obtained by Ballard et al. (1992) in a molecular analysis. Fortey and Thomas (1993) cautioned that the latter results need confirmation and Wägele and Stanjek (1995) questioned the sequence alignments in the study. Until more stable results are available, the choice of outgroup for an analysis of onychophorans remains contentious. Since the choice of outgroup will strongly affect the topology of the trees obtained (Ramsköld 1991), the more so the more different the outgroup codes from the ingroup taxa and the less support there is for the ingroup nodes, no outgroup (in the form of an actual taxon) was used in this analysis.

It may be noted that in an analysis that included most of the taxa treated herein (Monge-Nájera 1995), the Annelida was selected as outgroup, a choice based mainly on the analysis by Brusca and Brusca (1990). An attempt to code the characters used in this study for annelids failed, partly because tentative homologies are very uncertain and partly because the majority of characters used herein relate to legs and plates. Such features are absent in annelids, and the characters are thus inapplicable, a problem for which there is as yet no generally agreed solution (see discussions in Platnick et al. 1991; Wheeler and Nixon 1994).

When no outgroup can be chosen, one option is to construct a hypothetical ancestor (hypanc), for which the primitive state of each character is coded. In this study it proved difficult to construct a hypanc. To code character states for a hypanc it is necessary to be able to polarize the characters. There are different ways to polarize characters, primarily by using outgroup data or ontogenetic changes, but no such information is yet available for lobopodians. Good ontogenetic information is available for extant onychophorans, but the information is not applicable to most of the characters used in this analysis. A hypanc was eventually constructed, and the results obtained from its use are discussed below.

The problems with finding outgroups and determining character polarity in a phylogenetic analysis of lobopodians were anticipated in a preliminary analysis (Ramsköld 1992b:457), and up to now no acceptable solutions have presented themselves. However, bearing in mind the noise introduced in the analysis by using outgroups or hypancs, we do not regard it as a disadvantage to analyze only the actual taxa under study. Such an analysis will produce unrooted networks. A network lacks direction, and all branching points are triramous. The network is based on parsimony analysis and includes all information on character state changes, node support, and sister group relationships that can be found in a rooted tree. The difference is that until a network is rooted, it is not a hypothesis of a particular phylogeny, but of all the possible phylogenies. A network with $N$ terminal taxa will have $N$-2 nodes and 2$N$-3 branches. Thus, as in the present study, a network with twelve species will contain ten branching points and twenty-one branches. This means that the network can be rooted in twenty-one places, e.g., it presents twenty-one alternative phylogenies. The purpose of outgroups, whether real or hypancs, is above all to determine the correct placing of the root, that is, which of these twenty-one alternatives is the best phylogenetic hypothesis.

In this study, the data matrix (table 3.1) was analyzed with PAUP, version 3.0r. All characters were treated as unordered and of equal weight. The branch and bound algorithm was used, which means that the networks found are considered to be the shortest.

Three shortest networks were found, each with a length of thirty steps, consistency index (CI) of 0.833, and retention index (RI) of 0.865. One of the networks is shown in figure 3.12. The two others differ only in the relative positions of the two *Aysheaia* species and *Onychodictyon* distal to node 8 (fig. 3.12), due to changes in characters 2, 7, and 9 on branch 6–8 (between nodes 6 and 8), and 8–9 (between nodes 8 and 9).

The networks were then rooted in two different ways; first by using midpoint rooting, e.g., rooting at the longest branch. In this method the branch with the highest number of state changes is arbitrarily considered to be the one best suited for positioning of the root. In the three networks, there are two such branches: 4–5 (the one between nodes 4 and 5), and 5–6 (the one between nodes 5 and 6; fig. 3.12). The results are shown in figure 3.13. Tree *A* is rooted on branch 5–6, whereas trees *B* and *C* are rooted on branch 4–5 (fig. 3.12). The three trees differ only in the position of the clade (*Helenodora* + Recent Onychophora) and in the relative positions within a clade composed of the two *Aysheaia* species and *Onychodictyon*. A strict consensus is shown in figure 3.13D.

In all three trees the Cambrian lobopodians fall out as two distinct clades. The clade (*Helenodora* + Recent Onychophora) is sister group to either of these. The fossil taxa, not being sister taxa in any tree, would thus form a paraphyletic group that cannot be defined as a taxon distinct from extant Onychophora (including *Helenodora*). This result supports the suggestion that "the Recent Onychophora must be derived from marine ancestors, and could even turn out to be an ingroup taxon" (Ramsköld 1992b:458). If this is so, we must reject the use (Hou and Bergström 1995) of the consequently paraphyletic class Xenusia Dzik and Krumbiegel, 1989, as a taxon embracing all marine lobopodians.

In spite of the difficulties with ordering characters, it was decided to also try outgroup rooting using a hypanc. As mentioned above, outgroup rooting affects the topology of the tree, something that may or may not be advantageous, depending on

TABLE 3.1
Data matrix for cladistic analysis of Cambrian and post-Cambrian lobopodians

| *Taxon* | | | | |
|---|---|---|---|---|
| Minimal hypanc | 0???? | ????? | ?0??? | ??0? |
| Maximal hypanc | 00000 | 1?01? | 10000 | 0100 |
| *Aysheaia pedunculata* | 01000 | 01002 | 10000 | 1111 |
| *Aysheaia prolata* | 01000 | ????? | ?0?00 | 1??? |
| *Cardiodictyon catenulum* | 10200 | 10015 | 00?0? | 0000 |
| *Hallucigenia fortis* | 10210 | 10011 | 00111 | 0100 |
| *Hallucigenia sparsa* | 10210 | ?0011 | 00111 | 0100 |
| *Helenodora inopinata* | ??000 | 11??? | 1100? | ??00 |
| *Luolishania longicruris* | 0?100 | 0?114 | 10000 | 00?1 |
| *Microdictyon sinicum* | 10201 | 10011 | 00110 | 0100 |
| *Onychodictyon ferox* | 0?201 | 01003 | 10000 | 1110 |
| *Paucipodia inermis* | 10000 | 100?0 | 00?00 | 0100 |
| *Xenusion auerswaldae* | ??100 | 001?? | 10?0? | ?01? |
| Recent Onychophora | 00000 | 11016 | 11000 | ??00 |

See text for explanation of characters and hypancs; see figures 3.12–3.14 for results of analyses. Missing data are indicated by a question mark.

the quality of the broader phylogenetic hypothesis forming the basis for the choice of the outgroup.

The hypanc character states were inferred from general comparisons with several possible outgroups (annelids, tardigrades, and to some extent euarthropods) on the basis of generality of character distribution. Some characters could be coded with a certain degree of confidence (e.g., characters 1, 14–16). For others, the arguments

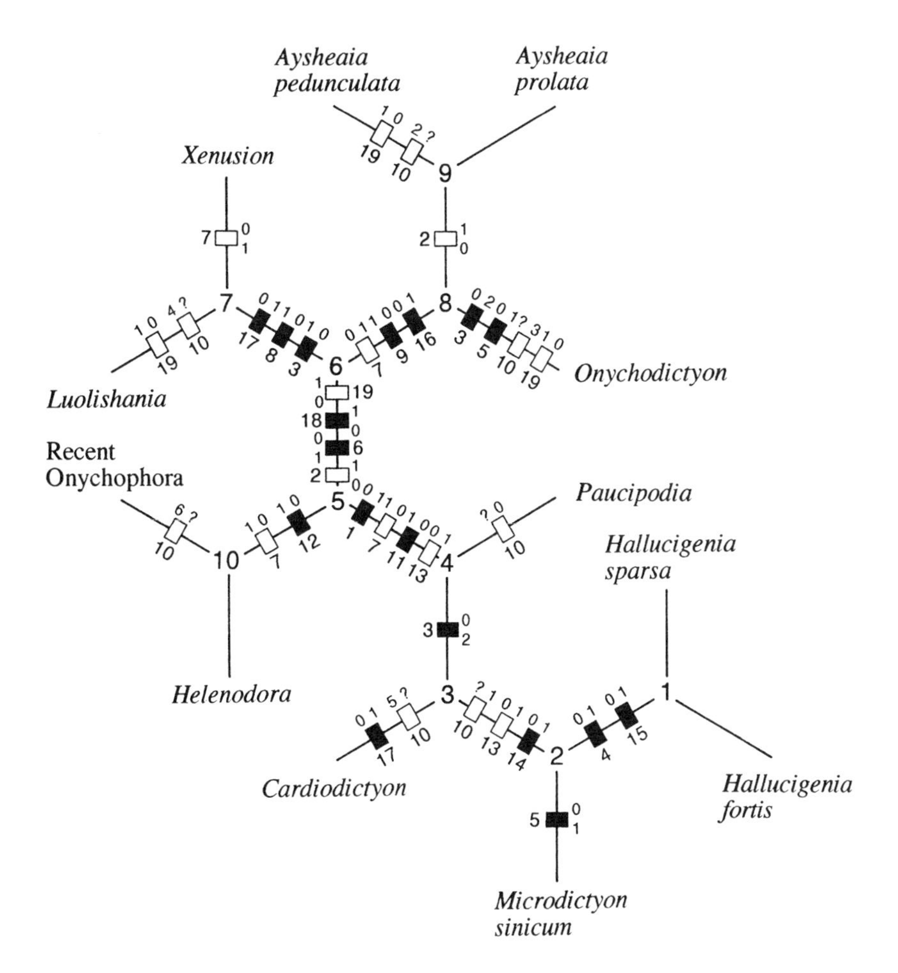


FIGURE 3.12

Network yielded by analysis of the twelve ingroup taxa but without a hypothetical ancestor. One of the three networks obtained by PAUP is shown with a complete presentation of all character state changes. The two other networks differ only in state changes in characters 2, 7, and 9 on branches 6–8 and 8–9 (see fig. 3.13 for these networks after rooting). Nodes are numbered 1 to 10 (as in fig. 3.13). Bars indicate character state changes on branch; solid bar = state change on this branch; open bar = state change either on this branch or on path to other position of bar showing state change in this character. Bars are given with character number on one side and the state change on the other, showing the state on each side of the bar (e.g., between nodes 4 and 3, character 3 state 3:0 supports node 4 and state 3:2 supports node 3). See text for discussion of the use and implications of the networks.

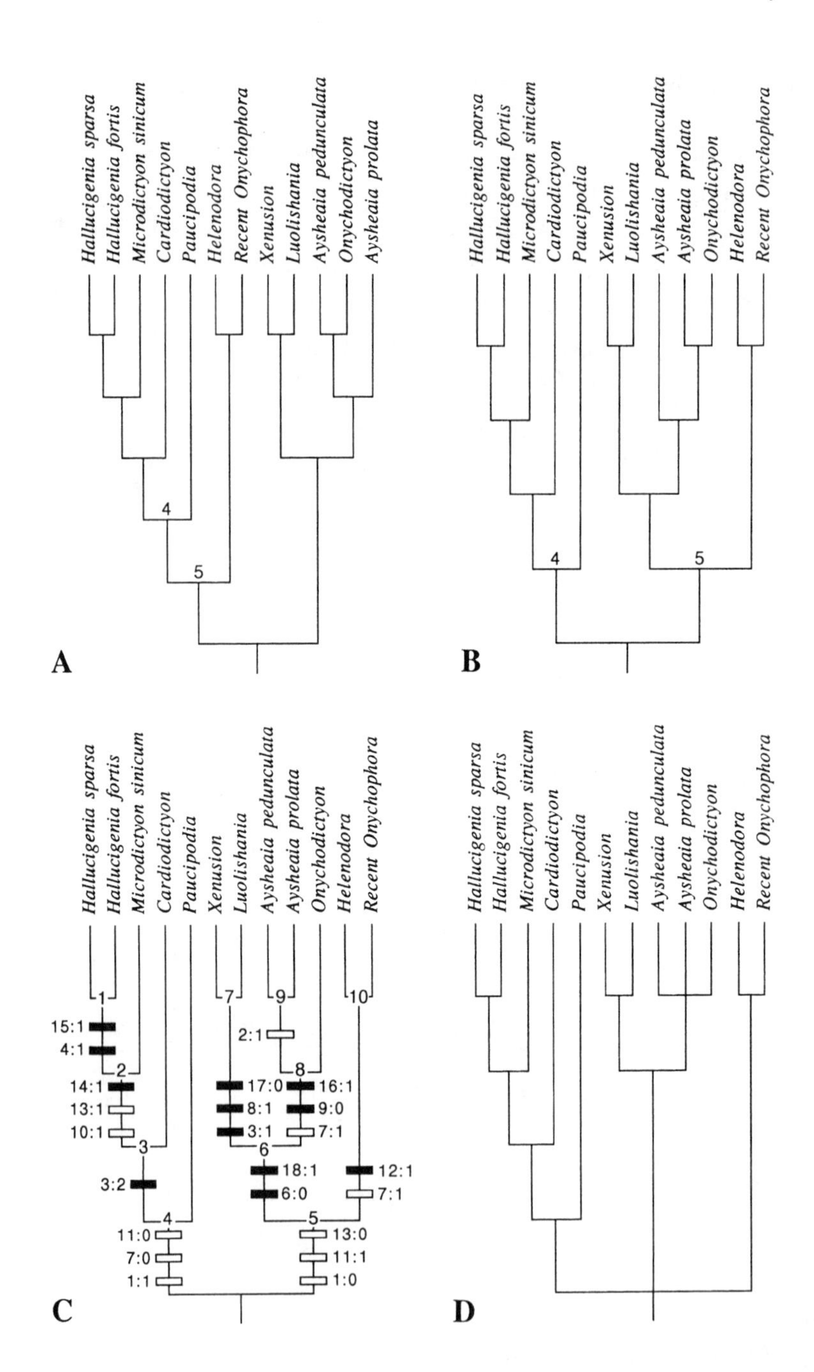


FIGURE 3.13

*(A–C)* The three networks obtained by the cladistic analysis with midpoint rooting. Length is 30 steps, consistency index 0.83, retention index 0.86. Nodes are numbered as in figure 3.12. The trees differ only in the position of the clade (*Helenodora* + Recent Onychophora) and in the position of *Onychodictyon* relative to the two *Aysheaia* species. In C, the node support by characters 1–19 (see text) is plotted on the branches. Filled bars indicate definite synapomorphies, open bars indicate the highest level for synapomorphies that may be more general (e.g., information about state is lacking for the branch/branches down tree); not all of the latter are plotted (see fig. 3.12 for complete presentation). *(D)* Strict consensus of cladograms in *A–C*.

for the coding range from reasonably well based to only weakly indicated by outgroup data. Due to this uncertainty in character state codings, the hypanc was coded both with few and with many characters (minimal and maximal hypanc, respectively, in table 3.1), and the data matrix was then analyzed with PAUP.

In general, the hypanc supported the results obtained from midpoint rooting. Runs with the minimal hypanc, with only three characters coded (1:0, 12:0, 18:0) yielded the exact same results as the maximally coded hypanc, with seventeen of the nineteen characters coded. The recoding of any, one, or several, of these seventeen characters, except 1, 12, or 18, to a question mark (e.g., missing data) had no effect on the results. Recoding of one or more of characters 1, 12, and 18 as missing data resulted in a less resolved tree.

The results from the analyses with a hypanc used as outgroup differ little from those where the networks were rooted with midpoint rooting. The shortest trees have the same length (thirty steps), C.I. (0.833), and R.I. (0.865). There are nine such trees (see strict consensus in fig. 3.14), compared to the three shortest networks found. The reason for the increase in the number of trees is that the hypanc roots not only on branches 5–4 and 5–6 (as in the midpoint rooting) but also on branch 5–10 (see fig. 3.12). With each of these rootings there are three trees differing in the

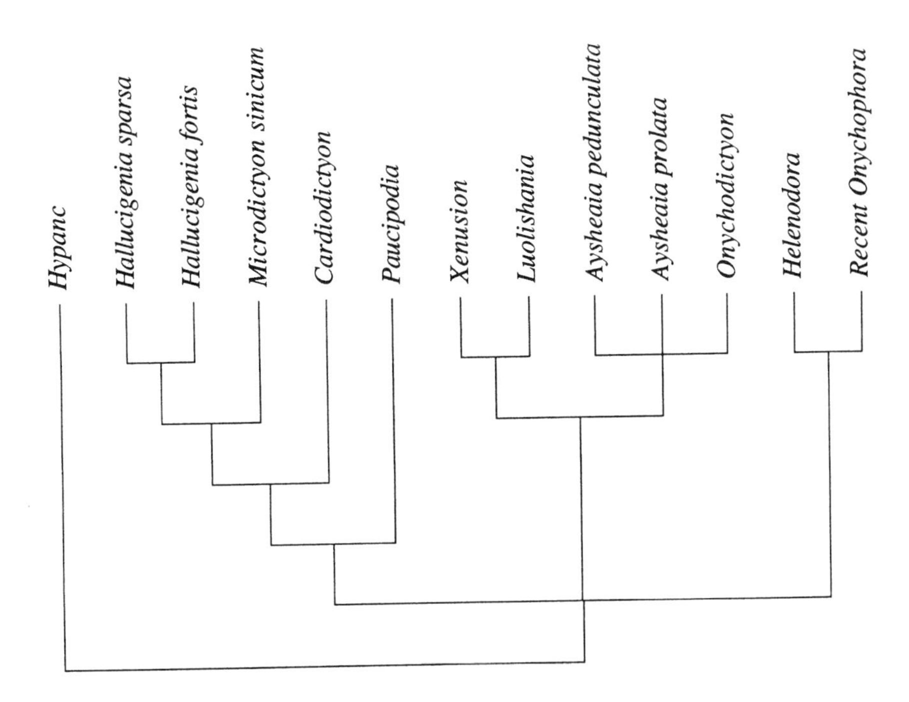


FIGURE 3.14

Result obtained by adding a hypothetical ancestor (hypanc) as outgroup to the twelve ingroup taxa. Strict consensus of nine trees. Length is thirty steps, consistency index 0.83, retention index 0.86. See text for details.

above-mentioned way within a clade composed of the two *Aysheaia* species and *Onychodictyon*.

The results with hypanc differ in one important respect from the midpoint rooted networks: in one topology (when the hypanc roots on branch 5–10, e.g., in three of the nine trees) the clade (Recent Onychophora + *Helenodora*) forms the sister group to a clade including all Cambrian lobopodians. This means that the paraphyly of the Cambrian lobopodians suggested above is not definite, although the topology of the consensus tree in figure 3.13D remains the same (cf. fig. 3.14).

In conclusion, the results obtained from networks rooted with midpoint rooting and the results yielded by inclusion of a hypanc corroborate each other. The question of monophyly or paraphyly of the Cambrian lobopodians is not finally settled, but the phylogeny of the these animals is otherwise well established.

## Taxonomic Considerations

The probability of paraphyly of the Cambrian lobopodians with regard to extant Onychophora indicates that the ancestry of the living, terrestrial forms is to be sought among the marine, Cambrian forms. The taxonomic conclusion to draw from this is that the Onychophora should also include these fossil forms, as suggested already by Ramsköld and Hou (1991). This more inclusive group constitutes the thus expanded Onychophora.

Previous taxonomy above the family level of the Cambrian lobopodians does not reflect phylogenetic relationships. Hou and Bergström (1995) introduced new names at the order level for the following forms: Archonychophora for *Luolishania*, Scleronychophora for *Microdictyon, Hallucigenia,* and *Cardiodictyon*, and Paronychophora for *Onychodictyon*. The available name Protonychophora Hutchinson, 1930, was used for *Aysheaia* and, tentatively, *Xenusion*. These four orders are rather randomly distributed in the cladograms presented herein (figs. 3.13, 3.14), and only one of the new taxa (Scleronychophora) coincides with a monophyletic group (it may be noted that a single species hardly constitutes a group, and we reject the practice to erect new orders for single species). For the extant forms, plus apparently *Helenodora,* the name Euonychophora Hutchinson, 1930, is available. As discussed below, we have chosen another route than to redefine existing names or to introduce new ones at specified levels in the taxonomical hierarchy.

Our results show three distinct, monophyletic groups within Onychophora (figs. 3.13, 3.14). None of these coincide with a previously established taxon, but with some modification, the names Euonychophora, Protonychophora, and Scleronychophora could perhaps be used for the three clades. Alternatively, new names could be established at the ordinal, class, or subphylum level, and/or any intermediate level(s). However, the literature already abounds in names applied to various combinations of the taxa under study here (see for example Hutchinson 1930; Dzik and Krumbiegel 1989; Anderson 1992; Hou and Bergström 1995). Each new analysis yielding results different from previous ones will provide opportunity for "erecting" new taxa above the family level, because names at higher hierarchical levels are not governed by the priority rules of the ICZN (International Code of Zoological Nomenclature). The already existing instability and confusion would then only get worse, something which is of benefit to no one. We have therefore decided to use an approach that will promote stability in names. This is the phylogenetic system of nomenclature proposed by Kevin de Queiroz and Jacques Gauthier (see de Queiroz and Gauthier 1994). The system of phylogenetic nomenclature is the logical extension of a cladistic approach to taxonomy. Its basis is the tenet of common descent, and its objectives are to make possible a taxonomy that accurately reflects phylogeny and is maximally stable. The system requires neither agreement on, nor detailed knowledge of, phylogenetic relationships. It leaves taxonomists free to change and differ in their conclusions about relationships, contents, and diagnoses of taxa in the light of new data and improved analytical methods. In this system, a new phylogenetic interpretation does not necessitate changes in the taxonomic names, it merely changes the composition of the clades covered by the names. For further insight into this system we refer the readers to the discussion by de Queiroz and Gauthier (1994).

For one of the three clades found in our analysis (figs. 3.13, 3.14), we use the existing name Euonychophora, and for the two others, we introduce the new names Alphonychophora and Betonychophora. We employ these names with an explicit phylogenetic definition (de Queiroz and Gauthier 1994). Stem-based definitions are used in order to secure maximal stability under all foreseeable circumstances, such as the discovery of paraphyly in the present analysis or of new forms that do not have an ingroup position in any of the clades found so far. This procedure means that each name denotes a clade that is defined as including all species sharing a more recent common ancestor with a specified species or clade than with another. Once defined, the name does not change.

Euonychophora Hutchinson, 1930, is redefined here as that monophyletic group composed of the clade (Peripatidae + Peripatopsidae) and all species (e.g., *Helenodora inopinata*) more closely related to it than to Alphonychophora or Betonychophora.

Alphonychophora *taxon novum* is that monophyletic group that includes *Aysheaia pedunculata* and *Aysheaia prolata* and all species (at present those of *Xenusion, Luolishania*, and *Onychodictyon*) more closely related to these than to Euonychophora or Betonychophora.

Betonychophora *taxon novum,* finally, is that monophyletic group that includes *Hallucigenia sparsa* and *Hallucigenia fortis* and all species (at present those of *Microdictyon, Cardiodictyon,* and *Paucipodia*) more closely related to these than to Euonychophora or Alphonychophora.

Note that Alphonychophora and Betonychophora assume different relative hierarchical positions depending on which of the three trees in figure 3.13 is selected.

## Comparison with Extant Groups

As long as only a few of the Cambrian animals discussed herein were known, it was necessary to make comparisons with a large number of groups whenever assessing affinities. Some of these exercises, such as the interpretation of *Xenusion* as a pennatulacean (Tarlo 1967) or of *Aysheaia* as an elasipod echinoderm (McKenzie 1983), may today seem far-fetched. They were, however, legitimate at the time. In this study, comparisons are restricted to the two main groups that are still of relevance when evaluating the affinities of the Cambrian lobopodians: the Onychophora and the Tardigrada. In our view, a comparison with euarthropods is hampered by the problems in homologizing more than a few features, and no such attempt is made here.

### Comparison with Onychophora

Many of the characters used to define extant Onychophora are histological or biochemical, whereas others are widely accepted as related to the shift from a marine to a terrestrial habitat. The latter characters include, for example, the oral papillae with slime glands and the tracheal system. The remaining anatomical characters regarded as onychophoran synapomorphies are few in number. Of the ten characters listed by Brusca and Brusca (1990) as synapomorphic for onychophorans, only four characters are potentially observable in marine forms preserved as fossils: (1) suppression of external segmentation; (2) subcutaneous hemal channels; (3) body papillae and scales; and (4) lobopods with pads and claws. If we accept these characters (and we know of no better ones) and if the Cambrian lobopodians form a monophyletic group together with the extant Onychophora, one or more of these characters must be identified in the fossils.

1. Suppression of external segmentation. The Cambrian lobopodians lack externally visible segmentation. The presence of regularly spaced legs and plates implies that segmentation is present, but forms lacking plates (*Aysheaia* and *Paucipodia*) show no trace on the trunk of external segmentation. In these forms the annulation is continuous and uniform, without any hint of segmental breaks. Suppression of external segmentation is therefore regarded as present in all known Cambrian lobopodians.

2. Subcutaneous hemal channels. These channels form part of the hydrostatic skeleton and are expressed on the cuticular surface as ridges, conventionally termed annuli. Each annulus contains a central, hemal channel. The annulation is present on both trunk and legs. In extant onychophorans, early embryos lack annuli, and annulation is gradually developed during ontogeny (Walker and Campiglia 1988). In the Cambrian lobopodians, annulation is present in all taxa (not observed only in the relatively poorly preserved *Hallucigenia sparsa* and *Aysheaia prolata*).

3. Body papillae and scales. These are minute structures, set on the annular ridges and may only be preserved in taxa where they are relatively large. Among the Cambrian lobopodians, such structures are observed with certainty only in *Aysheaia pedunculata* (fig. 3.1A). The raised patches on the annuli of *Onychodictyon* (fig. 3.6C) represent attachment sites for long papillae, and although these appear to be functional analogues to the papillae in extant onychophorans (both structures basically being sensory), the case for homology is only tentative. No small scales have been observed in Cambrian lobopodians, but the capacity to form strongly sclerotized cuticular structures may be homologous, and onychophoran scales and lobopodian plates may differ primarily in size.

4. Lobopods with pads and claws. All Cambrian lobopodians possess lobopods of a structure similar to those in extant onychophorans. Most Cambrian forms have one pair of talon-shaped claws on each leg, as do the extant onychophorans, but some have more. There are no pads like those in Recent taxa, and the claws are usually relatively larger. Forms with a soft tip of the leg protruding beyond the claws, as in *Aysheaia*, may have been able to walk with the claws held up from the substratum, whereas in most Cambrian taxa the claws were apparently terminal. It appears likely that the pads in extant onychophorans are an adaptation to a terrestrial, ground-living habitat, whereas many Cambrian forms were adapted to climbing. The marine forms would have had a near neutral buoyancy, with no need for pads on their legs.

We conclude that these four synapomorphies used to define the extant Onychophora are present also in the Cambrian lobopodians. Together these forms thus constitute a monophyletic group, defined by these four characters.

The question if the Recent terrestrial and the Cambrian marine lobopodians form two sister taxa, or if the Cambrian group is paraphyletic (that is, if it has given rise to the Recent forms), is tentatively answered by the phylogenetic analysis herein. As yet, a sister group to extant Onychophora (and forms like *Helenodora*) within the Cambrian taxa cannot be identified. Hou and Bergström (1995) regarded *Onychodictyon* as closer to modern onychophorans than any of the other Cambrian lobopodians. This was due to the supposed presence of antennae and jaws, features suggested to be present by Ramsköld (1992b), but now known to be absent. Monge-Nájera (1995) regarded *Aysheaia* as closest to extant Onychophora. However, in the cladogram (Monge-Nájera 1995, fig. 2) the crown group *(Aysheaia (Helenodora* + Onychophora)) is defined not by any features present but by three loss characters (loss of proboscis, loss of trunk spines, and loss of plate armoring). These "characters" are a priori assumptions about phylogeny used as characters and are thus invalid, leaving the clade without support. The present study refutes the position of either *Onychodictyon* or *Aysheaia* as sister taxon to post-Cambrian onychophorans.

It should be mentioned here that the reports of extant, marine "Pro-Onychophora" have been regarded as suspect (Jayaraman 1989; Monge-Nájera 1995), and they are not further commented upon here.

## Comparison with Tardigrada

The Tardigrada have repeatedly figured in discussions on the affinities of *Aysheaia* and other Cambrian lobopodians. The present analysis rejects hypotheses that the Cambrian lobopodians are intermediate between Onychophora and Tardigrada, and the comparison below indicates that they are not ancestral to both.

At present there is no consensus regarding the phylogenetic position of the Tardigrada. Kinchin (1994) recently summarized previous views on their affinities. Brusca and Brusca (1990) presented evidence for a position of Tardigrada above that of Onychophora, on the line leading to Euarthropoda. Monge-Nájera (1995) reviewed these and other characters and found only three characters that may unite Tardigrada with Arthropoda. Further, he gave reasons for doubting the homology in tardigrades and arthropods of these three characters. Analyses of 18S rDNA and rRNA sequences unite tardigrades with arthropods (Garey et al. 1996; Giribet et al. 1996), although onychophorans were not included in these works.

Fossil, true tardigrades have recently been discovered in Middle Cambrian rocks from Siberia (Müller et al. 1995). These phosphatized larvae are exceedingly similar to extant tardigrades, and we accept their assignment to Tardigrada. The fossil specimens have only three pairs of legs, but in the largest of the four individuals found so far there is what appears to be a rudiment of a fourth, posterior, leg pair. Therefore, adults did perhaps possess four completely developed leg pairs. As might be expected for a marine species, this fossil tardigrade lacks any cuticular plates. The claws are paired, a condition plesiomorphic for tardigrades, in which extant taxa with multiple claws in adults have a reduced number earlier in ontogeny (Kinchin 1994). This fossil tardigrade, by far the earliest known representative of the group, shows that the miniaturization of tardigrades (if we accept this general assumption) had taken place already during or before the Middle Cambrian.

Many of the characters conventionally regarded as apomorphic for tardigrades are histological and cannot be observed in fossils. Several of the anatomical features typically present in tardigrades, as well as the absence of many others, are usually considered as correlated with miniaturization. For both these reasons, it is likely that the ancestor of tardigrades before miniaturization may be difficult to recognize as such. This, unfortunately, leaves the field wide open to speculation, with little chance to either prove or disprove hypotheses.

*Aysheaia* shows some tardigrade similarities, such as the terminal mouth surrounded by papillae, the multiple claws, the lack of a trunk extension between the posterior legs, and the lack of true jaws and antennae. These similarities led Delle Cave and Simonetta (1975) to suggest that *Aysheaia* was intermediate between Onychophora and Tardigrada. Robison (1985), however, showed that these characters appear to be either plesiomorphic or convergent, and we accept his conclusions.

*Onychodictyon* shares some of the tardigrade similarities present in *Aysheaia*. The discussion by Robison (1985) applies equally well to these characters, and we reject them as possible synapomorphies. An additional character is the head shield in *Onychodictyon*. This structure has no counterpart in known fossil lobopodians or Recent onychophorans. The report of a bivalved head shield in *Cardiodictyon* and *Hallucigenia* (Hou and Bergström 1995) is not supported by our material (see above). In echiniscid tardigrades, however, there are multiple plates covering the trunk, including a frontal plate covering the head (e.g., Kinchin 1994, figs. 1.1, 2.5a, 3.7). Only limno-terrestrial tardigrades have plates, whereas marine forms lack them. The absence of any close similarity in topological relationships or structure of the plates in *Onychodictyon* and tardigrades indicates non-homology of the structures.

We conclude that, so far, none of the Cambrian lobopodians has been found to possess features that can be regarded with any confidence as derived characters uniquely shared with tardigrades.

## ACKNOWLEDGMENTS

This study was supported by the Wenner-Gren Foundation, Magn. Bergvalls Stiftelse, and the Royal Swedish Academy of Sciences (L. R.); the Chinese Academy of Sciences (L. R. and C. J.-y.) and the National Geographic Society (grants no. 4760–92, 5165–94) (C. J.-y.). L. R. thanks F. J. Collier for access to the Burgess Shale material in the Smithsonian Institution, and G. Budd for showing undescribed material from the Sirius Passet fauna. D. Collins generously provided information and photographs of unpublished Burgess Shale material, which catalyzed the discovery that *Acinocricus stichus* is a lobopodian.

## REFERENCES

Anderson, C. 1992. *Classification of Organisms: Living & Fossil.* Lancaster, Ohio: Golden Crown Press.

Ballard, J. W. O., G. J. Olsen, D. P. Faith, W. A. Odgers, D. M. Rowell, and P. W. Atkinson. 1992. Evidence from 12S ribosomal RNA sequences that onychophorans are modified arthropods. *Science* 258:1345–1348.

Bengtson, S. 1991. Oddballs from the Cambrian start to get even. *Nature* 351:184–185.

Bengtson, S., S. C. Matthews, and V. V. Missarzhevsky. 1986. The Cambrian netlike fossil *Microdictyon*. In A. Hoffman and M. H. Nitecki, eds., *Problematic Fossil Taxa*, pp. 97–115. New York: Oxford University Press.

Brusca, R. C. and G. J. Brusca. 1990. *Invertebrates*. Sunderland, Mass.: Sinauer.

Carlson, S. J. 1994. Investigating brachiopod phylogeny and classification: Response to Popov et al. 1993. *Lethaia* 26 [for 1993]: 383–384.

Chen, J.-Y. 1991. Report. In O. H. Walliser, ed., *IGCP 216, Bio-Events: Supplement to the Annual Report, 1990,* p. 5. Göttingen.

Chen, J.-Y., X.-G. Hou, and H.-Z. Lu. 1989. Early Cambrian netted scale-bearing, worm-like sea animal. *Acta palaeontologica sinica* 28:1–16.

Chen, J.-Y., G.-Q. Zhou, and L. Ramsköld. 1995a. The Cambrian lobopodian *Microdictyon sinicum. Bulletin of the National Museum of Natural Science (Taichung, Taiwan)* 5:1–93.

Chen, J.-Y., G.-Q. Zhou, and L. Ramsköld. 1995b. A new Early Cambrian onychophoran-like animal, *Paucipodia* gen. nov., from the Chengjiang fauna, China. *Transactions of the Royal Society of Edinburgh, Earth Sciences* 85:275–282.

Collins, D. 1986. Paradise revisited. *Rotunda, Royal Ontario Museum* 19:30–39.

Collins, D. H., D. E. G. Briggs, and S. Conway Morris. 1983. New Burgess Shale fossil sites reveal Middle Cambrian fossil complex. *Science* 222:163–167.

Conway Morris, S. 1977. A new metazoan from the Burgess Shale of British Columbia. *Palaeontology* 20:623–640.

Conway Morris, S. and R. A. Robison. 1982. The enigmatic medusoid *Peytoia* and a comparison of some Cambrian biotas. *Journal of Paleontology* 56:116–122.

Conway Morris, S. and R. A. Robison. 1988. More soft-bodied animals and algae from the Middle Cambrian of Utah and British Columbia. *University of Kansas Paleontological Contributions* 122:1–48.

de Queiroz, K. and J. Gauthier. 1994. Toward a phylogenetic system of biological nomenclature. *Trends in Ecology and Evolution* 9:27–31.

Delle Cave, L. and A. M. Simonetta. 1975. Notes on the morphology and taxonomic position of *Aysheaia* (Onychophora?) and of *Skania* (undetermined phylum). *Monitore Zoologico Italiano,* n.s., 9:67–81.

Delle Cave, L. and A. M. Simonetta. 1991. Early Palaeozoic arthropods and problems of arthropod phylogeny: With some notes on taxa of doubtful affinities. In A. M. Simonetta and S. Conway Morris, eds., *The Early Evolution of Metazoa and the Significance of Problematic Taxa,* pp. 189–244. Cambridge: Cambridge University Press.

Dzik, J. and G. Krumbiegel. 1989. The oldest "onychophoran" *Xenusion:* A link connecting phyla. *Lethaia* 22:169–181.

Fortey, R. A. and R. H. Thomas. 1993. The case of the velvet worm. *Nature* 361:205–206.

Garey, J. R., M. Krotec, D. R. Nelson, and J. Brooks. 1996. Molecular analysis supports a tardigrade-arthropod association. *Invertebrate Biology* 112:79–88.

Giribet, G., S. Carranza, J. Baguñà, M. Riutort, and C. Ribera. 1996. First molecular evidence for the existence of a Tardigrada + Arthropoda clade. *Molecular Biology and Evolution* 13:76–84.

Gore, R. 1993. Explosion of life: The Cambrian period. *National Geographic* 184: 120–136.

Hao, Y.-C. and D.-G. Shu. 1987. The oldest known well-preserved Phaeodaria (Radiolaria) from southern Shaanxi. *Geoscience* 1:301–310.

Hennig, W. 1966. *Phylogenetic Systematics.* Urbana: University of Illinois Press.

Hillis, D. M., J. P. Huelsenbeck, and C. W. Cunningham. 1994. Application and accuracy of molecular phylogenies. *Science* 264:671–677.

Hou, X.-G. and J. Bergström. 1995. Cambrian lobopodians: Ancestors of extant onychophorans? *Zoological Journal of the Linnean Society* 114:3–19.

Hou, X.-G. and J.-Y. Chen. 1989. Early Cambrian arthropod-annelid intermediate animal, *Luolishania* gen. nov., from Chengjiang, Yunnan. *Acta palaeontologica sinica* 28: 208–213.

Hou, X.-G., L. Ramsköld, and J. Bergström. 1991. Composition and preservation of the Chengjiang fauna: A Lower Cambrian soft-bodied biota. *Zoologica Scripta* 20:395–411.

Hutchinson, G. E. 1930. Restudy of some Burgess Shale fossils. *Proceedings of the United States National Museum* 78 (11): 1–24.

Hutchinson, G. E. 1969. *Aysheaia* and the general morphology of the Onychophora. *American Journal of Science* 267:1062–1066.

Jaeger, H. and A. Martinsson. 1967. Remarks on the problematic fossil *Xenusion auerswaldae. Geologiska Föreningens i Stockholm Förhandlingar* 88:435–452.

Jayaraman, K. S. 1989. Indian zoologist suspected. *Nature* 342:333.

Kinchin, I. M. 1994. *The Biology of Tardigrades.* London: Portland Press.

Maddison, W. P. 1993. Missing data versus missing characters in phylogenetic analysis. *Systematic Biology* 42:576–581.

McKenzie, K. G. 1983. On the origin of the Crustacea. *Memoirs of the Australian Museum* 18:21–43.

Monge-Nájera, J. 1995. Phylogeny, biogeography, and reproductive trends in the Onychophora. *Zoological Journal of the Linnean Society* 114:21–60.

Müller, K. J., D. Walossek, and A. Zakharov. 1995. "Orsten"-type phosphatized soft-integument preservation and a new record from the Middle Cambrian Kuonamka Formation in Siberia. *Neues Jahrbuch für Geologie und Paläontologie, Abhandlungen* 197:101–118.

Nash, M. 1995. When life exploded. *Time International* 146:62–70.

Platnick, N. I., C. E. Griswold, and J. A. Coddington. 1991. On missing entries in cladistic analysis. *Cladistics* 7:337–343.

Pompeckj, J. F. 1927. Ein neues Zeugnis uralten Lebens. *Paläontologische Zeitschrift* 9:287–313.

Ramsköld, L. 1991. Pattern and process in the evolution of the Odontopleuridae (Trilobita): The Selenopeltinae and Ceratocephalinae. *Transactions of the Royal Society of Edinburgh, Earth Sciences* 82:143–181.

Ramsköld, L. 1992a. The second leg row of *Hallucigenia* discovered. *Lethaia* 25:221–224.

Ramsköld, L. 1992b. Homologies in Cambrian Onychophora. *Lethaia* 25:443–460.

Ramsköld, L. and G. D. Edgecombe. 1991. Trilobite monophyly revisited. *Historical Biology* 4:267–283.

Ramsköld, L. and X.-G. Hou. 1991. New Early Cambrian animal and onychophoran affinities of enigmatic metazoans. *Nature* 351:225–228.

Ramsköld, L. and L. Werdelin. 1991. The phylogeny and evolution of some phacopid trilobites. *Cladistics* 7:29–74.

Rhebergen, F. 1990. Een raadselachtig fossiel: ?Onychophora? (An enigmatic fossil: Onychophora?). *Grondboor en Hamer* 44:130–131.

Rhebergen, F. and S. K. Donovan. 1994. A Lower Palaeozoic "onychophoran" reinterpreted as a pelmatozoan (stalked echinoderm) column. *Atlantic Geology* 30:19–23.

Robison, R. A. 1985. Affinities of *Aysheaia* (Onychophora), with description of a new Cambrian species. *Journal of Paleontology* 59:226–235.

Rolfe, W. D. I., F. R. Schram, G. Pacaud, D. Sotty, and S. Secretan. 1982. A remarkable Stephanian biota from Montceau-les-Mines, France. *Journal of Paleontology* 56:426–428.

Ruhberg, H. 1985. Die Peripatopsidae (Onychophora). *Zoologica* 137:1–183.

Scudder, S. H. 1890. New Carboniferous Myriapoda from Illinois. *Boston Society of Natural History, Memoir* 4:417–442.

Simonetta, A. M. 1976. Remarks on the origin of the Arthropoda. *Atti della Società Toscana di Scienze Naturali,* memoir series B, 82 [for 1975]:112–134.

Tarlo, L. B. H. 1967. *Xenusion:* Onychophoran or coelenterate? *The Mercian Geologist* 2:97–99.

Thompson, I. and D. S. Jones. 1980. A possible onychophoran from the middle Pennsylvanian Mazon Creek beds of northern Illinois. *Journal of Paleontology* 54:588–596.

Wägele, J. W. and G. Stanjek. 1995. Arthropod phylogeny inferred from partial 12S rRNA revisited: Monophyly of the Tracheata depends on sequence alignment. *Journal of Zoological Systematics and Evolutionary Research* 33:75–80.

Walcott, C. D. 1911. Cambrian Geology and Paleontology, II.5: Middle Cambrian Annelids. *Smithsonian Miscellaneous Collections* 57:109–144.

Walcott, C. D. 1931. Addenda to descriptions of Burgess Shale fossils (with explanatory notes by Charles E. Resser). *Smithsonian Miscellaneous Collections* 85:1–46.

Walker, M. and S. Campiglia. 1988. Some aspects of segment formation and post-placental development in *Peripatus acacioi* Marcus and Marcus (Onychophora). *Journal of Morphology* 195:123–140.

Walossek, D., K. J. Müller, and R. M. Kristensen. 1994. A more than half-a-billion-years old stem-group tardigrade from Siberia. *Abstract, Sixth International Symposium on Tardigrada.* Cambridge.

Wheeler, W. C., P. Cartwright, and C. Y. Hayashi. 1993. Arthropod phylogeny: A combined approach. *Cladistics* 9:1–39.

Wheeler, W. C. and K. Nixon. 1994. A novel method for economical diagnosis of cladograms under Sankoff optimization. *Cladistics* 10:207–213.

Whittington, H. B. 1978. The lobopod animal *Aysheaia pedunculata* Walcott, Middle Cambrian, Burgess Shale, British Columbia. *Philosophical Transactions of the Royal Society of London,* series B, 284:165–197.

Wills, M. A., D. E. G. Briggs, and R. A. Fortey. 1994. Disparity as an evolutionary index: A comparison of Cambrian and Recent arthropods. *Paleobiology* 20:93–130.

# CHAPTER 4

## Chengjiang Arthropods and Their Bearing on Early Arthropod Evolution

*Jan Bergström and Hou Xianguang*

## Abstract

The Lower Cambrian Chengjiang fauna from Yunnan, southwest China, is notably similar to the Middle Cambrian Burgess Shale fauna. The Chengjiang fossils are generally better preserved and more easily prepared than those from the Burgess Shale and other similar occurrences. A preliminary study of the arthropods has revealed a wealth of new information, which may challenge our views on the earliest morphologies and evolution of arthropods. An appendage that appears to represent a primordial type has a main stem of some fifteen to twenty very short and undifferentiated podomeres. The exopod is a simple rounded flap extending from the proximal part of the pediform stem. In the next steps, setae on the terminal leg podomere were superseded by spines, and setae were developed on the exopod.

From this level, there are at least three main evolutionary lineages. In one, the first post-antennal limb became greatly enlarged and branching. These are the "Great Appendage Arthropods." In the second, there was a successive evolution toward the crustacean state, as outlined by Walossek and Müller (1990). This included the development of a proximal endite, which ultimately formed a new podomere, the coxa, an invention exclusive for the crustaceans. It also included the separation of the first three limb pairs as a functional unit, and the associated development of the nauplius larva. Regarding the third lineage, the new understanding of the coxa makes it necessary to identify the trilobite-type outer ramus not as an enigmatic epipoditic "gill" but as an exopod homologous with the crustacean exopod. The exopod setae evolved into flat, trilobite-type lamellar setae. In most members of this group, the compound eyes had invaded the dorsal side of the head, and the legs were bent strongly outward at their base. Accordingly, the animals were usually notably broad, with a wide pleural fold. Here belong trilobites and similar groups, and probably the chelicerates.

In the primordial schizoramians, the segments were all similar behind the antennal segment. There usually were legs on all segments, as still in most trilobitomorphs. No part of the tergum behind the rostrum was involved in any formation of a head shield. Even this condition is traced up to the trilobitomorphs, in which the compound eyes moved up on the dorsal side behind the rostrum and were trapped there when a head shield developed, which it apparently had not yet done in *Sidneyia.*

The fossils invalidate Kukalová-Peck's (1992) hypothesis of a shared polyramous origin of an arthropod limb. They do not substantiate Lauterbach's (1980, 1983) daring speculations on primordial arthropods. They also do not support the Mandibulata theory, which is also made unlikely by strong embryological and molecular evidence.

The "soft-bodied" Lower Cambrian Chengjiang fauna was found by one of us (H. X.-g.) in 1984. The geographic location of the find is south of Kunming, Yunnan, in southwest China. It is remarkably similar to the Middle Cambrian Burgess Shale fauna from British Columbia, Canada, and must be considered as a representative of the same type of faunal community (Hou et al. 1991). However, it is probably some 15 million years older. In Siberian terms, it is probably of Atdabanian age. The "Cambrian explosion" is considered to have embraced the Tommotian and Atdabanian and to have lasted for about five million years (Bowring et al. 1993). This means that the Chengjiang fauna probably lived within the Cambrian explosion interval of most intense evolution. Gould's (1989) idea that most Cambrian animals belong to phyla that became extinct at a later date has no foundation in the systematic composition of faunas. In fact, both the fauna of the Burgess Shale and the Chengjiang fauna consist of roughly one-third sponges, one-third arthropods, and one-third lobopodians, brachiopods, worms, etc. (Whittington 1985a; Hou et al. 1991). Only about 5–10% of the species are not easily referable to extant phyla, which is a matter of uncertainty rather than proof of the presence of phyla that later became extinct.

The Chengjiang fauna from Yunnan is preserved in a mudstone that is much softer than the Burgess Shale from British Columbia. The animals are compressed, but not as flattened as in the Burgess Shale. These two factors make it much easier to prepare and interpret the fossils from Yunnan than those from British Columbia. The similarity between the faunas, with some genera and families shared, makes it possible for us to make a comparison with results from the Burgess Shale. It becomes increasingly apparent that many Burgess Shale species have been subject to misinterpretation in various respects.

This has been demonstrated regarding *Hallucigenia.* After studying Chinese material of the genus, Ramsköld and Hou (1991) realized that the previous reconstruction by Conway Morris (1977) showed the animal upside down. As noted by Bengtson (1991), this removed the enigmatic stamp from one more Cambrian group. However, turning again to the Burgess Shale material, Ramsköld (1992) introduced a new error by omitting the head from his reconstruction and interpreting the tail as the head. New Chinese material demonstrates the presence of a head at the end where Conway Morris (1977) had originally placed it (Hou and Bergström 1995). A misleading article on life in the Cambrian in *National Geographic* (Gore 1993:127) shows our reconstruction.

Similarly in Burgess Shale *Naraoia*, the proximal leg segment was tilted some 45° to fit against the body, and the number of cephalic legs, the number of endopod segments, the length and nature of the basal articulation as well as the insertion angle of the exopod, and the nature of the exopod constituents were mistaken (Whittington 1977). Notably, these cases are just examples. The kind of preservation of the Burgess Shale material misleads its interpreters to "enigmatize" the animals in their reconstructions. Conclusions on morphology, relationships, and evolution based on these reconstructions are hence unreliable at best and cannot serve for far-reaching evolutionary or phylogenetic interpretations (fig. 4.1).

It is our opinion that Størmer's (1944) famous study of the "soft-bodied" arthropods from the Burgess Shale and other lagerstätten on the whole gave us a more realistic view of the morphological variation and of general relationships than the later restudies by Whittington and his group (see papers by Briggs, Bruton, and Whittington in the list of references) or Simonetta and Delle Cave (1975). Some of Størmer's observations and conclusions no doubt were incorrect, but this does not change the positive main impression of his study. Later attempts to analyze the relationships (e.g., Briggs and Whittington 1981; Briggs and Fortey 1989; Briggs et al. 1992) have neglected Størmer's important observation on the unique shape of the setae (e.g., the lamellar setae) in trilobitomorphs and added some erroneous data. They have also ignored the overwhelmingly detailed and important information on the origination of arthropod structures resulting from studies of the Upper Cambrian alum shale (e.g., Walossek and Müller 1990, 1992; Walossek 1993). In addition, they have made use of the number of segments in the head although this appears to have no bearing on relationships between the groups. All this has resulted in analytical instability and thus strikingly variable phylogenetic trees.

The study of the Chengjiang arthropods is not completed. We wish to refrain from speculations on species that are as yet incompletely known and therefore base the following account on a limited number of species not yet formally classified in a hierarchy. However, a few observations must be made in order to relate the Chengjiang forms to the arthropods of other faunas.

*Anomalocaris* and allied taxa are not dealt with here (see Hou et al. 1995 for descriptions of Chengjiang anomalocarids). They have features in common with early schizoramian arthropods as well as features alien to arthropods, such as the circle of mouth plates.

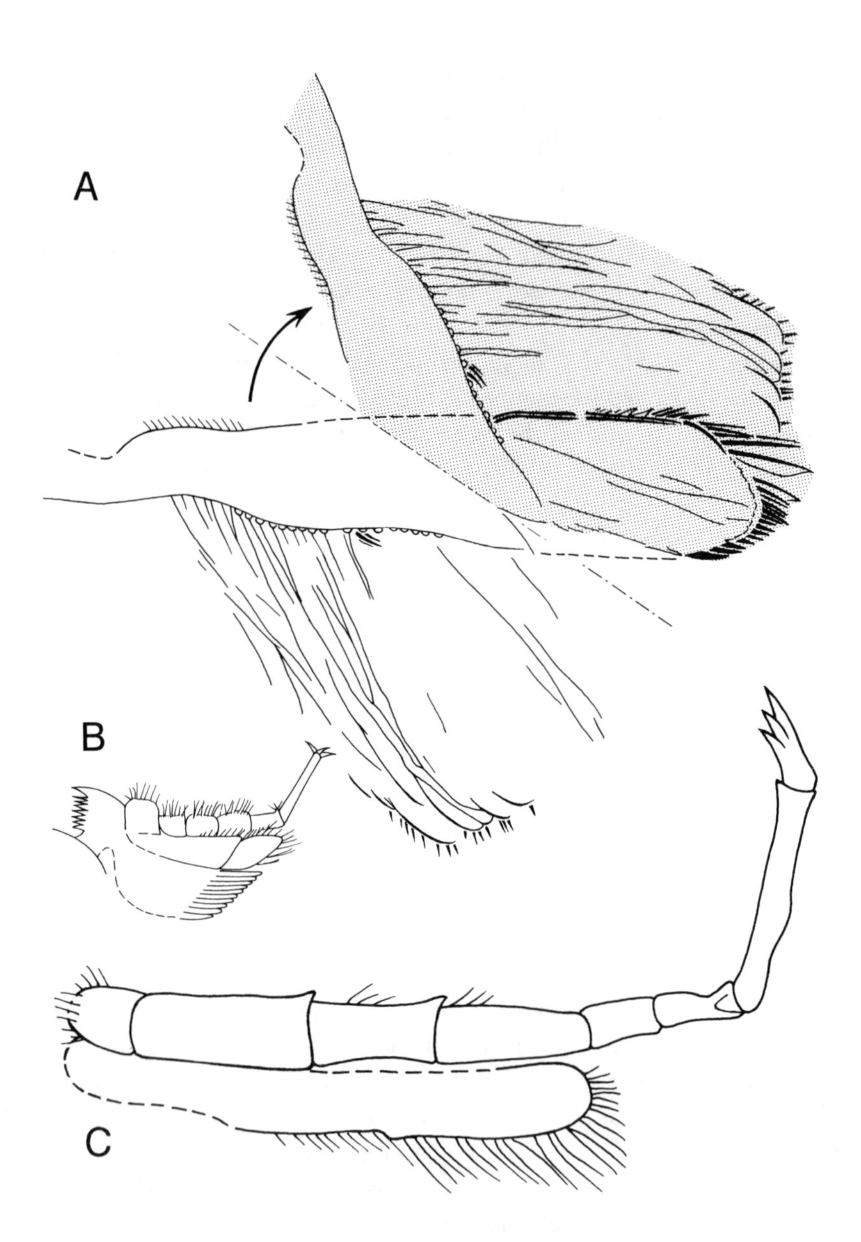


FIGURE 4.1

In light of new evidence, much of the confusion on early arthropod morphology and relationships caused by the Burgess Shale material is understood to be the result of poor preservation and mistaken interpretation. Thus, for instance, Bruton and Whittington (1983:563 and fig. 65) claimed that *Emeraldella brocki* had triramous limbs (fig. 4.1B herein, from their illustration). This was one reason for the Burgess Shale working group to reject Størmer's (1944) well-founded idea that many early arthropods had biramous "trilobite limbs" indicating affinity between them. A closer look at the specimen used for the reconstruction (shaded part of our fig. 4.1A, from Bruton and Whittington 1983, figs. 47, 49) shows that the supposed two rami consist of a single exopod ramus that is folded over on itself. Our figure 4.1C shows the *Emeraldella* limb as straightforwardly drawn from their figure 35, endopods 9 and 10 on the left side, and exopod 8 on the right side, with minor adjustments for size differences. *Emeraldella* thus had biramous "trilobite limbs," as noted by Størmer (1944).

A review of the stratigraphy and vertical distribution of the fauna was presented by Jin et al. (1993).

## Terminology

There is a considerable degree of terminological confusion in the literature. One reason is that terms are defined differently in crustaceans and trilobites. We therefore want to define our usage of some terms.

### *Tergites*

A tergite is a skeletal element on the dorsal side of the animal. Tergites can be confined to segments and are then called *segmental tergites*, *thoracic tergites*, etc. In this case, they are sometimes erroneously referred to as "segments" in the literature. The tergites can have pleural folds.

There is commonly a small tergite at the anterior end. This tergite is called the *rostral plate*. It can extend over to the ventral side. The (main part of the) head is covered by a tergite that in trilobites appropriately is called a head shield. Less appropriately, it has been given the name of the entire head (including), viz. "cephalon" (Moore 1959:O119). In crustaceans, it is called a cephalic shield. Here we use the term *head shield*. The head shield can extend into a pleural fold. A much more troublesome situation occurs when the alleged head shield is extended backward to cover one or more thoracic tergites. In crustaceans, such an extension is called a carapace. It has been defined as a fold extending from the maxillary segment. Dahl (1991) has demonstrated that this is an illusory definition, since no crustacean appears to have such a structure. Instead, the fold attaches to a thoracic segment in both branchiopods and malacostracans. In trilobites, the term *carapace* is synonymous with *dorsal exoskeleton* (Moore 1959:O119). We use the descriptive term *posterior extension of the head shield*. The implication is that the fold may attach directly to the head or to a thoracic extension of the head shield.

### *Limbs*

In trilobites, the first appendage is the antenna; it is succeeded by legs. In crustaceans, the first and second appendages are called antennula and antenna, respectively, or first and second antenna. Most arthropod groups conform with the trilobites as far as the use of the term *antenna* is concerned. To avoid confusion, we use *antenna* in animals with one pair of antennae, and *first* and *second antenna* in crustaceans with two pairs of antennae.

We use the terms *endopod* and *exopod* for the two branches of the schizoramian limb, even though we do not know the exact mode of branching in several forms. The shared stem "should" be a *basis* or *basipod*; this is difficult to define in forms such as *Fuxianhuia*, in which the leg clearly consists of a single pediform stem comprising presumptive basipod and endopod, while the exopod is a simple lateral fold.

Stiff outgrowths of the appendages are called *setae* if they are articulating at their attachment, *spines* if they are not. For trilobite-type stout lamellae we abandon the term *lamellar spine* in favor of *lamellar seta*, considering the fact that they are not spinous and that they have a joint at their base.

The stance of the appendages is subject to variation. In most early arthropods they extend downward and outward from the body, which means that they can support the body steadily when on the bottom. This stance can be called *semipendent*. Some crustaceans do not walk but use the legs for feeding, and they can hold them closely together along the midline. This is a *pendent* stance. It can be difficult to distinguish between these two types. A third type is the *laterally deflexed* stance. This is seen, e.g., in trilobites, in which the legs are attached to the underside of the body but extend laterally rather than downward. Usually the distal part is bent downward.

## Arthropods in the Chengjiang Fauna

The following list summarizes the arthropods described so far from the Chengjiang fauna. The bradoriids in the Lower Cambrian of China must be revised before a reliable number of species can be given. As the situation is now, compactional deformation features have been interpreted as biological features. As a result, one species of *Kunmingella* found together with the Chengjiang fauna at Maotianshan has been described as fifteen species distributed in three genera and three subgenera (Huo and Shu 1985). Arthropods that are treated in more detail here are marked with an asterisk (*).

Arthropoda incertae sedis

*Urokodia aequalis* Hou, Chen, and Lu, 1989
*Acanthomeridion serratum* Hou, Chen, and Lu, 1989
*Vetulicola cuneatus* Hou, 1987

Schizoramian arthropods devoid of exopod setae

*Fuxianhuia protensa* Hou, 1987*
*Chengjiangocaris longiformis* Hou and Bergström, 1991

Great Appendage Arthropods

*Leanchoilia illecebrosus* (Hou, 1987)
*Jianfengia multisegmentalis* Hou, 1987

Unassigned schizoramian arthropods with exopod setae

*Canadaspislaevigata laevigata* (Hou and Bergström, 1991)
*Chuandianella ovata* (Li, 1975)
*Isoxys auritus* (Jiang, 1982)

*Isoxys paradoxus* Hou, 1987
*Branchiocaris? yunnanensis* Hou, 1987
*Odaraia? eurypetala* Hou and Sun, 1988
*Combinivalvula chengjiangensis* Hou, 1987

Bradoriids

*Kunmingella maotianshanensis* Huo and Shu, 1983
*Kunyangella cheni* Huo, 1965
*Tsunyiella zhijinensis* Yin, 1978
*Jiucunella paulula* Hou and Bergström, 1991

Non-trilobite trilobitomorphs

*Naraoia longicaudata* Zhang and Hou, 1985*
*Naraoia spinosa* Zhang and Hou, 1985
*Kuamaia lata* Hou, 1987*
*Rhombicalvaria acantha* Hou, 1987
*Saperion glumaceum* Hou et al., 1991
*Retifacies abnormalis* Hou, Chen, and Lu, 1989
*Xandarella spectaculum* Hou et al., 1991*
*Sinoburius lunaris* Hou et al., 1991

Trilobites

*Eoredlichia intermedia* (Lu, 1940)
*Yunnanocephalus yunnanensis* (Mansuy, 1912)

## The Fossil Material

### Fuxianhuia

*Fuxianhuia protensa* is a fairly large arthropod, reaching a length of some 10 cm. Its general habitus resembles that of a eurypterid or euthycarcinoid (figs. 4.2, 4.3). A thorax of eighteen segments is succeeded by an abdomen devoid of legs and consisting of thirteen segments and a supposed telson. The entire thorax and the anterior part of the abdomen carry long pleural folds. In the main part of the abdomen there are only short folds. The supposed telson is extended into a median spine. In side view, there appears to be a slightly shorter spine below the median spine. This may be one of a pair of telson spines.

The head region is difficult to understand in all its details. The most anterior part is a separate rostrum. This is a short but wide tergite that ends laterally in a pair of clublike structures, the eyes. The central part of the head is rounded, but the head shield is extended laterally and posteriorly into a fold that covers two or three

FIGURE 4.2

*Fuxianhuia protensa*, no. NIGPAS 115319 in the archive of Academia Sinica, Nanjing. Photograph in dorsal view (see fig. 4.3 for drawing). Multisegmented endopods and round, flat exopods have been exposed by preparation. Note the crowding of the legs: two or three pairs of them appear to correspond to a single tergite. The specimen is 88 mm long.

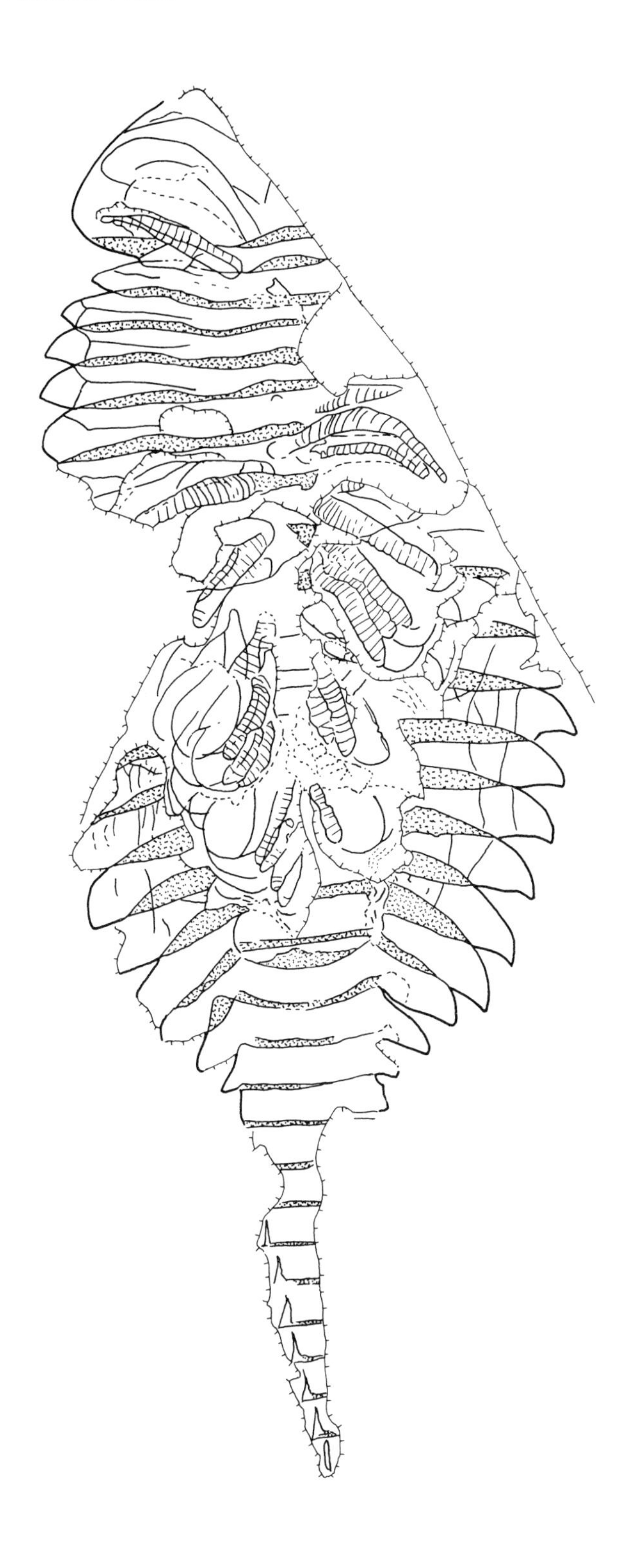

FIGURE 4.3
*Fuxianhuia protensa*, no. NIGPAS 115319. Drawing of specimen in figure 4.2.

successive tergites. A pair of short antennae extend from the lower anterior part of the head, from just behind the supposed acron. Behind the head follow three thoracic tergites with successively increasing width, but they are narrower than the successive part of the thorax.

This animal possesses features that appear to be of primordial arthropod design. In particular, the thoracic appendages exhibit a combination of undifferentiated characters (figs 4.2, 4.3). Thus, the pediform endopod is clearly the main stem of the limb, while the exopod is a lateral branch. Furthermore, there is a large number of short undifferentiated podomeres (up to twenty-one) in the main stem, and the terminal member is a simple, roundly conical element without a claw but with a number of short setae. The exopod is a very simple rounded flap devoid of visible setae. It simply extends from the side of the main stem. The exact mode of branching is not known. Even more surprising is that each tergite appears to cover more than one pair of legs, possibly around three pairs.

### *Canadaspidids*

So far, *Canadaspis perfecta* (Walcott, 1912) has been the best known canadaspidid (fig. 4.11b). This species was long thought to belong to the genus *Hymenocaris*. Now, the two genera are usually placed in two different orders, the Canadaspidida Novozhilov in Orlov, 1960, and the Hymenostraca Rolfe, 1969 (= Prophyllocarida Simonetta and Delle Cave, 1975). However, *Canadaspis perfecta* and *Hymenocaris vermicauda* Salter, 1853, are similar to one another in the shape of the posterior extension of the head shield, in the number and appearance of the abdominal segments, and in the split character of the furca. A new species in the Chengjiang fauna shares these characters and exposes the flat oval exopods typical of *Canadaspis perfecta*. As in *H. vermicauda*, the furca has two pairs of lateral spines. Without knowing the appendages of *Hymenocaris*, its exact relationships are in doubt, but it may be fairly close to *Canadaspis*.

*Perspicaris dictynna* (Simonetta and Delle Cave, 1975) is fairly poorly known, but one specimen appears to show similar flat oval exopods as *Canadaspis perfecta* (Briggs 1977, fig. 4; pl. 67, fig. 4). An appendage found in association with *P. recondita* (Briggs 1977, fig. 22; pl. 72, fig. 1, 3) is of unique appearance. The furca of *Perspicaris* consists of a single pair of unbranched spines, which is a difference between this genus and *Canadaspis* and *Hymenocaris*. The new species from the Chengjiang fauna appears to be morphologically intermediate in this respect.

### *Great Appendage Arthropods*

The Chengjiang fauna includes one species each of *Leanchoilia* and *Jianfengia*. They have not yet been closely studied. They have a shrimplike appearance, but have unsegmented exopod flaps fringed with setae. As in *Fuxianhuia* and *Canadaspis*, there is a pediform main stem, on which the simple exopod flap is just a lateral outgrowth. A uniting character is the so called great appendage, which is the most anterior appendage. Some podomeres of this appendage are produced into long extensions, making the appendage look branched. A small antenna is present in *Jianfengia*. Despite the lack of an exopod, the great appendage therefore is the second appendage. The Burgess Shale genera *Leanchoilia*, *Actaeus*, and *Yohoia* belong also in this group. *Sanctacaris*, also from the Burgess Shale, has similar exopods and may be related. We fail to see any similarity between this genus and the chelicerates, with which it has been associated (Briggs and Collins 1988).

Dieter Walossek has pointed out to us that there must be a clear functional difference between great appendages extending into long, slender, and annulated processes (*Alalcomenaeus*, *Leanchoilia*) and those with strong, sharp spines (*Yohoia*, *Jianfengia*). The latter probably speared their prey in much the same way as the stomatopod crustaceans. *Sanctacaris* probably represents a similar adaptation, but instead of a single large appendage there is a whole series of anterior grasping legs.

### Naraoia

Two species of *Naraoia* occur in the Chengjiang fauna. They are closely comparable with the two *Naraoia* species recorded from the Middle Cambrian of North America (cf. Whittington 1977). The preservation of the Chengjiang specimens is superior to that of the Burgess Shale specimens. This has made it possible to check the validity of observations made on the Canadian species. It turns out that a few general observations are correct, such as the observation of two large tergites, a gut with diverticula, one pair of antennae, and biramous appendages (figs. 4.4, 4.5, 4.11c–d). However, many other observations are incorrect, as is the conclusion that *Naraoia* is a trilobite. Thus, there are three pairs of post-antennal biramous limbs in the head, the supposed "coxa" is the basis and is in line with the endopod rather than tilted 45°, the attachment of the leg to the body is not constricted but large and roundly oval, the number of endopod segments is one more than depicted, the outer branch does not articulate only narrowly with the basis ("coxa") but with the full length of the basis and probably with part of the successive segment, and the supposed gill filaments are stiff lamellar setae with flattened cross section and distinct proximal articulation against the exopod axis. The basal part of the appendage is a long unsclerotized shaft. The shaft attaches to the body lateral to a medial sternite. Thus the appendage was not turned strongly laterally in the basis, like it was in trilobites, but was basically pendent. The variation in the fossil specimens indicates that the limb could be shifted between pendent and more laterally flexed postures and rotated in various ways. Our reinterpretation makes the *Naraoia* appendage closely comparable with the crustacean appendage, as was also realized by Walossek (1993, fig. 54).

*Naraoia* is obviously a trilobitomorph. The legs have the pendent character of *Marrella*, rather than the deflexed shape found, e.g., in trilobites, *Cheloniellon*, and chelicerates (fig. 4.11c). The lack of a limbless abdomen is a plesiomorphic character of no significance for the judgment on the naraoiid-trilobite relationship, and the development of a wide pleural fold may be a convergent phenomenon.

FIGURE 4.4

*Naraoia longicaudata*, isolated appendages without the soft cormus connecting them with the body. *(A)* NIGPAS 115315, unfolded specimen exposing segmented endopod and setose exopod. *(B)* NIGPAS 115316, specimen in which one branch is folded over the other. Length of exopod 14 mm. The difference clarifies the existence of a longitudinal joint between the two branches, which were probably held at an angle to one another.

There are no dorsal eyes in *Naraoia*, but a pair of fairly small rounded protrusions in front of the hypostome may represent paired ventral eyes.

The smallest specimens referable to *Naraoia* are only about 3 mm long (Hou et al. 1991, fig. 5). They have one pair of antennae and at least eight pairs of legs. The dorsum is covered by an undivided tergite, so no head can be recognized. Because of the undivided dorsal shield, the larvae were recognized as protaspids. This kind of larva may have occurred early in the evolution of schizoramian arthropods, and something similar is still present in the xiphosurids. The three first appendages are not in any way set off from the successive ones, which makes it clearly different from the nauplius larva of crustaceans.

### Xandarella

*Xandarella* is a genus so far found only in the Chengjiang fauna. Its only close relative appears to be another Chengjiang arthropod, which is so far undescribed. Although unlike all other trilobite-like arthropods in several respects, it is clearly of trilobitomorph design (figs. 4.6, 4.7).

FIGURE 4.5
*Naraoia longicaudata*, NIGPAS 115317, fragmentary exopods revealing the lamellar character of the setae. In dorsal view, the lamellae are regularly tilted toward the body midline in all specimens studied. Long side of photo 10.4 mm.

FIGURE 4.6

*Xandarella spectaculum*, NIGPAS 115286. Photographed in dorsal view (see fig. 4.7 for line drawing). The endopods and exopods of the limbs have been exposed by preparation. Scale as for figure 4.7.

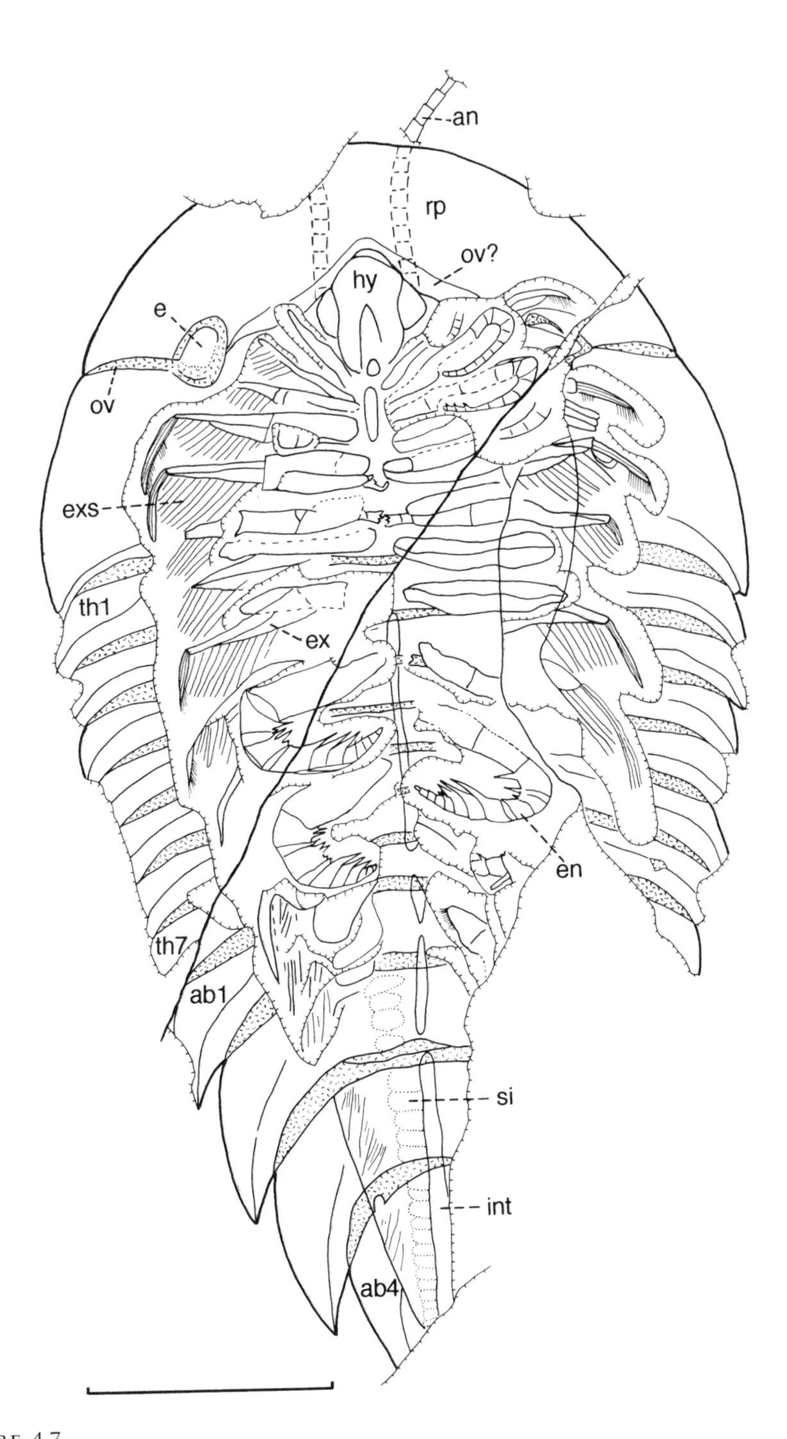


FIGURE 4.7

*Xandarella spectaculum*, NIGPAS 115286. Drawing of specimen in figure 4.6. Labels indicate the following: *ab1–ab4*, abdominal tergites; *an*, antenna; *e*, eye; *en*, endopod; *ex*, exopod; *exs*, exopod setae; *hy*, hypostome; *int*, intestine; *ov*, overlap between tergites; *rp*, rostral plate; *si*, segmental impression; *th1–th7*, thoracic tergites. Scale bar 1 cm.

The cephalon is large and semicircular because of a wide pleural fold. A presumed rostral plate is weakly defined by a thin line extending between the eyes and curving in front of the hypostome area. The paired compound eyes appear to lie on separate small tergites. Between the eye and the lateral margin of the head the boundary between the rostral and successive plate is a simple overlap obviously without fusion. This boundary probably marks the path of the eye moving from the ventral to the dorsal side of the head. In an embayment behind the head shield is a small tergite that is much smaller than the succeeding thoracic tergites. It seems to be a question of convenience whether this tergite is counted with the head or with the thorax. Behind it follow seven short thoracic tergites and four successively longer abdominal tergites.

One pair of uniramous antennae and six pairs of biramous appendages are found under the complex head shield. The small successive tergite and the seven thoracic tergites each correspond to one pair of legs. The abdominal tergites each cover more than one pair of legs (and thus segments), from two pairs in front to at least twelve pairs in the last part. The legs are of trilobitomorph design, being extended laterally and provided with endopod and exopod branches. The exopod has two segments, the proximal one with two rows of lamellar setae, the distal one with one row. Each seta is attached to the axis by a joint. All lamellar setae are probably directed posteriorly above/outside the next succeeding limb. Each lamella has a characteristic tilt (cf. figs. 4.5, 4.11d).

*Helmetiids*

*Kuamaia* is closely related to another Chengjiang arthropod, *Rhombicalvaria*, and also to the Burgess Shale genus *Helmetia*. *Kuamaia lata* is the best known species (figs. 4.8, 4.9). They share the general shape, with a spiny, trianguloid tail and a distinctive rostral plate. They have also a tendency to fuse the tergites, particularly laterally. The lamellar setae are flat and broadly lanceolate in shape (figs. 4.8, 4.9).

This tendency has gone much further in *Saperion glumaceum* (see Hou et al. 1991), whose tergum forms an oval shield that must have been virtually rigid. The rostral plate is similar to that of typical helmetiids. The segmental boundaries are still distinct in the middle part of the body but are effaced both along the margins and to a great extent in the anterior and posterior parts of the body. Therefore there are no distinct boundaries between head, thorax, and tail, if these tagmata can at all be recognized. This situation is similar to that found in *Tegopelte*, only that the effacement of the boundaries has proceeded still further in that genus. Contrary to Whittington's (1985b) judgment, there is no indication that *Tegopelte* had more than a single large tergite, which makes it difficult to define a head, because the legs indicate no tagmosis. The supposed tergite boundaries are neither regularly arranged nor strictly transverse and appear to be tectonically induced cracks. *Tegopelte* and *Saperion* are trilobitomorphs, but no trilobites.

### Sidneyia *and* Cheloniellon

The Burgess Shale genus *Sidneyia* and the Hunsrück Slate genus *Cheloniellon* are added for comparison and perspective. Both are trilobitomorphs with specializations indicating how chelicerates could have arisen from trilobitomorphs. In *Sidneyia*, the proximal leg segment has great similarities with the state found in the highly derived modern xiphosurids (Bruton 1981). This state cannot have been present in the earliest chelicerates, not even in the earliest merostomes. In *Cheloniellon*, the basis of the head legs is similar to that of the eurypterids, and the first post-antennal limb has become a pre-oral manipulating organ like the chelicera of chelicerates (Stürmer and Bergström 1978).

### *Trilobites and Chelicerates*

Trilobites are represented in the Chengjiang fauna, but we have not yet studied their appendages in detail. As seen from the cladogram (fig. 4.10), trilobites and chelicerates belong to the arachnomorph radiation, but share no characters they do not also share with other arachnomorphs. Lauterbach's (1980, 1983) speculation on a trilobite origin of chelicerates, or in his view the chelicerate nature of some Early Cambrian olenellid trilobites, is based on a single character in a paper animal constructed by him, while he ignored a number of characters present in the actual animals. The supposed observation to be seen only in the paper animal is that there are fifteen wide anterior thoracic segments and additional narrow posterior segments in *Olenellus*. This magical count would make the number of segments identical to that in the chelicerates. However, *Olenellus* has only fourteen wide segments, *Elliptocephala* and *Nephrolenellus* each have thirteen. In Lauterbach's drawings of *Olenellus*, the fifteenth thoracic tergite has the width of the fourteenth but the axial spine of the fifteenth tergite. Lauterbach's definition of the prothorax and his phylogenetic discussion are based on this paper animal. Ramsköld and Edgecombe (1991) convincingly showed that his conclusions are incorrect: *Olenellus* is a fairly late olenellid trilobite, sharing trilobite synapomorphies, and has nothing to do with the origin of the chelicerates.

Lauterbach's belief that the axial spine of the fifteenth thoracic segment in *Olenellus* corresponds to the telson of merostomes is also unrealistic, first because the mentioned axial spine is segmental, the telson post-segmental, and second because the spinous telson is not a synapomorphy of the Chelicerata, but only of the Merostomata.

The chelicerates appear to form one of several branches in the "trilobitomorph" tree. They do not appear to be represented in the Chengjiang fauna or in any other Burgess Shale-type fauna and are therefore not dealt with in any detail here. As demonstrated by the list of characters and the cladogram, the chelicerate lineage probably arose as one of the top shoots in that tree.

## Key Evolutionary Characters

Schram (1993) recently offered a negative review of a valuable volume resulting from a successful symposium on crustaceans. A main point of his is that there is too little of a cladistic approach. However, his comments indicate that he has not read the volume carefully. In the case of a paper by Bergström (1992), which concerns us because it is on the type of material dealt with here, he claims that twelve characters

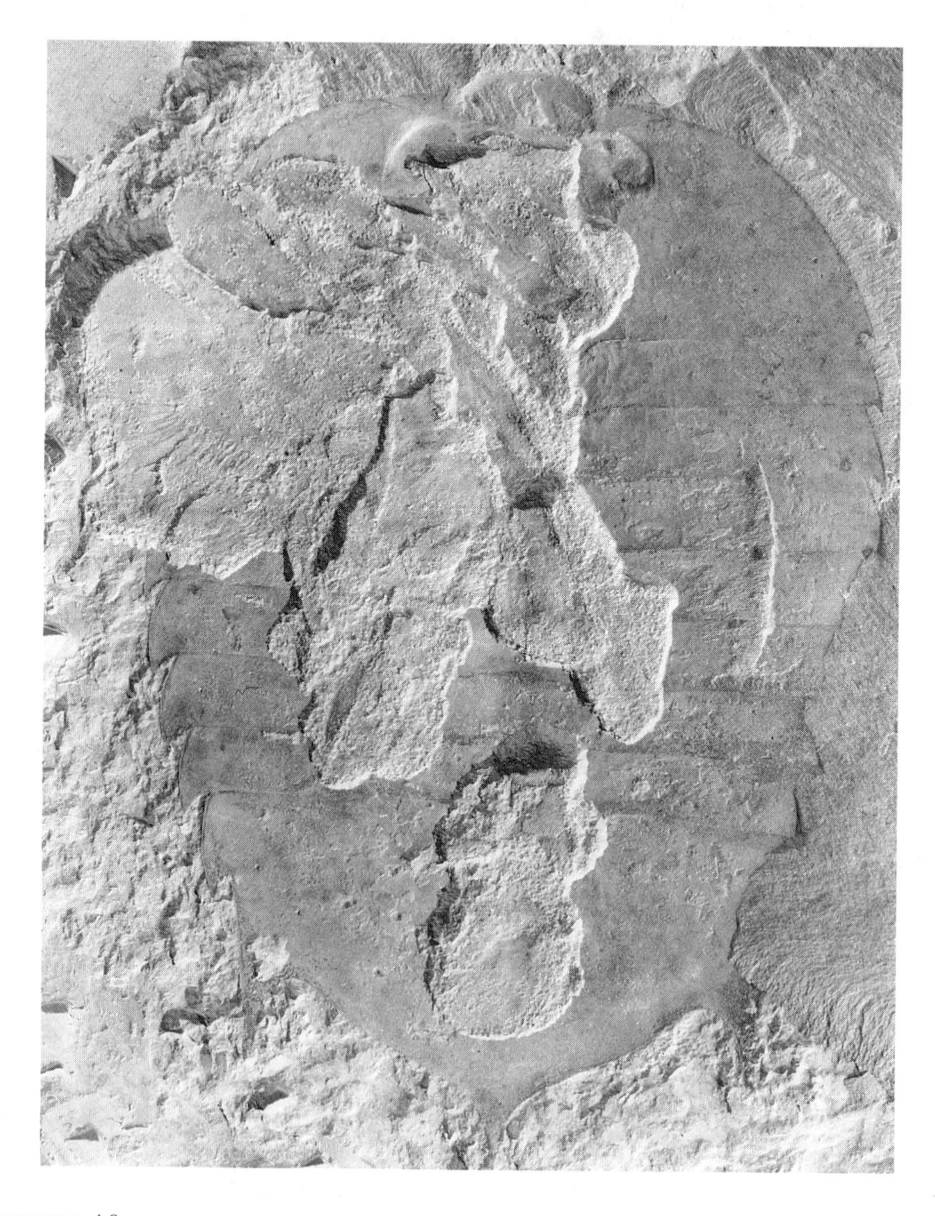

FIGURE 4.8

*Kuamaia lata*, NIGPAS 115318, photographed in dorsal view (see fig. 4.9 for line drawing). Endopods and exopods exposed by preparation. Letters as in figure 4.7. Scale as for figure 4.9.

have been used to sort thirteen arachnomorph and twenty-four crustacean-like genera. In his statement, Schram made three fundamental mistakes. First, the twelve characters of the character matrix are basic ones used for the separation of the arachnomorphs from other schizoramian arthropods, but not for the subdivision into further subgroups. Thus they are used for differentiating two groups and not thirty-seven as claimed by Schram. Second, a long list of additional characters was used for the subdivision in the cladogram, and those characters are actually listed in

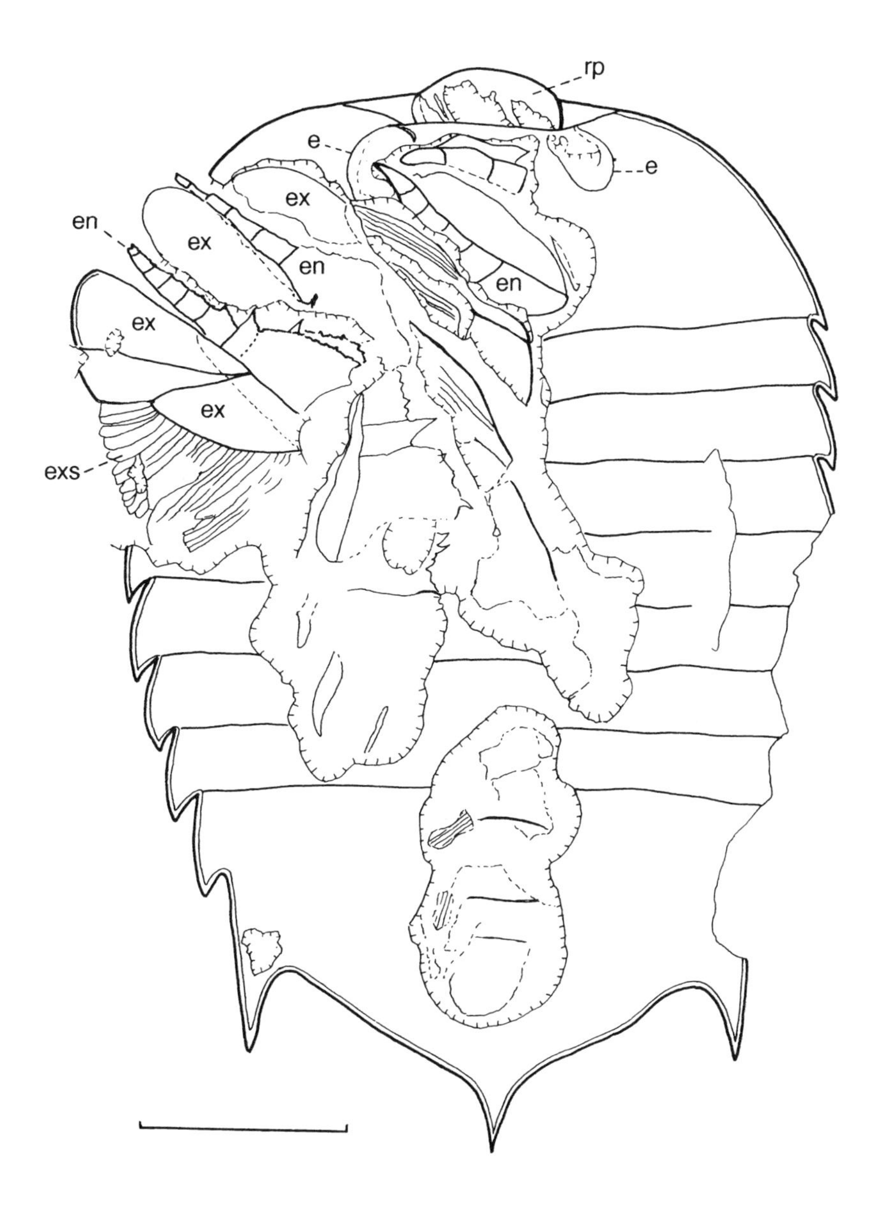


FIGURE 4.9
*Kuamaia lata*, NIGPAS 115318. Drawing of specimen in figure 4.8. Letters as in figure 4.7. Scale bar 1 cm.

the figure caption (Bergström 1992, fig. 3). Third, and most fundamental, it is not the number of characters but their relevance that is important. For instance, an echinoderm can be reliably identified as an echinoderm virtually on a single character, for instance, the crystal structure of its skeletal elements. It is not only superfluous but simply incorrect to list characters such as color, size, presence and number of arms and cirri, presence or absence of stalk, state of symmetry, etc., to classify it as an echinoderm. Furthermore, selecting protostomian-type features such as the trochophora-like doliolaria larva or the schizocoel of some ophiuroids for comparison would lead to placement at the wrong end of the animal kingdom. Characters used without an understanding of their historical order will most probably distort any cladogram. This is why computer programs for cladograms are very dangerous tools for those who think they have found a shortcut to map phylogeny.

Schram's (e.g., 1986) idea that remipedians are particularly primitive crustaceans is at odds with the very detailed observations made by Walossek and Müller (1990) on Cambrian "stem-group crustaceans." Thus, Schram (1993:820) finds it convenient to regard the observations and ideas of Walossek and Müller (1992) as controversial. However, alignment of rRNA sequences indicates that remipedians are maxillipods closely related to copepods (Abele in Walossek 1993:64). Schram has overlooked the close similarity in the limbs between remipedians and copepods and instead speculated that the remipedian body shape is primitive rather than derived. Similarly, although the malacostracan character of phyllocarids is well known among crustacean specialists, Schram has moved them to the phyllopods based on the presence of "filter-feeding" although the method of feeding is quite different from that of phyllopods.

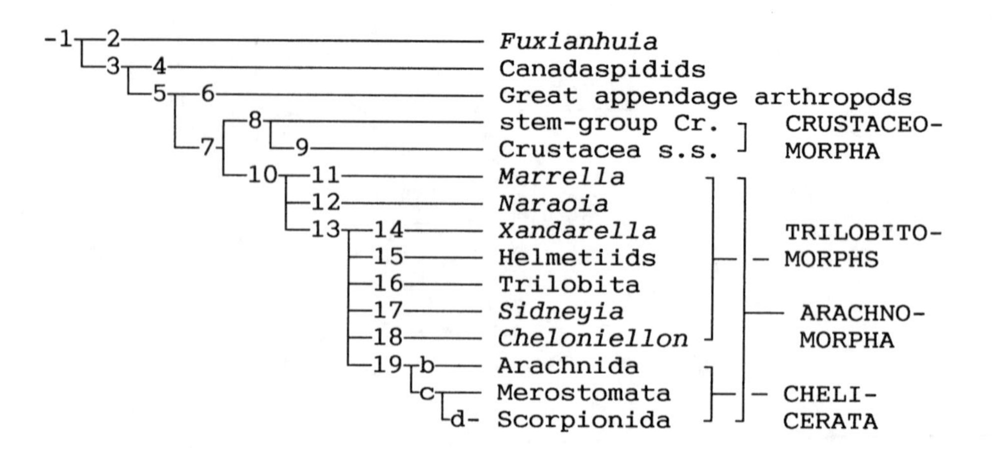


FIGURE 4.10
Suggested phylogeny of schizoramian arthropods. For characters, see text.

One very important lesson is that characters are not equally important in all situations (e.g., at all systematic levels) for the recognition of phyletic relationships. Evolution has a history, and if the historical events are ordered incorrectly, they are simply misleading.

The following characters have been used for the construction of the cladogram (fig. 4.10). For convenience, the cladogram has only one number in each internode even if there are two or more apomorphic characters, distinguished below by letters. Derived characters are as follows:

1a. Acquisition of lobopodian legs, which were semipendent to support the body safely on the substrate.

1b. Segmentation. This could easily be derived from the pseudo-segmentation characteristic of acoelomates and pseudocoelomates. It is worth remembering that most early bilaterian branches, including most acoelomates and pseudocoelomates, are pseudomeric and that pseudomery lingers on even in some coelomate phyla, namely among the nemertines and in primitive mollusks (Bergström 1991). The evidence for a pseudomeric origin of the Coelomata as a whole is thus very strong. The apparent misfit between tergite and appendage numbers in *Fuxianhuia* should be noted. Leg multiplication is known from branchiopods, but it cannot be taken for granted that the situation in *Fuxianhuia* is identical. It could also represent a primordial condition in which the metamery was not yet perfect.

1c. Hemocoel. A secondary body cavity or coelom may have developed more than once among the so called coelomates and also independently in the "pseudocoelomate" phyla Tardigrada, Priapulida, and Chaetognatha. It is utilized as a hydroskeleton in burrowing and sessile filter-feeding phyla. It is found in the embryo of many arthropods, where it may have a function in embryogenesis. It then degenerates, and the ultimate cavity is formed mainly by the blastocoel and is called a hemocoel because it houses the blood. It is formed in some arthropods by splits in the mesoderm, in others by mesoderm marginal folds that incorporate blastocoelic space.

1d. Acquisition of an articulated exoskeleton covering body and legs and associated loss of ciliation.

1e. Legs multisegmented, with a simple exopod flap (except in the antenna, which was plesiomorphically kept uniramous). Multisegmentation can be secondary, but the very simple and homonomic appearance of the multisegmented legs of several very old arthropods indicates that it is a synapomorphy for the (schizoramian) arthropods. In legs of this morphology, the exopod is clearly a simple unsegmented and non-setigerous, lateral extension from the segmented pediform leg (fig. 4.11a).

1f. Pair of telson spines. These spines are widely distributed among the arthropods, and are found even in *Fuxianhuia* and *Canadaspis*. In the latter, the spines appear merely as a prolongation of the surface ornament.

1g. Compound eyes on the ventral side of the head (Schizoramia). Among the Early Paleozoic arthropods, only chelicerates and some trilobitomorphs had their eyes dorsally on the head shield. When eyes are known from other Paleozoic forms,

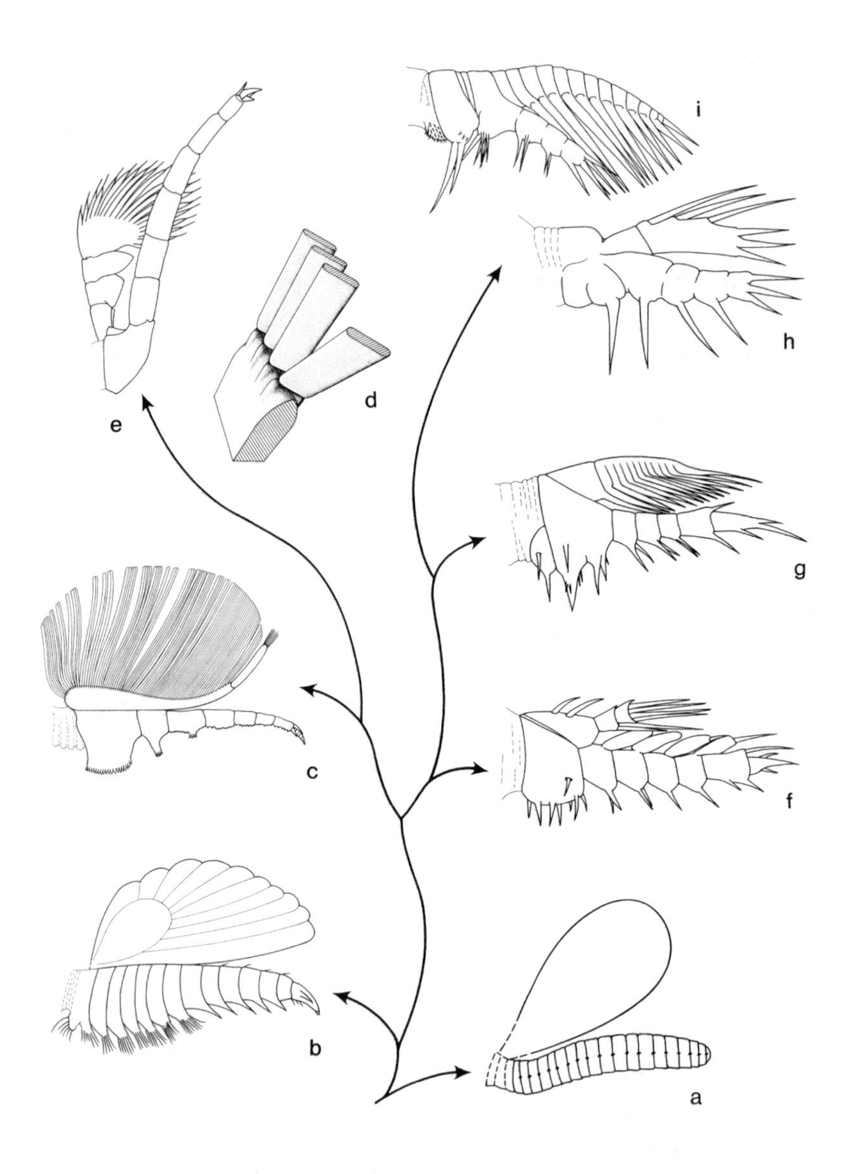


FIGURE 4.11

Suggested evolution of schizoramian limbs. *(a) Fuxianhuia protensa* Hou, 1987; *(b) Canadaspis perfecta* (Walcott, 1912); *(c–d) Naraoia longicaudata* Zhang and Hou, 1985, *(d)* detail showing insertion with proximal articulation typical of setae (but not of gills; compare probably marrellomorph setae illustrated by Walossek et al. 1993, fig. 5A–B); *(e)* trilobite *Ceraurus pleurexanthemus* Green, with laterally deflexed limb (modified from Størmer 1939); *(f)* "trilobite" *Agnostus pisiformis* (Wahlenberg, 1818) (redrawn from Walossek 1993); *(g)* stem-lineage crustacean *Martinssonia elongata* Müller and Walossek, 1986 (redrawn from Walossek 1993), with "incipient coxa" in the shape of an endite on the cormus; *(h, i) Skara anulata* Müller and Walossek, 1985. *(a, c, d)* Lower Cambrian Chengjiang fauna, *(b)* Middle Cambrian Burgess Shale fauna, *(e)* Middle Ordovician Trenton Limestone, *(f–i)* Upper Cambrian "Orsten" fauna. *(a and b)*, primitive limbs with multiarticulate endopod, non-differentiated podomeres, and simple exopod flap without strong setae. Note lamellar setae (and flexed basipod in part) in trilobitomorph lineage *(c–e)* and successive development of coxa in crustacean lineage *(f–i)*. Note the primitive nature of the alleged crustacean *Canadaspis (b)*.

they are situated on the underside of the head, or at least beneath the front margin of the head shield.

2. Formation of limbless abdomen; "head" tergite with posterior fold.

3. Spine(s) on tip of leg.

4. Formation of limbless abdomen; large bivalved tergal fold formed.

5. Setation on exopod (fig. 4.11c–i). Advanced end groups such as naraoiids, trilobites, agnostids, and crustaceans have exopods richly provided with setae. Setae appear to be absent in *Fuxianhuia*, as well as in the Burgess Shale *Canadaspis*. In both genera, the exopod is a simple unsegmented rounded flap, while the endopod is stout and multisegmented, with undifferentiated podomeres. We suggest that this morphology including the lack of exopod setae is primitive.

6a. Antenna reduced. Second limb forms "great appendage," which lacks an exopod. Despite some variation, it appears quite characteristic of a group represented in the Chengjiang fauna by *Alalcomenaeus* and *Jianfengia*.

6b. Loss of paired telson spines. The great appendage arthropods have a telson produced into a single median spine or plate.

7a. Segmentation of exopod. The exopod is unsegmented in *Fuxianhuia* and *Canadaspis*, which for other reasons appear to be early offshoots. Segmentation is also lacking in the Great Appendage Arthropods, which are also morphologically distant from the advanced trilobite-like and crustacean-like forms.

7b. A derived egg cleavage pattern, with a unique cell fate map not seen in any other spiral group (Anderson 1973:268–288). Whereas, for instance, in annelids, mollusks, myriapods, and hexapods the mesoderm is formed by the 4d cell, in crustaceans it is formed by the 3A, 3B, and 3C cells. Lack of spiral cleavage in chelicerates makes a direct cell-to-cell comparison impossible, but the space relationship between mesoderm and endoderm is like that of crustaceans, e.g., inverted in comparison with the other groups. This apomorphy may have been introduced still earlier, but the lack of living representatives of earlier offshoots makes it impossible to place it lower than at the divergence between the lineages leading ultimately to crustaceans and chelicerates.

8. "Stem-group crustacean" characters. The successive acquisition of character states in this lineage was discussed at length by Walossek and Müller (1990) and Walossek (1993): first antenna modified from purely sensory to swimming and feeding function, endite formed on all post-antennal legs proximal to the basis podomere, and the exopod became flagelliform, with inwardly directed setae at least in the second antenna and mandible.

9. "Crown-group crustacean characters" (see Walossek and Müller 1990; Walossek 1993): The apomorphic states for the true crustaceans include a nauplius larva with enhancement of the "proximal endite" and of the basis of the antennae and mandibles to form two distinct podomeres (fig. 4.11h–i), a large, fleshy labrum projecting over a funnel-shaped atrium oris, a post-oral sternum with paragnaths, which originate from the mandibular sternite, and a special setation and particularly setules on all parts concerned with feeding; associated structures of the adult, including a second antenna succeeded by a mandible.

10. Exopod setae develop into flat and closely set lamellar setae (fig. 4.11c–e). As was demonstrated by Bergström (1992, figs. 1–2), flat and closely set lamellar setae (trilobite lamellae, lamellar spines) occur exclusively in a group held together also by a set of additional characters: hypostome present, legs present to end of body (plesiomorphy), typically dorsoventrally flattened body, typically no paired appendages from the telson. In most of the forms the legs are laterally deflexed, there are large pleural folds, and the eyes are dorsal. The whole group constitutes the trilobitomorphs. The morphology and arrangement is clearly derived. Loss of furca.

11. Segmental pleura lacking. This feature is characteristic of the Middle Cambrian *Marrella* and *Burgessia* and the Devonian *Mimetaster*, in which the only pleural extensions are horns of the cephalon. The Devonian *Vachonisia*, which may be related, has an extensive shield.

11b. Multisegmentation of the exopod, each segment carrying only one lamellar seta of a row. There may be one row of lamellar setae, as in *Mimetaster* and *Marrella*, or two rows as in an exopod from the Middle Cambrian of the Georgina Basin, Queensland, Australia, illustrated by Walossek et al. (1993, fig. 5A, B).

12. Two dorsal tergites with wide pleural folds. The separation of the dorsum into two tergites is uncommon among arthropods. Head with five limb-bearing segments.

13a. Appendages laterally deflexed (fig. 4.11e). The proximal podomere, the basis, is strongly curved. This is a character of the advanced trilobitomorphs. In *Marrella*, *Mimetaster*, *Burgessia*, and *Naraoia* the appendages are still semipendent, as in many Cambrian arthropods (the reconstruction by Hughes 1975 of the marrellomorph *Burgessia* shows strongly laterally deflexed and curved limbs, but in the actual specimens the limbs tend to be almost straight and ventrolaterally directed).

13b. Eyes located dorsally on shield. This is a character exclusively found in trilobitomorphs and chelicerates (Bergström 1992) as long as we deal with Lower Paleozoic schizoramians (but also in peracarids and tracheates among younger arthropods). All other early arthropods with eyes have them on the underside of the head, and they never penetrate the dorsal shield.

14. Abdomen with only four tergites, each covering more than one segment. Head with antennae and probably six leg pairs.

15a. Broad exopod axis with broad lamellar setae.

15b. Rostral plate small, and eyes situated close to the anterior end.

15c. Head includes antennae and probably three leg pairs.

15d. Segmental tergites becoming more or less fused. There was probably still some flexibility in *Kuamaia*, whereas in *Saperion* the entire dorsum must have acted as a stiff shield.

16. Calcification of dorsum, doublure, and hypostome; submarginal ecdysial suture; circum-ocular ecdysial suture; trilobation; eye penetrates through slit in head shield (rather than through a round opening as in other groups); eye-ridge that was originally double. In most post-Cambrian trilobites, the lower part of the circum-

ocular suture had been lost, the eye surface being fused to the free cheek. The head has antennae and three to four pairs of legs; the glabellar segmentation indicates that four leg pairs is the original number.

17. Large rostral plate. No fusion of post-rostral tergites to form a head. Tail fan.

18. Rostral sutures lost. First post-antennal limb modified to form a pre-oral uniramous "second antenna." Basis segment of successive four leg pairs with strong medially directed enditic spines. Large head tergite covers anterior five segments, while last leg of head is under a separate segmental tergite.

19. Loss of antennae; tagmosis affecting limbs (probably with loss of exopod in prosoma, of endopod in opisthosoma); eye forms rounded opening in tergite (rather than slit); prosoma with seven to eight segments (including antennal).

## The Schizoramians

The cladogram presents a picture that is consistent in having very few reversals. The first bifurcation is at a level with appendages being present along the entire body. The post-antennal limbs all consist of a multiarticulate (about twenty segments) pediform leg, with a simple exopod flap proximally on the outer side. In several lineages the legs disappear from the posterior part of the body, which will then form an abdomen. The number of podomeres decreases in different lineages. Furthermore, the exopod tends to develop marginal setae and a more complex shape.

We know from crustaceans that it is very difficult to use the presence or absence of a posterior fold ("carapace") in any phylogenetic analyses. Such a fold seems to come and go (Dahl 1983; Walossek 1993). The presence in *Fuxianhuia* and *Canadaspis* indicates that it is a very old character, although it is also absent in many early forms.

The Chengjiang arthropods provide us with new insight into the early evolution of schizoramian arthropods. *Canadaspis* and *Fuxianhuia* appear to represent an original stage with uniform legs, consisting of a multisegmented, pediform main branch and an exopod formed by a simple, rounded, lateral flap devoid of setae. *Fuxianhuia* may not yet have achieved true segmentation, and the termination of the leg is very clumsy and lacks claws.

Setae were added to the exopod in the next step. Groups on this level include the Great Appendage Arthropods such as *Alalcomenaeus*, the alleged chelicerate *Sanctacaris*, and probably *Habelia* and *Molaria*.

Thereafter, two main evolutionary lineages can be discerned. In one of them, the appendages stayed semipendent or even became distinctly pendent. This branch consists of agnostids, "stem-lineage crustaceans," and true crustaceans. The successive acquisition of character states was ably described by Walossek and Müller (1990, 1992) and Walossek (1993).

In the other lineage, the exopod setae became very long and flattened to be accommodated in a very tight row. The elements have commonly been regarded as "gill filaments." The basal constriction and the close similarity between the trilobite and crustacean type appendages make it obvious that they correspond to the crustacean setae. This is also clear from their striking rigidity, borne out both by the character of their preservation (they are held at an angle with the bedding plane without being distorted or compressed) and the fact that they made distinctive scratches preserved in trace fossils (Bergström 1973, 1976; Seilacher 1994).

This evolutionary lineage is made up of the trilobitomorphs and, apparently, the chelicerates. After the marrellomorphs and naraoiids were branched off, the lineage leading to the remaining trilobitomorphs and chelicerates developed a strong flexure in the basis of the limbs and moved the compound eyes to the dorsal side. The eyes appear to have moved up between the rostral tergite and the successive tergite. This is where it is actually seen in *Sidneyia*. If this scenario is right, the crustacean-like arthropods and the trilobitomorph-chelicerate lineage thus form two advanced sister groups within a monophyletic group Schizoramia.

Our results thus support Størmer's (1944) conclusion that the Arachnomorpha (e.g., Chelicerata plus Trilobitomorpha including Trilobita) are united by distinct synapomorphies. They also give a strong indication that the arachnomorphs and "pan-crustaceans" form sister groups within a major schizoramian group. This challenges Manton's (1977) speculation that there are four major arthropod groups, namely, the Uniramia, Crustacea, Chelicerata, and Trilobita.

## The Origin of the Uniramian Arthropods and the Mandibulata Concept

It is more difficult to make definite statements on the origin of the Uniramia (Myriapoda and Hexapoda = Atelocerata) and therefore also on the reality of the old unit Mandibulata (e.g., the Myriapoda, Hexapoda, and Crustacea). We have not seen any limbs in the Chengjiang fauna that could be interpreted as uniramian limbs or precursors of such. It may be said that the limb of *Fuxianhuia* brings schizoramians one step closer to uniramians from a purely morphological point of view. However, the morphology of the *Fuxianhuia* endopod is one that might be expected in a primeval arthropod. It does not prove or even indicate a shared origin of schizoramians and uniramians within the Arthropoda.

Budd (1993, fig. 4) developed a scheme in which *Anomalocaris*, *Opabinia*, and his genus *Kerygmachela* form a sister group of schizoramian arthropods. The list of characters includes a number of mistakes that are fatal for the result. For instance, the uniting characters are said to be a "biramous" lobopodium, with gills borne on the lateral branch (Budd's apomorphy number 8). However, Budd illustrates *Kerygmachela* as clearly uniramous, with the supposed outer branch being a pleural fold forming a discrete extension from the body, not from the leg. Furthermore, no gills are known from either *Opabinia* or *Anomalocaris* (Bergström 1987; Dzik and Lendzion 1988), and gills in schizoramians are not on the lateral branch, the exopod (except for xiphosurids; the related eurypterids had gills directly on the underside of the body, see Wills 1966), but may be newly formed as epipodites in some crustaceans.

The Carboniferous (Mississippian) *Tesnusocaris* is too young to be of much value and too poorly preserved to be safely interpreted. Still, Emerson and Schram (1990, 1991) used their interpretation of it to suggest that biramous appendages arose through fusion of uniramous appendages. There is thus some similarity between this case and the one presented by Budd. *Tesnusocaris* is controversial even on the descriptive level, and it shows no obvious morphological similarity with the primitive Cambrian arthropods.

Are mandibulates a monophyletic group, or did mandibles develop more than once? There is a striking scarcity of fossil uniramians, or at least of arthropods that can be positively identified as uniramians, in the Cambrian. We have no evidence of any mandibulate schizoramians in the Lower Cambrian. The Burgess Shale arthropod that is most often considered as a crustacean is *Canadaspis* (e.g., Briggs 1978, 1983, 1992). Results from studies on the arthropods from the Upper Cambrian "Orsten" (a bituminous limestone, usually translated as "stinkstone") indicate that *Canadaspis* lacks a series of characters that are typical of true crustaceans (Walossek and Müller 1990), and our study of Chengjiang arthropods indicates that it is a very primitive non-crustacean schizoramian. The oldest definite crustaceans are from the "Orsten" of Sweden. However, *Branchiocaris* and its allies may represent Middle Cambrian phyllopod crustaceans.

A true coxa was not developed in the stem-group crustaceans (Walossek and Müller 1990), and as a consequence they can not have had any mandible, as this structure is formed by the coxa of the mandibular segment. Three main alternatives appear available regarding the origin of the Uniramia (myriapods and hexapods) and the fate of the Mandibulata concept: (1) the uniramians evolved from early stem-group crustaceans, which would make the Mandibulata a monophyletic group; (2) the uniramians share an origin with schizoramians; this origin might have been fairly close to *Fuxianhuia*, and mandibulates would then have originated twice; (3) uniramians and schizoramians do not share an arthropod ancestor, e.g., the Arthropoda are not monophyletic. A fourth alternative, in which crustaceans evolved from uniramians, is distinctly in conflict with the paleontological evidence on the origin of the crustaceans (Walossek and Müller 1990 and the evidence presented herein).

As to the first alternative, it would seem that the crustaceans and their appendages are too specialized to have given rise to uniramians, and there is no obvious place for uniramian origins in the "pan-crustacean" lineage as delineated by Walossek and Müller (1990). However, lacking a useful record of Cambrian, Ordovician, and Silurian uniramians, this is difficult to judge. If parallel-sided myriapod-like forms such as the Lower Cambrian *Urokodia* (Hou et al. 1989) are uniramians, the chance of a derivation from late stem-group crustaceans decreases. However, as summarized by Wägele (1993), the greatest similarities with uniramian arthropods are found in the true crustaceans, not in their predecessors.

Wägele (1993) has collected a series of arguments in favor of the idea of the Mandibulata as a monophyletic end group of arthropods. What makes it particularly difficult to judge, however, is that he does not discuss the arguments against this idea, except for some that are probably incorrect. We know from living crustaceans (Cladocera, Cirripedia, Copepoda, Malacostraca) that their embryology (cell fates in egg cleavage) is strongly derived, with the positions for mesoderm and endoderm being virtually shifted in comparison with all other living animals except for chelicerates. The uniramian embryology is primitive and similar to that of other spiralian coelomates such as bryozoans, mollusks, annelids (Anderson 1973; see cladogram and comment, point 7b). From an embryological point of view, crustaceans and chelicerates therefore could be derived from uniramians, but not vice versa. As noted earlier (Bergström 1986), crustaceans share a distinctive cytochrome C mutation in an extremely stable N-terminus position with mollusks and deuterostomians, whereas uniramians are primitive in this respect. It is clear that the embryological and molecular evidence is in conflict with the Mandibulata hypothesis, which is convincing only if presented without this counterevidence. The fossils give no definite clue, but the embryological data alone make it highly improbable that uniramians evolved from the lineage leading to the crustaceans after the trilobite-chelicerate lineage was split off.

The fossil material does not permit us to choose between alternatives (2) and (3). The legs of primitive schizoramians such as *Fuxianhuia* and *Canadaspis* would appear to form also a good starting-point for uniramian legs. If either of these two alternatives is the right one, however, the Mandibulata would be diphyletic.

Kukalová-Peck (1992) found fossil evidence that some Permian insects had several exites on their legs and claimed that this is a more primitive situation than a more "derived" biramous state in trilobite-like arthropods and many crustaceans. She regarded this speculation as proof that all arthropods were originally polyramous, and that they therefore form a single phylum, in which biramous and uniramous groups are secondary developments. Her claim (Kukalová-Peck 1992:254) that primitive fossil crustaceans and chelicerates had additional exites on "several and various leg segments" finds no support in the fossil material. On the contrary, the Cambrian material shows that there was originally one, and only one, large exopod, which was unsegmented, contrary to the exites in the Permian insects. Her claim (same page) that chelicerates and trilobites had jaws is probably a slip of the pen. Her claim that the jaws (or corresponding appendages) of chelicerates, trilobites, crustaceans, insects, and myriapods have the same identical segments is false: Walossek and Müller (1990) clearly demonstrated how the crustacean limb, and particularly the second antenna and mandible, came into existence through the addition of a proximal segment to the limb found in trilobites and other primitive biramian arthropods. Our material shows that Kukalová-Peck's speculations based on insects and incorrect information on Lower Paleozoic marine arthropods have no bearing either on the formation of legs and jaws or on the phylogeny and interrelationships on the main arthropod groups.

Shear (1992) believed in Kukalová-Peck's speculations and, like her, regarded the Uniramia as a nonexisting phylum. We certainly agree that the Onychophora do not belong with the Myriapoda and Hexapoda, but unlike Kukalová-Peck and Shear, we do not regard that as proof that the two latter are related to other arthropods. The hard facts are lacking. Shear's speculation on another supposed phylum containing the Onychophora, Tardigrada, and Pentastomida demonstrates the weakness of the beliefs about relationships on the phylum level when they are based only on morphology. The pentastomids are now regarded as close relatives of branchiuran crustaceans (Wingstrand 1972; Abele et al. 1992), and they are utterly unlike tardigrades; onychophorans belong with the coelomates but are certainly no crustaceans; tardigrades have a number of features (such as cloaca, radial musculary pharynx with stylets, lack of circular muscles, lack of circulatory system) indicating that they are most probably aschelminths related to gastrotrichs and nematodes. Segmentation and legs obviously evolved independently in the three groups thought by Shear (1992) to belong in the same phylum, so why not in uniramians and schizoramians as well?

The Middle Cambrian *Cambropodus* was described as a uniramian arthropod (Robison 1990). We agree with Wägele (1993:275) that this interpretation is highly speculative. We do not feel convinced that the organism is an arthropod. The supposed legs are smoothly curved, without indication of segments, and extend from onion-shaped bases with no obvious counterpart in arthropods. Anyway, *Cambropodus* has no obvious connection with the arthropod lineages preserved in the Chengjiang fauna.

The mosaic pattern of character distribution makes it an objective fact that parallelism is extremely common in the animal kingdom. If that were not the case, the relationships between, say, the phyla would have been known long ago. If we still want to judge parallelism as unlikely in the specific case of mandible formation in the Arthropoda, then how do we reconcile the concept of the Mandibulata with the fact that uniramians are distinctly plesiomorphic in their cell fates, while crustaceans and chelicerates have a strongly derived pattern? And how do we explain that crustaceans share with mollusks and deuterostomians, but not with uniramians, a distinctive and entirely unique mutation in the cytochrome C molecule? As long as there is not even an attempt to answer these and similar questions, the theory that the Mandibulata form a monophyletic group is neither proven nor particularly reasonable.

## The Origin of Arthropods

Charles Darwin found the sudden appearance of the Cambrian faunas disturbingly incompatible with his ideas of evolution based on natural selection. He was unaware of genetics and knew of no mechanism that could bring about fast changes. He therefore supposed that more primitive arthropods and other animals had been evolving for long times before they were preserved in Cambrian strata. Likewise, Dawkins (1986) in his influential *The Blind Watchmaker* believes in long Precambrian prehistories of the phyla we see in the Cambrian because he believes neither in macromutations nor in sudden divine genesis of the phyla. In the absence of knowledge, Darwin could believe. In the days of Dawkins, however, science had evolved further. We now know of many mechanisms that may move forward evolution very quickly and even in what appears as steps. Such mechanisms include endosymbiosis, heterochrony, and several other types that are not to be considered as macromutations. We also know that there are no major sedimentary gaps in the Late Precambrian, that there are trace and body fossils, but that there are virtually no signs of coelomate phyla. If soft-bodied arthropods did exist, they ought at least to have left their tracks in the sediment, and they have not. We can only conclude that they were not there (e.g., Bergström 1990, 1991).

Others, including Glaessner (1984), have thought that arthropods are represented among the Ediacara life-forms from the Late Precambrian. However, the similarities between the Ediacaran organisms and arthropods are very superficial; in their basic organization they have nothing in common (e.g., Seilacher 1989).

The first arthropods appear to have evolved less than five million years before the Chengjiang fauna lived. We believe that *Chengjiangocaris* and other arthropods retaining primordial features give us an indication of how the arthropods came into existence from pseudo-segmented, wormlike animals through development of appendages and segmentation. A gene system capable of forming tagmosis was already in existence. Neither macromutation nor special creation is needed for an explanation.

## ACKNOWLEDGMENTS

Per Ahlberg, Lund, criticized a version of our manuscript in a constructive way. Dieter Walossek, Ulm, critically studied our specimens and gave us many invaluable suggestions and detailed information on "Orsten" and other early arthropods. He also thoroughly criticized our manuscript. We are deeply indebted for the constructive discussions and support, even if we have not always accepted the suggestions given to us.

## REFERENCES

Abele, L. G., T. Spears, W. Kim, and M. Applegate. 1992. Phylogeny of selected maxillopodan and other crustacean taxa based on 18S ribosomal nucleotide sequences: A preliminary analysis. *Acta Zoologica* 73:373–382.

Anderson, D. T. 1973. *Embryology and Phylogeny in Annelids and Arthropods*. Oxford: Pergamon Press.

Bengtson, S. 1991. Oddballs from the Cambrian start to get even. *Nature* 351:184.

Bergström, J. 1973. Organization, life, and systematics of trilobites. *Fossils and Strata* 2:1–69.

Bergström, J. 1976. Lower Paleozoic trace fossils from eastern Newfoundland. *Canadian Journal of Earth Sciences* 13:1613–1633.

Bergström, J. 1986. Metazoan evolution: A new model. *Zoologica Scripta* 15:189–200.

Bergström, J. 1987. The Cambrian *Opabinia* and *Anomalocaris*. *Lethaia* 20:187–188.

Bergström, J. 1990. Precambrian trace fossils and the rise of bilaterian animals. *Ichnos* 1:3–13.

Bergström, J. 1991. Metazoan evolution around the Precambrian-Cambrian transition. In A. M. Simonetta and S. Conway Morris, eds., *The Early Evolution of Metazoa and the Significance of Problematic Taxa*, pp. 25–34. Cambridge: Cambridge University Press.

Bergström, J. 1992. The oldest arthropods and the origin of the Crustacea. *Acta Zoologica* 73:287–291.

Bowring, S. A., J. P. Grotzinger, C. E. Isachsen, A. H. Knoll, S. M. Pelechaty, and P. Kolosov. 1993. Calibrating rates of Early Cambrian evolution. *Science* 261:1293–1298.

Briggs, D. E. G. 1977. Bivalved arthropods from the Cambrian Burgess Shale of British Columbia. *Palaeontology* 20:595–621.

Briggs, D. E. G. 1978. The morphology, mode of life, and affinities of *Canadaspis perfecta* (Crustacea; Phyllocarida), Middle Cambrian, Burgess Shale, British Columbia. *Philosophical Transactions of the Royal Society of London*, series B, 281:439–487.

Briggs, D. E. G. 1983. Affinities and early evolution of the Crustacea: The evidence of the Cambrian fossils. In F. R. Schram, ed., *Crustacean Issues*. Vol. 1, *Crustacean Phylogeny*, pp. 1–22. Rotterdam: Balkema.

Briggs, D. E. G. 1992. Phylogenetic significance of the Burgess Shale crustacean *Canadaspis*. *Acta Zoologica* 73:293–300.

Briggs, D. E. G. and D. Collins. 1988. A Middle Cambrian chelicerate from Mount Stephen, British Columbia. *Palaeontology* 31:779–798.

Briggs, D. E. G. and R. A. Fortey. 1989. The early radiation and relationships of the major arthropod groups. *Science* 246:241–243.

Briggs, D. E. G., R. A. Fortey, and M. A. Wills. 1992. Morphological disparity in the Cambrian. *Science* 256:1670–1673.

Briggs, D. E. G. and H. B. Whittington. 1981. Relationships of arthropods from the Burgess Shale and other Cambrian sequences. In M. E. Taylor, ed., *Short Papers for the Second International Symposium on the Cambrian System 1981: United States Department of the Interior, Geological Survey, Open-File Report* 81–743:38–41.

Bruton, D. L. 1981. The arthropod *Sidneyia inexpectans*, Middle Cambrian, Burgess Shale, British Columbia. *Philosophical Transactions of the Royal Society of London*, series B, 295:619–656.

Bruton, D. L. and H. B. Whittington 1983. *Emeraldella* and *Leanchoilia*, two arthropods from the Burgess Shale, Middle Cambrian, British Columbia. *Philosophical Transactions of the Royal Society of London*, series B, 300:553–585.

Budd, G. 1993. A Cambrian gilled lobopod from Greenland. *Nature* 364:709–711.

Conway Morris, S. 1977. A new metazoan from the Cambrian Burgess Shale, British Columbia. *Palaeontology* 20:623–640.

Dahl, E. 1983. Malacostracan phylogeny and evolution. In F. R. Schram, ed., *Crustacean Issues*. Vol. 1, *Crustacean Phylogeny*, pp. 189–212. Rotterdam: Balkema.

Dahl, E. 1991. Crustacea Phyllopoda and Malacostraca: A reappraisal of cephalic and thoracic shield and fold systems and their evolutionary significance. *Philosophical Transactions of the Royal Society of London*, series B, 334:1–26.

Dawkins, R. 1986. *The Blind Watchmaker*. London: Longman.

Dzik, J. and K. Lendzion. 1988. The oldest arthropods of the East European platform. *Lethaia* 21:29–38.

Emerson, M. J. and F. R. Schram. 1990. The origin of crustacean biramous appendages and the evolution of Arthropoda. *Science* 250:667–669.

Emerson, M. J. and F. R. Schram. 1991. Remipedia, part 2: Paleontology. *Proceedings of the San Diego Society of Natural History*, 1991, no. 7:1–52.

Glaessner, M. F. 1984. *The Dawn of Animal Life: A Biohistorical Study*. Cambridge: Cambridge University Press.

Gore, R. 1993. Explosion of life: The Cambrian period. *National Geographic* 184:120–136.

Gould, S. J. 1989. *Wonderful Life: The Burgess Shale and the Nature of History*. New York: Norton.

Hou, X.-G. and J. Bergström. 1995. Cambrian lobopodians: Ancestors of extant onychophorans? *Zoological Journal of the Linnaean Society* 114:3–19.

Hou, X.-G., J. Bergström, and P. Ahlberg. 1995. *Anomalocaris* and other large animals in the Lower Cambrian Chengjiang fauna of southwest China. *Geologiska Föreningens i Stockholm Förhandlingar* 117:163–183.

Hou, X.-G., J.-Y. Chen, and H.-Z Lu. 1989. Early Cambrian new arthropods from Chengjiang, Yunnan. *Acta Palaeontologica Sinica* 28:42–57.

Hou, X.-G., L. Ramsköld, and J. Bergström. 1991. Composition and preservation of the Chengjiang fauna: A Lower Cambrian soft-bodied biota. *Zoologica Scripta* 20:395–411.

Hughes, C. P. 1975. Redescription of *Burgessia bella* from the Middle Cambrian Burgess Shale, British Columbia. *Fossils and Strata* 4:415–435.

Huo, S.-C. and D.-G. Shu. 1985. *Cambrian Bradoriida of South China*. Xi'an: Northwest University Press.

Jin Y.-G., X.-G. Hou, and H.-Y. Wang. 1993. Lower Cambrian pediculate lingulids from Yunnan, China. *Journal of Paleontology* 67:788–798.

Kukalová-Peck, J. 1992. The "Uniramia" do not exist: The ground plan of the Pterygota as revealed by Permian Diaphanopterodea from Russia (Insecta: Paleodictyopteroidea). *Canadian Journal of Zoology* 70:236–255.

Lauterbach, K.-E. 1980. Schlüsselereignisse in der Evolution des Grundplans der Arachnata (Arthropoda). *Abhandlungen des naturwissenschaftlichen Vereins in Hamburg (NF)* 23:163–327.

Lauterbach, K.-E. 1983. Synapomorphien zwischen Trilobiten- und Cheliceratenzweig der Arachnata. *Zoologischer Anzeiger* 210:213–238.

Manton, S. M. 1977. *The Arthropoda: Habits, Functional Morphology, and Evolution*. Oxford: Clarendon Press.

Moore, R. C., ed. 1959. *Treatise on Invertebrate Paleontology, Part O, Arthropoda 1*. Boulder, Lawrence: Geological Society of America, University of Kansas Press.

Ramsköld, L. 1992. The second leg row of *Hallucigenia* discovered. *Lethaia* 25:221–224.

Ramsköld, L. and G. D. Edgecombe. 1991. Trilobite monophyly revisited. *Historical Biology* 4:267–283.

Ramsköld, L. and X.-G. Hou. 1991. New Early Cambrian animal and onychophoran affinities of enigmatic metazoans. *Nature* 351:225–228.

Robison, R. A. 1990. Earliest-known uniramous arthropod. *Nature* 343:163–164.

Schram, F. R. 1986. *Crustacea.* New York: Oxford University Press.

Schram, F. R. 1993. Review of Boxshall, Strömberg, and Dahl, *The Crustacea: Origin and Evolution. Journal of Crustacean Biology* 13:820–822.

Seilacher, A. 1989. Vendozoa: Organismic construction in the Proterozoic biosphere. *Lethaia* 22:229–239.

Seilacher, A. 1994. How valid is *Cruziana* stratigraphy? *Geologische Rundschau, Special Volume: NE African Geology*, 94–117.

Shear, W. A. 1992. End of the "Uniramia" taxon. *Nature* 359:477–478.

Simonetta, A. M. and L. Delle Cave. 1975. The Cambrian non-trilobite arthropods from the Burgess Shale of British Columbia: A study of their comparative morphology, taxonomy, and evolutionary significance. *Palaeontographia Italica* 69:1–37.

Størmer, L. 1939. Studies on trilobite morphology, part 1: The thoracic appendages and their phylogenetic significance. *Norsk Geologisk Tidsskrift* 19:143–273.

Størmer, L. 1944. On the relationships and phylogeny of fossil and recent Arachnomorpha: A comparative study on Arachnida, Xiphosurida, Eurypterida, Trilobita, and other fossil Arthropoda. *Skrifter Utgitt av Det Norske Videnskaps-Akademi i Oslo, I. Matematisk-Naturvidenskapelig Klasse* 5:1–158.

Stürmer, W. and J. Bergström. 1978. The arthropod *Cheloniellon* from the Devonian Hunsrück Shale. *Paläontologische Zeitschrift* 52:57–81.

Wägele, J. W. 1993. Rejection of the "Uniramia" hypothesis and implications of the Mandibulata concept. *Zoologische Jahrbücher: Abteilung für Systematik, Ökologie, und Geographie der Tiere* 120:253–288.

Walcott, C. D. 1912. Middle Cambrian Branchiopoda, Malacostraca, Trilobita, and Merostomata. *Smithsonian Miscellaneous Collections* 57:145–228.

Walossek, D. 1993. The Upper Cambrian *Rehbachiella* and the phylogeny of Branchiopoda and Crustacea. *Fossils and Strata* 32:1–202.

Walossek, D., L. Hintz-Schallreuter, J. H. Shergold, and K. J. Müller. 1993. Three-dimensional preservation of arthropod integument from the Middle Cambrian of Australia. *Lethaia* 26:7–15.

Walossek, D. and K. J. Müller. 1990. Upper Cambrian stem-lineage crustaceans and their bearing upon the monophyly of Crustacea and the position of *Agnostus. Lethaia* 23:409–427.

Walossek, D. and K. J. Müller. 1992. The "alum shale window": Contribution of "Orsten" arthropods to the phylogeny of Crustacea. *Acta Zoologica* 73:305–312.

Whittington, H. B. 1977. The Middle Cambrian trilobite *Naraoia,* Burgess Shale, British Columbia. *Philosophical Transactions of the Royal Society of London*, series B, 290: 409–443.

Whittington, H. B. 1985a. *The Burgess Shale.* New Haven: Yale University Press.

Whittington, H. B. 1985b. *Tegopelte gigas*: A second soft-bodied trilobite from the Burgess Shale, Middle Cambrian, British Columbia. *Journal of Paleontology* 59:1251–1274.

Wills, L. J. 1966. A supplement to Gerhard Holm's "Über die Organisation des *Eurypterus fischeri* Eichw." with special reference to the organs of sight, respiration, and reproduction. *Arkiv för Zoologi* 18:93–145.

Wingstrand, K. G. 1972. Comparative spermatology of a pentastomid, *Raillietiella hemidactyli*, and a branchiuran crustacean, *Argulus foliaceus*, with a discussion of pentastomid relationships. *Kongelige Danske Videnskabernes Selskab: Biologiske Skrifter* 19:1–72.

## NOTE ADDED IN PROOF

Since completion of the manuscript, *Naraoia longicaudata* has been reassigned to the new genus *Misszhouia* Chen et al., 1997, and a few species have been added to the list on pp. 156–157 (see Hou and Bergström 1997; Ramsköld et al. 1997).

Chen, J.-Y., G. D. Edgecombe, and L. Ramsköld. 1997. Morphological and ecological disparity in naraoiids (Arthropoda) from the Early Cambrian Chengjiang fauna, China. *Records of the Australian Museum* 49:1–24.

Hou, X.-G., and J. Bergström. 1997. Arthropods of the Lower Cambrian Chengjiang fauna, southwest China. *Fossils and Strata* 45:1–116.

Ramsköld, L., J.-Y. Chen, G. D. Edgecombe, and G.-Q. Zhou. 1997. *Cindarella* and the arachnate clade Xandarellida (Arthropoda, Early Cambrian) from China. *Transactions of the Royal Society of Edinburgh, Earth Sciences* 88:19–38.

# CHAPTER 5

## Early Arthropod Phylogeny in Light of the Cambrian "Orsten" Fossils

*Dieter Walossek and Klaus J. Müller*

## Abstract

The perfectly preserved small benthic fossils from the Swedish Upper Cambrian "Orsten" contribute significantly to the understanding of morphology, life habits, and phylogeny of Early Paleozoic representatives of Arthropoda, especially on the path toward the Crustacea. The recognition of primary and subsequent evolutionary novelties in the crustacean lineage leading toward its crown group is possible by attributing some of the forms to the stem lineage while others are considered representatives of crown-group taxa. The "Orsten" material serves particularly well for a better understanding of the derivation of crustacean types of limbs from a more primordial type, as known from trilobites and other early arthropods. According to this hypothesis, the limb portions of arthropods can now be homologized in that they comprise principally a joint socket and a basipod (*not coxa*) carrying the two rami, endopod and exopod. This design is also recognizable among extant chelicerates that have evolved stenopodial, uniramous walking legs: limulids have a rigid limb basis carrying the large endopod, but a vestigial exopod is still present on their last prosomal leg (flabellum). All of the flat opisthosomal legs comprise a basis, a short endopod, and a flat exopod. One of the innovations of Crustacea *s.l.* (Pan-Crustacea) is a separate, mobile and setigerous enditic process at the inner proximal edge of the basipod, developed most likely in accordance with new feeding and locomotory strategies. In certain limbs of the post-antennular series of the Eucrustacea this "proximal endite" grew out as a separate portion of the stem, the *coxa*, but in others it retained its original design as a small endite (for example in the trunk limbs of Entomostraca). It also becomes clear that the process of specialization of the last head limb, the so-called second maxilla, occurred independently after the branching event into the major lineages Malacostraca and Entomostraca. Accordingly, the shape of the maxillules and maxillae and their individual developmental fates can indeed serve as tools to distinguish between the major crustacean lineages. Examples are given of how particular "Orsten" forms contribute to the systematic status of particular crustacean taxa, such as the Branchiopoda and the Maxillopoda.

## Reconstructing Phylogeny

One of the major advantages of the concept of phylogenetic systematics (Hennig 1966) over other concepts for reconstructing phylogeny is that it demands testing hypotheses of relationship and explaining their consequences. As an example, the hypothesis of the polyphyletic origin of the Arthropoda implies that characters shared between its presumed members should have evolved independently. It has been repeatedly shown, however, that the huge number of characters shared by the members of Arthropoda is plausible only under the view of arthropod monophyly. This does not mean that we understand a great deal about early arthropod evolution. The earliest euarthropods, among them the earliest representatives of the future major lineages, were indeed genetically not yet very distinct, even from their precursors. Hence it is difficult to estimate what really characterized the euarthropod ground plan. Potential ground plan characters of Euarthropoda include:

1. metameric tergites with lateral extensions, pleurotergites, and membranous connections (*arthrodization*);
2. the extent to which head tagmosis had advanced, externally and internally (*cephalization*);
3. whether a head shield as the product of fusion of anterior tergites had developed;
4. how much the head shield extended to any side; and
5. whether sclerotization had also already affected the limbs (*arthropodization*).

With regard to the plasticity in the design of the arthropod cuticle to form articulations or fuse these again, we cannot even be sure that the "typical" arthropod limb, with the cuticle differentiated into articulated portions, evolved only once. Furthermore, morphology can be altered at any time, and evolving "lineages" may modify their original apomorphic characters (constitutive characters *sensu* Hennig) again (fig. 5.1). A particular example is demonstrated below with the Maxillopoda. In the light of the incompleteness of the fossil record, we can, at best, hypothesize evolutionary levels that certain fossils have achieved rather than create precocious cladograms including such fossils. Again, fossils in a chronological sequence do not necessarily express evolutionary transformation of a character (Willmann 1989). In our view, we have to wait for restudy of such early fossils as they are the real entities

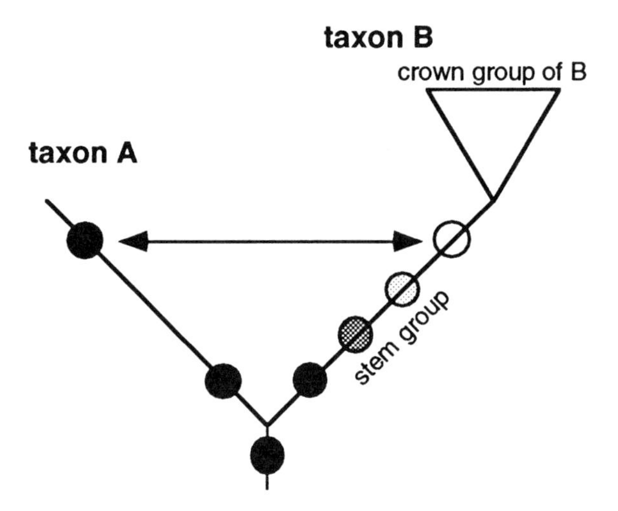


FIGURE 5.1
Scheme of a hypothesized sister group relationship of two taxa A and B, in which the synapomorphy of A and B (black dot) changes gradually along the stem lineage of B (into white dot; modified from fig. 3 of Willmann 1989).

that enlighten the evolutionary scenario leading to the Euarthropoda and its major lineages. This will be more helpful than developing hypotheses derived exclusively from living taxa or from molecular data, or from fossils alone.

## Ground Patterns and Fossils

There are two complementary ways for achieving a better understanding of the relationships of taxa: the reconstruction of ground patterns or ground plans and the recovery of well-preserved fossils.

### Ground Pattern

The ground pattern is the sum of all available data for a taxon (structures, functions, life habits), such as the maximum number of segments. It represents the reconstruction of an organism possessing that portion of the set of features that had been developed during the early history of the group to which it belongs, and those characters newly developed in its immediate stem species. The ground pattern is, however, devoid of any of the apomorphies developed in descendants of the ground plan representative. As an example, particular features of eumalacostracan crustaceans, and even Decapoda, as included in the "diagnosis" of Crustacea in many textbooks (e.g., Wehner and Gehring 1990:671–676), cannot have characterized the ground plan of Crustacea *s.l.*, not even at the crown-group level, the Eucrustacea. For any comparison between extant taxa above the species level it is necessary to work out ground patterns, either of the whole organisms (ground plan representatives) or of particular structures or functional systems.

## Fossils

Fossil organisms, if they are not stem species, can be expected to differ from the ground pattern of a certain evolutionary level of a taxon in having their own, specific adaptive modifications (autapomorphies). Not all and possibly not even many characters of the fossils are plesiomorphic if they are compared with their extant relatives. On the other hand, once fossils are recognized as members of a particular taxon by shared features (synapomorphies), their character set can assist to evaluate and to polarize the status of further features within that taxon. In other words, they help to determine the sequential development and evolutionary fate of characters (structures or functional systems) and also to recognize homoplasies, i.e., independent modifications of features with the result of similarity in structure, form, or function.

In the last ten to fifteen years, we have been able to present a detailed picture of three-dimensionally preserved, secondarily phosphatized "Orsten"-type assemblages from southern Sweden and related occurrences by the recognition of forms as follows:

1. non-crustaceans, such as a chelicerate larva (Müller and Walossek 1986b, 1988a); *Agnostus pisiformis* (Müller and Walossek 1987), stem-group Pentastomida (Walossek and Müller 1994, Walossek et al. 1994) and a stem-group tardigrade (Müller et al. 1995);
2. representatives of particular eucrustacean groups, such as *Bredocaris admirabilis* (Maxillopoda; Müller and Walossek 1988b), the Skaracarida (Maxillopoda; Müller and Walossek 1985a; Müller and Walossek 1988b; Walossek and Müller 1992) or *Rehbachiella kinnekullensis* (Branchiopoda; Walossek 1993, 1995), and;
3. a set of derivatives of the stem lineage of Eucrustacea, including *Martinssonia elongata* (Müller and Walossek 1986a) and three more taxa from Sweden (Walossek and Müller 1990, 1991, 1992) and one from Poland (Walossek and Szaniawski 1991).

This picture contrasts significantly with that of Lauterbach who repeatedly claimed (e.g., 1988) that the "Orsten" arthropods represent a cluster of stem-group mandibulates. It also contrasts with the cladistic approach of Wilson (1992) who in his computer-based analyses did not use any of the specific characters that have led us to the above-mentioned reconstructions of the phylogenetic relationships of these forms. In several other cases, even our published data were scored incorrectly, the consequences of which are evident.

The larval world contributes its own set of data for phylogenetic analyses (cf. Walossek 1993 for further references). It is obvious that reproduction (and adult life) becomes impossible if ontogeny fails at some stage, and adult features of the Arthropoda in particular are developed through a series of steps or molts, often with the need to develop new structures since larvae live in different regimes from the adults. Not only are the strategies for escaping from the high risk during molting important here but the specific adaptations of larvae to their life regime — as a new evolutionary strategy and a "decoupling" of larval and adult evolution — also provide us with further structures important in a reconstruction of phylogeny. The discovery of various larval series in the "Orsten" is of exceptional value, since this additional information is now also available as a part of the fossil record.

Any phylogenetic approach has to consider the rules of evolution and principles of parsimony to reduce the number of assumed character reversals. In striking contrast to computer-based cladistic analyses, however, the Hennigian method demands that the worker explain evolutionary changes (steps) and consider their phylogenetic consequences. Since features develop sequentially and hierarchically, the character set has to be evaluated again after each recognized branching node for changes (transition; fig. 5.1) as well as for the appearance of new features. Also, incompleteness of information must be stressed in phylograms derived from such reconstructions. Moreover, according to Fisher (1981), hypotheses can be tested at the level of evolutionary scenarios. This helps to revise interpretations of homology or polarity and to reduce the number of possible hypotheses; in other words, it increases the probability of a certain choice.

In all, there should be only one possible tree left, the one that reflects the natural system of diversification. Fryer's (1992) attempt to derive crustaceans directly from hemocoelic worms via phyllopod limb-bearing animals similar to Anostraca (in his view having unsegmented limbs), which contradicts the structural evidence from fossil and extant Arthropoda, is an example of non-consideration of consequences when hypothesizing evolutionary paths.

## "Orsten" Fauna

### Diversity of Shape and Lifestyle

The finest Paleozoic fossils providing the most minute details suitable for direct comparison with extant organisms are from the Swedish "Orsten" (fig. 5.2). In particular their morphological details have led us to reconstruct the environment in which these forms lived as a soft-bottom zone inhabited by a large variety of mostly

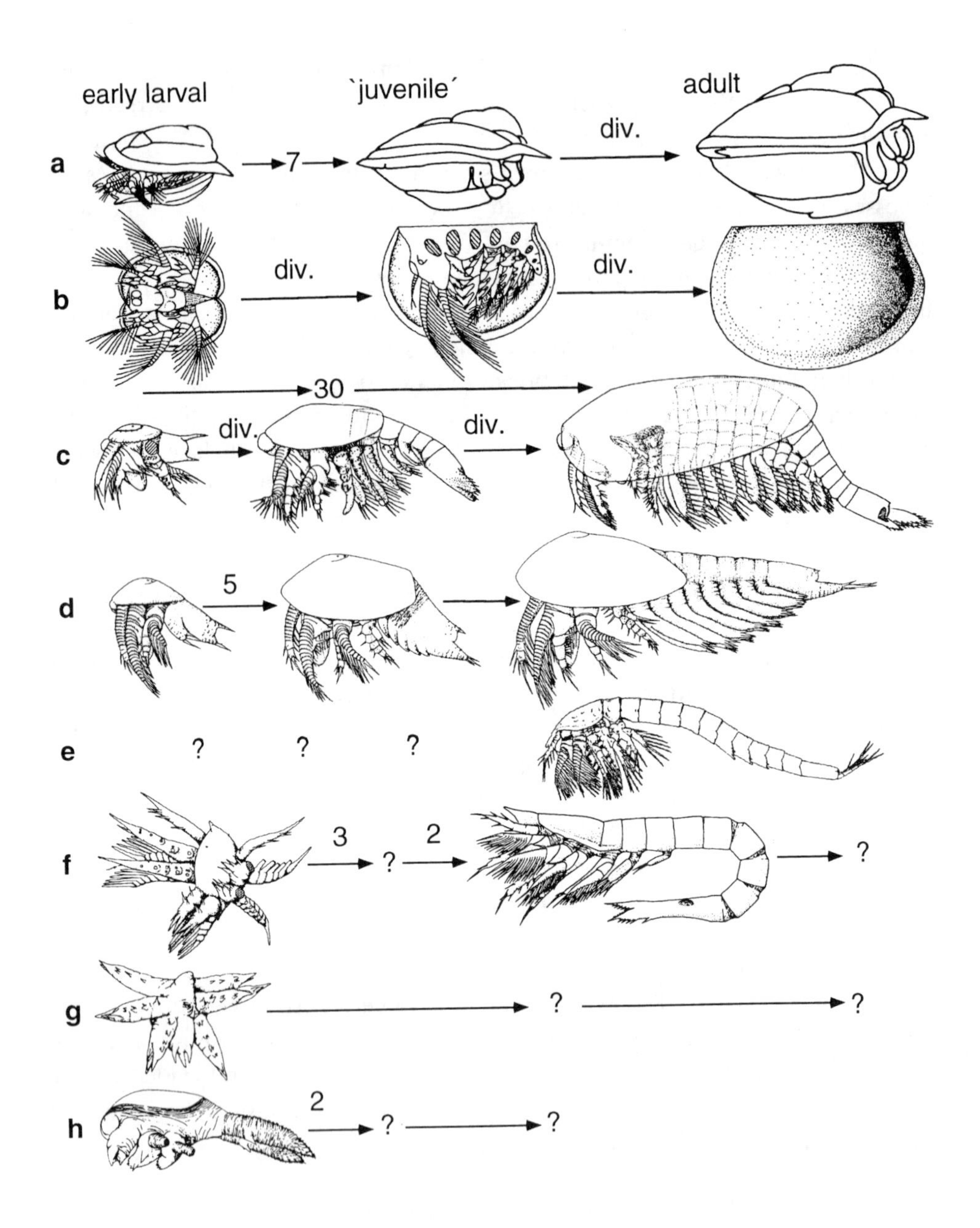


FIGURE 5.2

Examples of life-form types and ontogenetic series of "Orsten" arthropods (not to scale, appendages omitted in some cases [oval filled with lines]). *(a) Agnostus pisiformis* (Linnaeus, 1757) (shield length 5 mm). *(b)* Phosphatocopina (shield length 2–3 mm). *(c) Rehbachiella kinnekullensis* Müller, 1983 (total length of TS13 stage 1.5–17 mm). *(d) Bredocaris admirabilis* Müller, 1983 (length of adult 0.85 mm). *(e) Skara anulata* Müller, 1983 (total length 1.2 mm). *(f) Martinssonia elongata* Müller and Walossek, 1986 (total length 1.5 mm, juvenile?). *(g, h)* Isolated larvae; *(g)* type-A larva (from Müller and Walossek 1986a, body length 0.1 mm), *(h)* type-D chelicerate larva (from Müller and Walossek 1986a, body length 0.35 mm).

meiofaunal, hence small animals (e.g., Müller and Walossek 1991; Walossek 1993, fig. 2). Generally even the largest specimens preserved in this type of fossilization are less than 2 mm long, a restriction for which we still have no suitable explanation. A closer view of the ontogenetic series of available "Orsten" forms reveals that strategies were quite diverse and that not necessarily all were larvae, nor were all members of a permanent meiofauna.

For *Agnostus pisiformis,* eight instars were found with preserved ventral integument appendages, the smallest approximately 320 μm and the largest 800 μm long (fig. 5.2a). Growth continued to a stage with calcified shields and a length of about 5 mm, which can be found in huge masses in the rock. Growth of the phosphatocopines starts with small larvae having shield sizes of 400 μm, while adults generally have shield lengths of up to 3 mm (fig. 5.2b; Müller and Walossek 1987). After thirty instars *Rehbachiella kinnekullensis* attained a length of about 1.5–1.7 mm (fig. 5.2c). Yet even this stage was still immature, and growth may have continued toward the centimeter range. This organism progressively developed a complex filter apparatus and most probably left the benthic zone during growth, so that small instars had a much better preservation potential (= temporary member of the meiofauna community; Walossek 1993). The metanauplius larva of *Bredocaris admirabilis,* about 250 μm long, molted five times to metamorphose into an 850 μm long adult, which may have lived at or slightly below the sediment-water interface (fig. 5.2d; Müller and Walossek 1988b). So, *Bredocaris,* completing its life cycle within the flocculent layer, is interpreted as a permanent member of the meiofauna community.

No larvae are known for the two species of Skaracarida, *Skara anulata* and *S. minuta* (fig. 5.2e; Müller and Walossek 1985a), and *Dala peilertae* (Müller 1983; Müller and Walossek 1991). It is possible that these animals changed to a life at or near the bottom after a series of pelagic larval stages.

*Martinssonia elongata* (fig. 5.2f), a slim bottom dweller similar to Skaracarida, is known from three lecithotrophic early larvae, recognizable by their ill-developed limb armature and missing mouth and anus, and two stages with a segmented trunk (fig. 5.4). It remains uncertain if the largest stage, about 1.5 mm long, is already the adult (Müller and Walossek 1986a).

In some cases only early larvae have been encountered, such as in the case of the non-feeding 100–120 μm small type-A larvae (fig. 5.2g) and a possibly parasitic chelicerate larva of 350 μm, strikingly resembling the protonymphs of extant Pantopoda (fig. 5.2h; Müller and Walossek 1986b). In such cases, adults were not preserved, most likely due to life in a different regime.

## Stem-group and Crown-group Eucrustacea

Following the concept of Ax (1985), a monophylum can be seen as a compound (Pan-monophylum *sensu* Lauterbach 1989) consisting of a stem lineage and the crown group, which is the taxon that embraces all descendants of the last common ancestor of the extant species of the group. The usage of the stem-lineage concept permits the distinction of evolutionary plateaus, and this, prior to the possibility of a detailed phylogenetic approach, helps to distinguish between characters "already developed" and those "not yet developed." This may even aid in the recognition of whole functional systems that develop in accordance with functional constraints, as will be shown for the feeding apparatus of the crustacean head below.

A closer view of current "diagnoses" of Crustacea reveals that crustacean monophyly has never been questioned but also has never been validated. Traditional characters, such as "two pairs of sensorial antennae," "two pairs of maxillae," or "carapace" refer to particular ingroup taxa (autapomorphies, see above) rather than characterizing the ground plan. It was particularly Lauterbach (1983, 1986) who made an attempt on the basis of the concept of phylogenetic systematics to recognize autapomorphies of the Crustacea. His approach included, however, many deficiencies in detail, misunderstandings of structural and functional details, and erroneous interpretations of "Orsten" fossils in particular, which he never studied himself. For example, Lauterbach (1988) used a metanauplius of *Bredocaris* as the ground plan nauplius of his Mandibulata. Not realizing that the antennules in a particular specimen were simply broken off, he misinterpreted the second antennae as "biramous antennules," consequently transferring the mandibles into "mandible-like antennae," and the maxillules into "brush-like mandibles." He also did not recognize that this metanauplius already had buds of further limbs on its hind body — revealing itself to be an advanced larval stage — and that the larval sequence of *Bredocaris* is a specific feature (autapomorphy) of a particular taxon within the Maxillopoda (see below).

As a consequence, hitherto any phylogenetic approach and attempt to assign crustacean-like fossils to the Crustacea has been bedeviled from the beginning by the difficulty of knowing what Crustacea are — and their ingroup taxa, too.

## Evolutionary Trends in the Stem Lineage of Eucrustacea

### *Primordial innovations*

A set of "Orsten" fossils gives interesting insight into the principal innovations of the early eucrustacean lineage and the individual specific adaptations of its members (fig. 5.3). These fossils have permitted us, additionally, to uncover a set of characters for the ground plan of crown-group crustaceans, the Eucrustacea. These fossils are:

1. *Martinssonia elongata* Müller and Walossek, 1986a, a slender form of crustacean-like appearance but homonomous post-antennular head limbs (figs. 5.2f, 5.4);
2. *Goticaris spinulosa* Müller and Walossek, 1990, and *Cambropachycope clarksoni* Müller and Walossek, 1990 (fig. 5.3), both with a huge uniform compound eye at their forehead, and;

3. *Henningsmoenicaris scutula* (Walossek and Müller, 1991) with a bowl-shaped shield and stalked eyes.

A further form from Poland, *Cambrocaris baltica* Walossek and Szaniawski, 1991, apparently a mobile swimmer with a set of laterally directed paddling limbs, was later added to this assemblage.

In our view, the ground plan of Euarthropoda, as exhibited in various early arthropods known from Burgess Shale — type lagerstätten, included:

1. a head with *at most three pairs* of post-antennular biramous appendages of the type known from various trilobites and early arthropods (e.g., Chen et al. 1991, fig. 7 for *Misszhouia,* the earliest known arthropod limb at that time; see below), i.e., a basipod, a seven-segmented endopod, and a flap-like exopod;
2. a dorsal shield covering the same number of segments and representing the product of fusion of their tergites;
3. a hypostome with the mouth opening at its rear; and
4. a trunk consisting of a fairly limited number of tergite-bearing segments (approximately twenty) back to the non-somitic end piece bearing the anus (telson or pygidium).

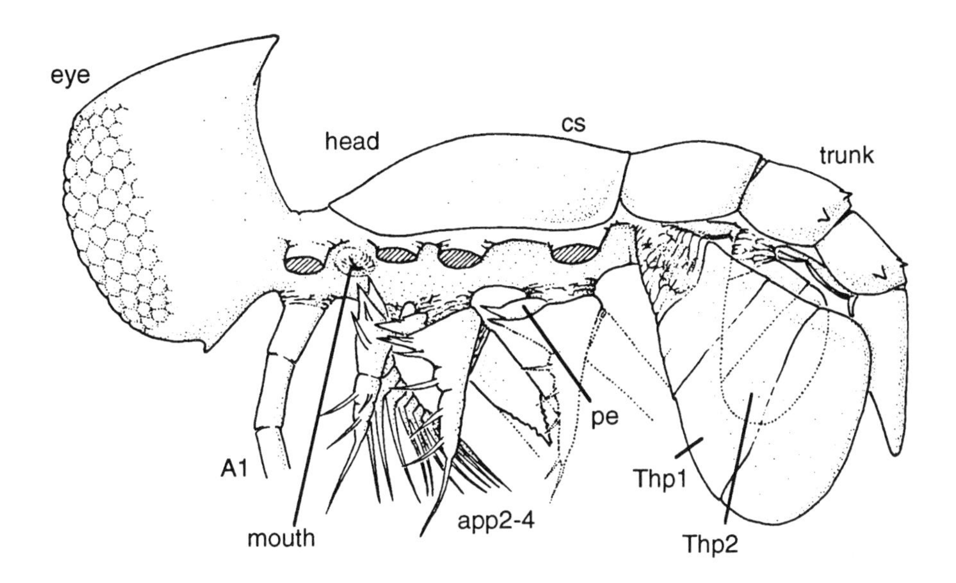


FIGURE 5.3

Reconstruction of the stem-group eucrustacean *Cambropachycope clarksoni* Walossek and Müller, 1990, from the Upper Cambrian "Orsten" of Sweden (from Walossek and Müller 1992, fig. 1A; length approx. 1.5 mm; uncertain structures added with stippled lines). The huge, faceted, unpaired frontal eye is shared with another "Orsten" form, *Goticaris longispinosa* Walossek and Müller, 1990.

We assume that crustacean evolution started with such a morphology but with a few significant morphological alterations that led to new locomotory and feeding habits, possibly in accordance with a reduction in size and a meiobenthic swimming mode of life. The morphological innovations at the ground plan level of Crustacea (Crustacea *s.l.*) we recognized in "Orsten" forms as named above mainly concern the appendages. These include:

1. the tilting of exopod setation toward the endopod in those post-antennular limbs that bear a multiannulated exopod. A medially directed exopod setation permits swimming coupled with sweeping food particles toward the mouth;
2. the development of a separate, setulate or spine-bearing endite at the inner proximal edge of the limb stem. This separate endite may have enabled the manipulation of food particles in the proximity of the ventral body surface while the distal parts could serve different functions, and it is a further indicator of a new life strategy of the crustacean lineage; and
3. the first antennae (antennules) have not only achieved a striking similarity to the limb stems and inner rami of the subsequent limbs but were most likely also involved in locomotion and feeding rather than purely carriers of sense organs (specific sweeping setation);

It is noteworthy that the latter feature is still recognizable today in the naupliar antennules particularly of extant Maxillopoda among Entomostraca and penaeid Decapoda among Malacostraca. Consequently, the antennules should have functioned that way at least in the ground pattern of the early larvae of Eucrustacea (= crown-group crustaceans).

## Plesiomorphies of the Stem-group Level of Eucrustacea

Early larvae are now known from all four Swedish stem-group forms but have been published in detail so far only for *Martinssonia* (fig. 5.4a–c; SEM examples of *Goticaris* and *Henningsmoenicaris* in Walossek and Müller 1990, figs. 2–4, reconstruction in fig. 5). These larvae have three post-antennular limbs, as known from *Agnostus pisiformis* and trilobites. Based on this, we have suggested that the head of the ground plan of Euarthropoda did not include four, as traditionally assumed, but only three post-antennular limb-bearing segments (Walossek and Müller 1990, 1992; Walossek 1993). This hypothesis confirms the concept Cisne (1982) had developed from his reinterpretation of the head segmentation of the Ordovician trilobite *Triarthrus eatoni* (more recently confirmed by Whittington and Almond 1987). This species had long remained the basis of discussions of arthropod head segmentation following the description of its external anatomy by Størmer (1939). In consequence, the early larvae of the "Orsten" stem-group crustaceans and those of the trilobites, the protaspids, are considered as "complete-headed" larvae, adding only trunk segments during ontogeny.

Besides the head tagma, we have observed structural retention from preceding (plesiomorphic) evolutionary stages in particular in the trilobitoid hypostome and in the larval development starting with a "complete-headed" larva.

Among the stem-group crustaceans the degree of specialization of post-antennular limbs differs between taxa. *Cambropachycope* and *Goticaris* have similarly structured, post-antennular head limbs and uniramous, post-cephalic, leaf-shaped limbs, while in *Martinssonia* and *Cambrocaris,* the post-antennular limbs are more or less identical, both having multiannulate exopods. In *Henningsmoenicaris* the anterior two post-antennular limbs bear multiannulate exopods, while all posterior limbs have leaf-shaped exopods, a character also known from most crown-group Crustacea.

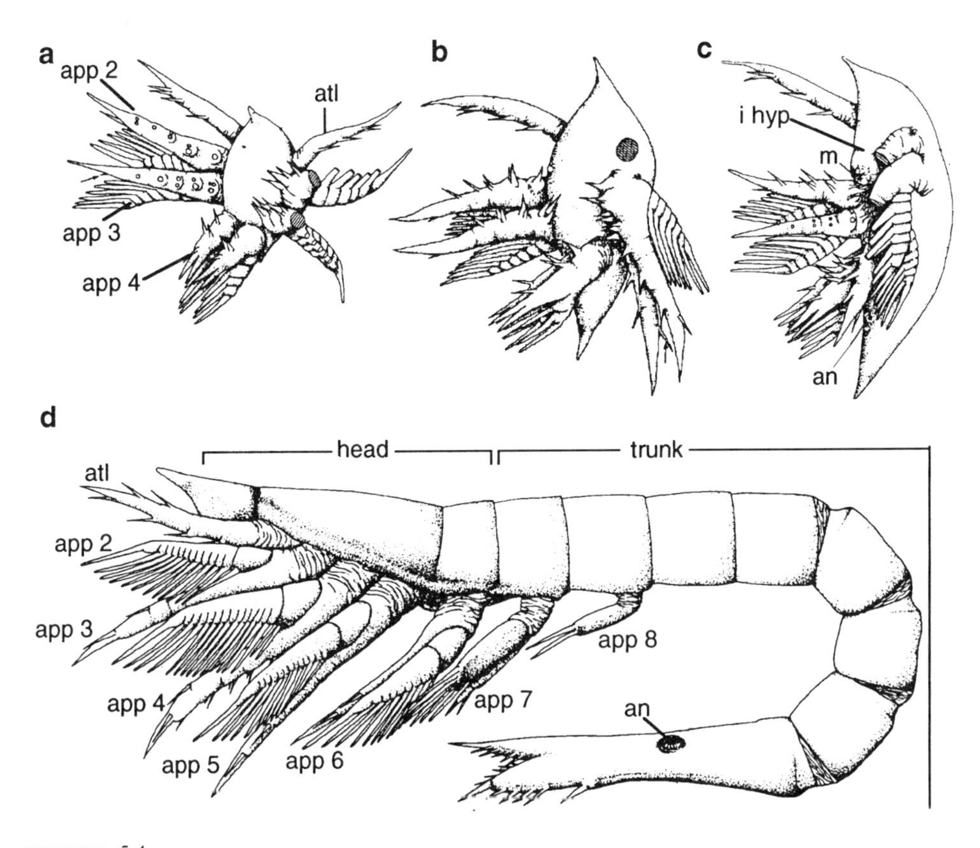


FIGURE 5.4

Larval development of *Martinssonia elongata* Müller and Walossek, 1986. *(a)* First larval stage with four functional pairs of limbs but lacking mouth and anus. *(b)* More spindle-shaped second stage, still non-feeding. *(c)* Third stage, possessing mouth and anus. *(d)* Possibly still immature stage 5, with segmented trunk terminating in a slender, bifurcate portion carrying the anus on its ventral side (from Müller and Walossek 1986a).

Thus, slightly differentiated head limbs occur only in *Henningsmoenicaris;* yet, none of these forms developed a specialized set of post-antennular head limbs in the sense of the second antennae, mandibles, and maxillules as present in crown-group crustaceans (for the maxilla see below). On the other hand, there is already a differentiation into forms having three post-antennular head limbs (a plesiomorphy), such as *Cambropachycope* and *Goticaris*, and those having four post-antennular head limbs, such as *Martinssonia, Henningsmoenicaris*, and *Cambrocaris*. It remains uncertain which state of exopod shape is original. In Phosphatocopina the exopods seem to develop from the multiannulate condition into a leaf shape. On the other hand, *Fuxianhuia* from the Lower Cambrian (Bergström and Hou, this volume) has leaf-shaped exopods, similar to other early arthropods from the same locality (D. W., pers. observation).

*Implications*

Our assumption that the formation of a head tagma had progressed during arthropod evolution only to the third pair of biramous legs at the ground plan level of Euarthropoda has important implications, because it contrasts with the traditional view that the euarthropod head already included a maxillary segment (i.e., four post-antennal legs) upheld in most recent literature (e.g., Kraus and Kraus 1994).

In our view, the chelicerate line should have started at the same level of head formation and included progressively three more segments to form their characteristic prosoma, and not just two, as traditionally assumed. Notably, such an original head tagma is still recognizable today in the early larvae, protonymphs, of the Pantopoda, which possess chelifores (post-antennular appendage 1) and two more limbs (appendages 2, 3; antennule lost in the chelicerate line, possibly even later, see Müller and Walossek 1986b, larva type D).

Larger "heads"occur in many subordinate taxa within the Arthropoda, and we consider them to be the results of convergent evolution. This implies again the necessity to reconstruct the ground pattern of taxa, because from any level ingroup taxa may well evolve the same number of head segments as could unrelated taxa, which is evident with regard to possible relationships between Crustacea and Tracheata ("myriapods" and hexapods). Both are said have two pairs of maxillae, but, as will be shown below, this is not true for the ground plan of Crustacea (Crustacea *s.l.*) and also not true in detail for the crown group (Eucrustacea; see below).

Again returning to chelicerates, they should have bypassed levels of four head limbs — as in tracheates and crustaceans — of five — as in various crustacean taxa (= cephalothorax of e.g., copepods, syncarids, peracarids) — and six — as also known from various crustacean taxa.

This interpretation gives new momentum to the search for fossil "intermediates" not only in the Paleozoic fossil record but also among "survivors" in later times. It also affects any discussion of head tagmosis of Euarthropoda and the development of the characteristic dorsal cover, the head shield. In the ground plan of Euarthropoda, such a shield should accordingly have included no more than three post-antennular limb-bearing tergites, rather than four, as traditionally assumed. The head shield reveals, on the other hand, an enormous plasticity in size already among Lower to Middle Cambrian arthropods, from being very short and narrow (e.g., *Jianfengia*) to very wide and long (e.g., *Canadaspis*). Its varying fusion with more body segments is also evident already in these early arthropods. Accordingly, the head shield's degree of lateral or posterior expansion appears much influenced by simple functional needs — too much for it to represent an appropriate character for phylogenetic approaches. Convergent evolution of a similar shape hence occurred repeatedly.

### *Innovations of the Crown-group Level: The Eucrustacea*

Major changes seem to have affected the ventral head region with its feeding structures in particular, such as the mouth region and associated structures on the limbs, and they will be described separately below. In the stem group, the mouth is located in an exposed position at the posterior end of a bulging structure, the hypostome, between the antennules and first biramous limbs, much as in *Agnostus* (Müller and Walossek 1987, e.g., fig. 24). A specific setulation around the hypostome is not developed, nor is an atrium oris, the funnel-shaped recession of the mouth, as in extant Crustacea, nor a labrum, containing many slime glands and forming the ceiling of the atrium oris. Likewise missing are the hairy sternum, at least as the product of fusion of the cephalic sternites as far back as the maxillules, and a pair of humps, paragnaths (products of the mandibular sternite), which form the slope into the atrium oris, as developed in all crown-group crustaceans.

Hence, while at the stem-lineage level feeding must still have operated in a similar fashion to the old arthropod mode of sucking in particles that were not ground up but simply pushed toward the mouth at the rear of the hypostome, all characters associated with this new complex feeding system can thus be attributed to the ground plan of the crown-group crustaceans, the Eucrustacea, such as, the labrum, atrium oris, formation of a rigid sternum, setulation of different parts of the mouth region, and specific armature of associated limbs.

It is noteworthy that the *hypostome* and the *labrum* are two different organs, which must not be confused terminologically. Yet the hypostome — as a character of the ground plan level of Euarthropoda — has never been lost completely in Eucrustacea. It still forms a short bridge at the anterior proximal edge of the labrum.

## Developmental Changes in the Appendages of Crustacea

*Martinssonia, Cambropachycope, Goticaris, Henningsmoenicaris,* and *Cambrocaris* all possess a separate setigerous endite at the proximal edges of the stems of the post-antennular limbs. We consider this endite, which split off from the original limb stem carrying the two rami, a feature unique to the Crustacea. Its development, recognized in all "Orsten" stem-group crustaceans as well as in *Cambrocaris* (fig. 5.5b), may have permitted manipulation of food particles more proximally, while the more distal limb parts became free for other purposes (Walossek and Müller 1990). Its recognition led to the establishment of a new hypothesis for the origin and further development of the different types of post-antennular limbs of Crustacea (Walossek 1993, fig. 54), which is slightly emended in the following discussion.

### *Evolution of the Crustacean Types of Post-antennular Limbs*

In our view, the ancient euarthropod limb was made up of a narrow, almost transversely oriented stem portion continuing into a inner ramus or endopod and a leaf-shaped outer ramus or exopod, which arose from the outward sloping edge of the stem. This original condition — only one stem and two rami (fig. 5.5a), as is well documented by various early arthropod fossils with preserved appendages (e.g., Müller and Walossek 1987 for *Agnostus;* Chen et al. 1991, fig. 7 for the leg of *Misszhouia* from the Lower Cambrian, see our fig. 5.6A) — is homologous in all Euarthropoda.

The stem is termed the *basipod* according to the situation in Crustacea, the only extant euarthropod group that has retained such biramous limbs (fig. 5.5).

In the stem lineage of the Crustacea, a new element appears at the inner proximal edge of the basipod in the form of a movable feeding device. This *proximal endite* may have developed primarily in the head and anterior trunk limbs, in accordance with a new locomotory and feeding system, as described above. How this structure developed is still uncertain, e.g., whether it is an entirely new element or merely a separation of the proximal set of grasping spines at the inner edge of the original basipod (see fig. 5.5a).

Subsequently, the post-antennular limbs of crustaceans underwent different changes, partly in sets and partly individually. The developmental fate of the proximal endite and basipod led to two extremes in the modification of the proximal endite:

1. its enlargement, leading to the formation of a separate, articulated limb portion, the "coxa," proximal to the basipod. Two states are recognized: (a) ring-shaped; (b) with bulged median edge; and
2. the retention of the ancestral condition (short, lobate endite), but enlargement of the basipod instead.

The situation of the mandible is an example of the extreme expansion of the coxal portion (state 1b). Here the proximal endite grows to enormous size during ontogeny (fig. 5.5c), while its "palp," comprising the basipod and rami, may either eventually disappear completely (ground plan of Entomostraca) or reappear, after gradual reduction and loss at a later stage of larval development as a tripartite element (ground plan of Malacostraca). Notably at this stage the mandibular coxa has taken over the function of a grasping element of the limb, located immediately distal to the body-limb joint — originally that of the basipod. The retention of the palp in adult Maxillopoda is considered to be a phenomenon in accordance with the paedomorphic origin of this group rather than a plesiomorphy.

At the other extreme, the proximal endite retains in one way its plesiomorphic shape as a, more or less, separately movable setigerous endite (= "gnathite," "arthrite," etc.), but the basipod elongates to form the major portion of the limb (often called "protopod"; see, for example, *Rehbachiella* in fig. 5.5d, 5.5e for a limb bud). This shape can be recognized in:

1. the maxillule of all entomostracan taxa;
2. the maxilla of Eucrustacea; and
3. all trunk limbs of Entomostraca, e.g., Cephalocarida, Maxillopoda, and Branchiopoda (ground plan design, further modifications possible).

It remains uncertain if the numerous enditic lobes along the median edge of the basipod were originally present. Their presence in the maxilla of Malacostraca and the maxilla and trunk limbs of Cephalocarida as well as, basically, Maxillopoda and Branchiopoda, seems to point in that direction. In fact, the proximal endite is enhanced as a separate limb portion proximal to the basipod only in certain postantennular limbs of Eucrustacea, and even there to a different degree and on different limbs in the major crustacean taxa. In addition to the extreme situation of the mandible, the intermediate situation of coxa and basipod being of about the same size, is characterized by the following:

1. the larval second antenna and mandible (example in fig. 5.5f and g, for cirripede nauplii);
2. the adult second antenna;
3. the anterior eight thoracopods of Malacostraca (Phyllocarida and Eumalacostraca); and
4. the maxillule of Malacostraca (fig. 5.7).

This differentiated evaluation of crustacean limb-types and the feeding system serves to identify several "Orsten" forms, such as *Bredocaris, Skara,* and *Rehbachiella,* at least as representatives of the crown group of Crustacea rather than as members of the stem lineage of "Mandibulata," as claimed by Lauterbach (e.g., 1986, 1988). All three taxa have not only specialized second antennae, mandibles, both with coxa and basipod, and maxillules, but also one more pair of limbs included with the head tagma. Their cephalic feeding system, again, includes the labrum, sternum back to the second maxilla, paragnaths, and the special armature of the limbs and setulation of the mouth area.

Considering the different development of the proximal endite/basis system, a clear distinction can now be made between "phyllopodous" limbs of phyllocarid (leptostracan) Malacostraca and of the Branchiopoda. In the former group, the stem consists of a flattened coxa plus basipod, while the limbs of the latter group have an elongated basis plus a proximal endite (examples in Walossek 1993, figs. 46, 47). Further differences in detail are in accordance with the completely different filter systems of both taxa (details in Walossek 1993:100–107).

According to our hypothesis, the original euarthropod limb should have consisted only of the three elements: *basipod, endopod,* and *exopod* (figs. 5.5a, 5.6A), while crustacean limbs embrace four elements: *basipod with a movable endite, the "proximal endite," and two rami.* Since all portions were subject to further modification

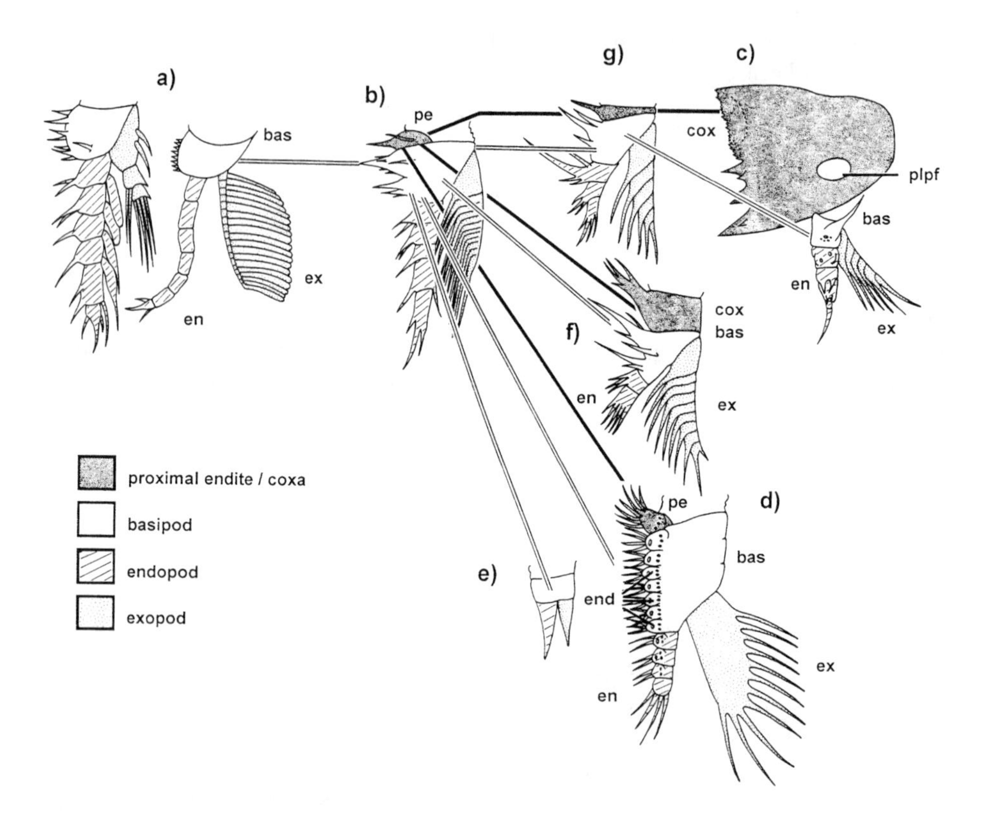


FIGURE 5.5

Evolution of the crustacean types of limbs (from Walossek 1995, fig. 7, modified from Walossek 1993, fig. 54; not to scale). *(a)* Schematic trilobitoid limbs with basipod, endo-, and exopod (setation outward oriented). *(b)* Limb of stem-group eucrustacean, exemplified by *Martinssonia elongata* Müller and Walossek 1986, with small "proximal endite" and distal part as in trilobitoid limb, except for inward orientation of exopod setation; right side: limbs of crown-group eucrustaceans, exemplified by their morphological extremes. *(c)* Branchiopod mandible, with enlarged coxa and palp comprising basipod, endopod, and exopod, which becomes reduced eventually. *(d)* Branchiopod phyllopodium, exemplified by the trunk limb of *Rehbachiella kinnekullensis* Müller, 1983, with enlarged basipod but small-sized "proximal endite" as the evolutionary equivalent of the coxa. *(e)* Initial stage of phyllopodium-type limb of *Rehbachiella. (f–g)* Developmental transition of limbs retaining the plesiomorphic design of distal limb parts exemplified by *(f)* the second antenna and *(g)* the mandible of a barnacle nauplius, both limbs having coxae instead of proximal endites.

on each leg due to functional requirements, this provides us with a large data set for phylogenetic analyses. Again, development on particular limbs of additional joints enhancing flexibility may be a further, complicating possibility, as is, for example, evident in the thoracopods of Calmanostraca (Notostraca and Kazacharthra; see below).

Traditionally, the Crustacea are generally understood to possess a limb stem consisting of a coxa and basipod. As presented above, this proves not to be true when considered in detail. In other words, Eucrustacea cannot be characterized by the possession of a coxa and basipod. It is noteworthy that Calman (1909:51) had already mentioned the presence of "gnathobases" on all trunk limbs as a "primitive character" of Branchiopoda — in fact representing our "proximal endite." We are currently investigating whether the pleopods of Malacostraca have a limb stem subdivided into coxa and basipod, or if they possess a proximal endite. If not, they would reflect the more ancestral, "pre-crustacean" design of a euarthropod limb.

The partial reflection of the shape of the original euarthropod limb shape in crustacean limbs can also be recognized in the oblique insertion area of the exopod along the outer edge of the basipod (fig. 5.6). Furthermore, in *Misszhouia* (fig. 5.6A) the proximal endopodal podomere is partly connected with the exopod. The same feature has been observed in *Agnostus* and the stem-group crustacean *Henningsmoenicaris* (Walossek and Müller 1990, fig. 5d, e). Remarkably, all major spines and setae on the median protrusions of the proximal endite and basipod are retained in their number and position from the stem-group level, as developed in *Martinssonia,* to the larval second antenna of a barnacle nauplius (fig. 5.6B, C).

In our view, arthropods with preserved limbs from the Lower Cambrian of China confirm this new hypothesis. Even more, they also provide us with a remarkable precursor type in the legs of *Fuxianhuia,* described in greater detail by Bergström and Hou (this volume). Of its approximately twenty annuli the proximal ones are very short, while those of the distal set are longer. There is in fact no distinction between a firm basipod and endopod. The exopod is preserved too, being a simple feeble plate lacking marginal setation. Its insertion is still uncertain, but may have been at one of the longer annuli immediately distal to the fine annuli.

We assume that such a "primordial" limb with short muscles connecting each of the annuli, as is deduced from the situation in crustacean endopods, functioned rather lobopodium-like, with a very limited range of action for each individual annulus.

The modification of such a limb type may have been paralleled by improvement of the muscle system running into the limb. Since this requires a firmer attachment area, it may have led to a compromise between the formation of a rigid basipod through the fusion of several longer annuli and the development of a membranous joint or shaft through a softening of the short proximal annuli to give the limb greater mobility. The endopod would, according to this hypothesis, include the remaining distal set of the original annuli. At this level of evolution, the number of

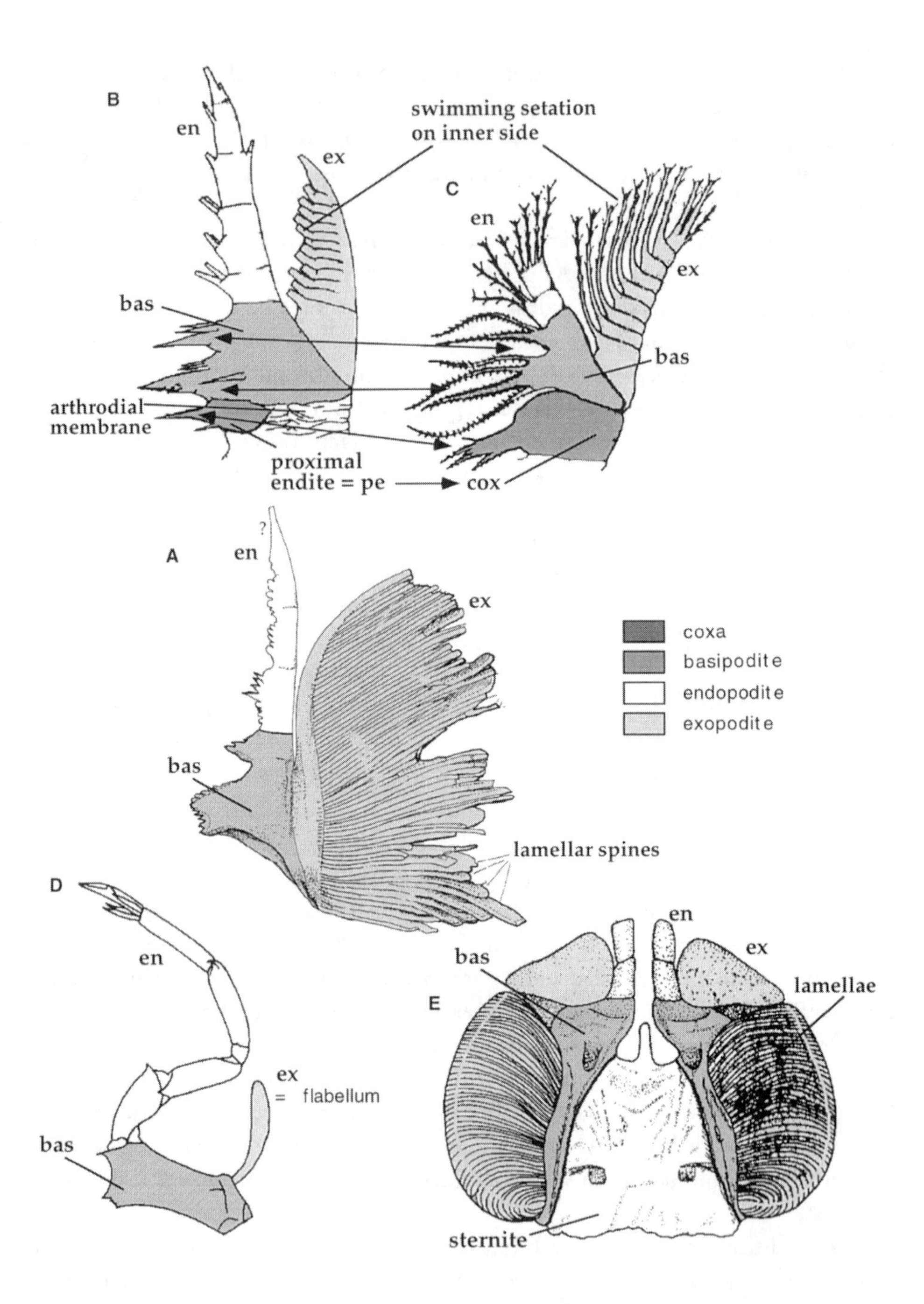


FIGURE 5.6

Comparison of *(A)* a *Misszhouia* leg (Lower Cambrian of China, drawn from Chen et al. 1991, fig. 7); *(B)* a post-antennal limb of the stem-group eucrustacean *Martinssonia* (Upper Cambrian of Sweden, from Müller and Walossek 1986a); *(C)* the second antenna of an extant cirriped nauplius (from Costlow and Bookhout 1957); *(D)* last prosomal leg of *Limulus;* and *(E)* opisthosomal limb of *Limulus*. Arrows connect corresponding setae/spines on the legs of the stem group and the maxillopod crustacean (not to scale).

endopodal podomeres seems to have become fixed at seven, as can be observed in the limbs of *Misszhouia,* in known appendages of trilobites, or in *Agnostus.*

Crustacea *s.l.* have at most five endopodal podomeres, and this number is retained as a plesiomorphy in two extant taxa of the crown group, the Malacostraca and Cephalocarida. Within the Eucrustacea, this basic number can be further reduced to four and fewer portions (not necessarily podomeres).

Our hypothesis on the evolution of the crustacean appendages, based on morphological data from both fossils and extant arthropods, has several implications for the homology and terminology of arthropod limbs. First of all, the limb stem of trilobites and other early arthropods, traditionally termed "coxa,"is in fact the equivalent of the basipod of Crustacea. Again, telopod and endopod are synonyms, and the so-called pre-epipod of Trilobita represents the exopod (Størmer 1939 used the name "pre-epipod" because in comparison with the crustacean limb a coxa does not carry a ramus but only an exite).

The hypothesis can be directly applied to the Chelicerata. In this group, analogous to several crustacean taxa, evolution has enforced walking on uniramous legs, clearly in the aquatic realm. Yet, *Limulus,* as a representative of xiphosuran chelicerates, nicely shows the evolutionary pathway toward the uniramous state in still having an exopod on the last prosomal leg, the so-called flabellum (fig. 5.6D), and on all opisthosomal legs, the operculum, and the gill-bearing limbs (fig. 5.6E). From this state the originally flattened setae seem to have broadened to form the gill blades of more derived chelicerates. Subsequently during evolution, these legs are generally assumed to have merged with the ventral body, including the gill blades, into concavities to lead over to the book gill/lung system of the arachnid lineage (cf. Siewing 1985, fig. 846).

Again, the original spine-bearing, blade-like inner edge of the basis, as developed in *Misszhouia,* trilobites, and *Agnostus,* is still recognizable in the prosomal legs of *Limulus,* being slightly set off from the major portion there (grinding edge). The same can also be seen in all those arachnids in which the anterior prosomal legs still take part in the feeding process, particularly the eurypterids and scorpions. These process-like enhanced inner edges of the basipods are typically anteriorly turned (e.g., Storch and Welsch 1991, figs. 159, 160). The same can be observed in the two post-cheliphoran limbs of the chelicerate larva with supposed pantopod affinities from the "Orsten" (larva type D in Müller and Walossek 1986b, fig. 9).

Interestingly, the basis of the prosomal legs became progressively more connected with the body in the arachnid lineage, so that derived arachnids, such as the spiders, walk exclusively on their endopods, with the major joint being between basis and ramus. This situation is in striking contrast to "walkers" among Crustacea (analogous), which retain the old body-limb joint for movement of the limbs (particularly Eumalacostraca: joint proximal to the coxa).

Regrettably, it is not yet possible to recognize either the original basipod-endopod system or a coxa-basipod-endopod system in the uniramous legs of tracheates. Since grinding edges can be developed on either basipod (trilobites, chelicerates) or coxa (crustaceans), it is important to know from which part the tracheate mandible originated; position and function alone can no longer serve as an argument for the assumption of relationships between crustaceans and tracheates on the basis of their so-called mouthparts, e.g., mandibles and two pairs of maxillae (= "Mandibulata"), as claimed even in most recent literature (e.g., Kraus and Kraus 1994). Again, land-living crustaceans conservatively retain all post-mandibular legs and the design of these. This is in striking contrast to tracheates, where the first post-antennular segment lacks limbs, the mandible is just a gnathobasic structure, and the so-called maxilla and labium (name used for the sternal plate in crustaceans) show no relicts of the ground plan shape as presented above. Detailed comparisons with the trilobitan and crustacean legs under the new hypothesis are urgently needed.

Possibly our hypothesis can also lead us toward a better understanding of the morphology and systematic status of the arthropods in the Burgess Shale. Some of these have been related to the Crustacea and even phyllocarid Malacostraca. For example, *Canadaspis perfecta*, described in detail by Briggs (1978), had trunk legs with a stem portion subdivided into seven annuli and a seven-segmented endopod, giving the impression of a limb with fourteen segments. There is no differentiation of post-"mandibular" limbs (Briggs 1992). Head structures (at least second post-mandibular leg not incorporated into head; post-mandibular legs identical in design), trunk tagmosis (total segment number uncertain, "abdomen" limbless and seven-segmented), and shield morphology (bivalved, including at most only the maxillulary segment) are not in accordance with the morphology of extant Phyllocarida (Leptostraca). The character set of *Canadaspis*, including its limb design, suggests that it represents a very early offshoot of the euarthropod lineage.

Previous misunderstanding of the fossil evidence for arthropod limbs (Cisne's reinvestigation of the *Triarthrus* legs dates back to 1975) and the failure to recognize the different fate of each of the crustacean limbs may have been the major reasons for the creation of "ur-type models," as can be found even in modern textbooks. For example, it has long been assumed that all limb portions should have had median "endites" and lateral "exites" originally, and either their reduction or retention on particular limb articles should have led to the different types of arthropod limbs. This idea presented, for example, in Storch and Welsch (1991, fig. 146) has never been supported by fossil evidence. In fact, endites occur as subdivisions of the inner edge of the basipod, if enlarged. Otherwise, there is only one median setiferous or spine-bearing edge developed on the basipod and/or coxa. The outer edge of the limb carries one exopod arising from the basipod and one or two epipodites ("exites") arising from the basis or coxa and basipod if two portions are present (for a more detailed discussion of limb structures of post-maxillary limbs, see Walossek 1993:118–120).

Again, in light of the recently discovered early arthropod fossils from the Chengjiang fauna especially, there is no longer a basis for hypothesizing a tripartite structure of the original arthropod limb stem divided into a "pre-coxa," coxa, and basipod. Indeed, Kaestner (1967) in his illustration of a copepod mandible (his fig. 661) misunderstood the articulating membrane between the coxa and basipod as a "coxa" (in contrast to the correct drawing of this limb in Calman [1909]), which led him to interpret the coxa as a "pre-coxa."

Hypotheses such as an origin of the rami from separate legs, as promoted by Schram and Emerson (1991), or that of Fryer (1992), who considers a phyllopodous limb the ancestral type, have no substantial foundation and are not discussed here in detail. The same must be said for the attempts of Averof and Akam (1995) who try to compare molecular data with the speculations of all these authors, neglecting the evidence presented for early arthropod fossils — in particular on limb morphology — subsequent to Cisne (1975).

## Characters of the Crown-group of Crustacea

The ground plan representative of the crown-group of Crustacea, the Eucrustacea, included, as far as we can now state, several external adult features:

1. a head made up of five limb-bearing segments (four in the ground pattern of Euarthropoda and still in the early stem lineage of Crustacea);
2. a head shield enclosing the same number of segments;
3. differentiated second and third head appendages, the second antennae and mandibles, both with a limb stem comprising a coxa and basipod;
4. an additional pair of differentiated, brush-shaped post-mandibular head limbs, the maxillules, serving in food transport toward the mouth;
5. a sternum resulting from fusion of all post-oral, cephalic sternites (mandible to maxillae);
6. a trunk made up of about fifteen to sixteen segments, fourteen or fifteen of which carry limbs;
7. a conical telson (possibly the condensed product of more posterior somites and the original non-metameric end piece);
8. the terminal anus; and
9. paddle-shaped furcal rami with marginal setation, most likely developed as steering devices.

Well-developed biramous natatory second antennae in the adult characterize in particular the Onychura (= Diplostraca) and, possibly, the Lower Triassic to Upper Jurassic Kazacharthra among phyllopod Branchiopoda, the Middle Devonian *Lepidocaris* among anostracan Branchiopoda, and the Ostracoda among Maxillopoda. A purely sensorial state of the second antennae is more or less restricted to the Malacostraca or special taxa, or even restricted to a certain sex. Accordingly, this situation cannot represent a ground plan character of Eucrustacea. Modifications from this state, such as the sexually differentiated antennae of Euanostraca, or large-scale reduction to even complete loss, as in various parasitic crustaceans, are simply in-group deviations from the basic plan.

Compound eyes are, as a plesiomorphic character, a further part of the ground plan of Crustacea. It remains unclear, however, whether the compound eyes were originally sessile or stalked. Our knowledge of the "Orsten" fauna cannot contribute much to this question (Walossek and Müller 1990; Walossek 1993:115). Eyes are only visible in a few taxa. In *Cambropachycope* and *Goticaris* they represent huge uniform organs. In *Henningsmoenicaris* they develop ontogenetically from small humps at the inner anterior rim of the head shield into stalked eyes positioned at the anterior edge of the hypostome (Walossek and Müller 1990, fig. 4b). *Bredocaris* possesses sessile eyes from the first larval stage (Müller and Walossek 1988b, pls. 3, 7, 8–12, 14), which is also the situation in Maxillopoda. The fate of the eyes of *Rehbachiella,* which are developed already in the orthonauplius, could not be monitored in later larval stages (Walossek 1993). On the other hand, the compound eyes of Crustacea emerge from the ventral side, i.e., below the head shield, and this holds true even for the large, calcified decapods. Only in certain taxa within the Peracarida are the eyes located dorsally on the shield (e.g., amphipods and isopods; the situation in phyllopod branchiopods is different, see below). This basic difference relative to various early arthropods, trilobites, chelicerates, and tracheates has never been seriously considered (only Bergström [1992] mentions this difference to some degree), so a conclusion cannot yet be presented.

Uncertainty also remains in regard to the original presence or absence of epipodites as flabby outgrowths of the outer edge of the stems of post-maxillary limbs serving as gills or osmoregulatory organs in several extant crustacean taxa. Larvae as well as small-sized crustaceans typically lack such organs. Hence their absence in "Orsten" forms cannot be used as an argument for one or the other assumption (Walossek 1993:119–120; for discussion of more morphological characters, particularly on head shield versus "carapace," see pp. 107–124).

As compared to the larvae of the stem-group crustaceans, the nauplius of all extant Crustacea — where still present as a free-living larva in the ontogenetic sequence — bears a special feeding system. This includes a differentiated limb setation, setulation of the oral region, labrum and atrium oris, but only three pairs of appendages, the antennules, antennae, and mandibles. The naupliar apparatus serves during development as a kind of locomotive for the growing larva until the post-naupliar limb apparatus is fully functional. It is known from extant Euanostraca that during later development both apparatuses are functioning in parallel, and the naupliar one is lost only shortly before the adult state (e.g., Anostraca; see Barlow and Sleigh 1980; Fryer 1983).

The second antennae and mandibles are virtually identical in the larval phase, with both serving in locomotion and feeding in extant crustacean feeding larvae (examples see figs. 5.5f, g, 5.6c). Hence, a subdivision of the euarthropod head into a prosencephalon and gnathencephalon between the second antenna and mandible (e.g., Siewing 1985:750) is at odds with this fact and that second antennae and mandibles turned into specialized head limbs not before the late stem-lineage level of Eucrustacea. Particularly in Copepoda and Cirripedia, the naupliar second antennae can possess a large coxa with an enditic process and armature much more elaborate than in the corresponding mandible (e.g., Dahms 1990, figs. 5, 6).

The development of a nauplius as the first free-living larva in the ontogenetic sequence, inclusive of its structural design, is considered here as a further apomorphy of the Eucrustacea:

1. a "short-headed" larva with only three pairs of limbs and adding further head segments and trunk segments during growth, in contrast to "complete-headed" larvae of earlier levels (still in the stem lineage of Eucrustacea);
2. having only two pairs of post-antennular limbs, the second antenna and mandible, which are identical in shape and function for locomotion and feeding, and comprising limb stems made of a coxa and basis;
3. having a labrum as a glandular organ that projects over an atrium oris leading to the recessed mouth opening;
4. possessing a hairy sternum resulting from fusion of the antennal and mandibular sternites and setulation around the mouth, and differentiated setal armature on specific parts of all appendages; and
5. carrying a small head shield, which covers the three limb-bearing naupliar head segments.

Table 5.1 lists a selection of external ground pattern characters of Eucrustacea. These can now be differentiated into characters new to the crown-group level (apomorphic) and plesiomorphies inherited either from the stem lineage (apomorphies of Crustacea *s.l.* or Pan-Crustacea) or from even earlier levels (for abbreviations, see table 5.2; for values in parentheses, see text above). The list emphasizes that there are indeed several apomorphic features that characterize Eucrustacea as a monophylum.

The Phosphatocopina seems to be a group that may require splitting of the data set in order to differentiate even more precisely. Their detailed description is under way, but we can state already now that this group possesses the labrum, a short atrium oris, and paragnaths on the sternum (Müller and Walossek 1985b, fig. 2f), but it has a serial set of post-mandibular limbs (Müller and Walossek 1985b, fig. 6b), e.g., no specialized first maxilla, and the sternum seems not to include the fourth post-antennular head segment, the maxillary segment. As a result, the Phosphatocopina cannot represent crown-group Crustacea, but may be the sister group of the Eucrustacea. Accordingly, they cannot belong to any crown-group taxon either, nor to any entomostracan taxon, nor to the Maxillopoda, and certainly not to Ostracoda. As a further consequence, the Phosphatocopina can no longer serve as evidence of the occurrence of Ostracoda and Eucrustacea as early as the Lower Cambrian.

In the Upper Cambrian "Orsten," the Phosphatocopina are the most abundant components of the secondarily phosphatized arthropod assemblages (e.g., Müller 1979, 1982). Representatives of them have been found in the Middle Cambrian of Australia, with forms that even lack exopods on all post-antennular limbs (Walossek et al. 1993), and even in the Lower Cambrian (Comley, Shropshire; Hinz 1987). Accordingly, stem-group Eucrustacea appear in the fossil record already almost simultaneously with the earliest arthropods known so far from the Lower Cambrian of China.

Head tagmosis advanced in the stem lineage of Eucrustacea from a state with three post-antennular segments to a stage with four such segments (table 5.1, column B). In parallel, the head shield also attained a stage at which it includes five limb-bearing segments. The progressive inclusion of segments, as we suggest for the evolution leading to Eucrustacea, is recapitulated during ontogeny, best recognizable in *Rehbachiella,* where the inclusion of the maxillary segment shows considerable delay. Also, in penaeid Malacostraca the head shield of the protozoëa still reaches back only to the maxillulary segment (Kaestner 1967, fig. 697), while in thecostracan Maxillopoda, the maxillary segment is not incorporated in the head during the whole naupliar phase (e.g., Cirripedia). Moreover, the large head shield of cirripede

TABLE 5.1
Ground plan features of Eucrustacea

| CHARACTER | A | B | C |
|---|---|---|---|
| head + shield embracing 5 limb-bearing somites | | +(*) | |
| uniramous antennule (A1) | | | + |
| A1 locomotory/feeding device, at least in larvae | | + | |
| mouth projected by glandular organ, the labrum | (+) | | |
| recessed mouth opening with atrium oris | (+) | | |
| sternum including Mx2-sternite | + | | |
| paragnaths as humps on the mandibular sternite | (+) | | |
| biramous post-antennular limbs | | | + |
| ex setation against en (multiramous status) | | + | |
| separate endite at inner proximal edge of basis | | + | |
| A2 stem with coxa + basipod | + | | |
| Md stem with coxa + basipod | (+) | | |
| Mx1 with proximal endite | | + | |
| Mx1 specialized as mouthparts | + | | |
| Mx2 with proximal endite | | + | |
| Mx2 as a morphological trunk limb | | | + |
| trunk comprising 15/16 somites 14 carrying legs | ? | ? | ? |
| tagmosis unspecialized | | | + |
| trunk end: conical telson, terminal anus and furca | (+) | | |
| larva (orthonauplius) with 3 limbs: A1, A2, Md | + | | |

Column A: apomorphy of Eucrustacea; column B: apomorphies known from representatives of the stem lineage, (*) developed later on stem lineage; column C: plesiomorphies of Crustacea *s.l.* (+) status still uncertain, since n part also known from phosphatocopines.

cyprids seems to include only the maxillulary segment (Høeg, pers. comm.). Inclusion of head segments, however, is a process that should be distinguished from the development of shield size and the range of extension of its margins to any side (for detailed discussion see Walossek 1993:110–115).

Dorsal and ventral structures did not evolve at the same rate. Rather, the ventral head structures reveal an apparent delay in their evolutionary speed of incorporating and modifying additional limbs relative to the dorsal situation. While the maxillule becomes specialized as a "mouthpart" at the crown-group level, this is not so for the fourth pair of post-antennular head limbs, as is particularly evident from the extant Cephalocarida. In this group, the maxillae are shaped like a trunk limb (e.g.,

TABLE 5.2
Abbreviations used in the figures and lists/tables

| | |
|---|---|
| Abd | abdomen |
| an | anus |
| app | appendages |
| A1 | first antenna or antennule (also atl) |
| A2 | second antenna or antenna |
| bas | basipod(ite) |
| cox | coxa |
| cs | cephalic or head shield |
| div. | diverse |
| en | endopod(ite) |
| ex | exopod(ite) |
| fr | furcal rami |
| hyp | hypostome |
| i | initial, incipient |
| j | joint |
| la | labrum |
| lsp | lamellar spines (e.g., *Misszhouia*) |
| L1–L4 | earliest larvae of *Rehbachiella* |
| Md | mandible |
| Mx1 | first maxilla or maxillule |
| Mx2 | second maxilla or maxilla |
| pe | proximal endite |
| pgn | paragnaths |
| ples | plesiomorphic |
| plpf | palp foramen, opening for palp on basipod |
| rop | rostral process (Skaracarida) |
| rud | rudimentary, of larval shape |
| s | seta |
| sen | sensillum (*Dala peilertae*) |
| tel | telson |
| Thp | thoracopods |
| Ts1–13 | instars of *Rehbachiella* with developing trunk segmentation |

Sanders 1963). The implication of a trunk limb–shaped maxilla in an extant crustacean taxon is that the Eucrustacea cannot be characterized as having two pairs of post-mandibular mouthparts or maxillae. Notably, the "Orsten" forms *Bredocaris admirabilis, Rehbachiella kinnekullensis,* and, possibly *Lepidocaris rhyniensis* Scourfield, 1926, from the Devonian, all of which have been identified as members of specific crustacean taxa on the basis of shared derived characters (see below) reveal the same situation as the Cephalocarida.

A further implication is that the fourth pair of head limbs specialized into "second maxillae" convergently within the Eucrustacea Accordingly, particularly the recognition of the different fate of the maxillules and maxillae may provide a useful tool to evaluate the phylogeny within Eucrustacea.

Indeed, the different shape of the maxillule may serve as a character to distinguish between the two major lineages, Malacostraca and Entomostraca (fig. 5.7). In Malacostraca the stem of the maxillule comprises a coxa and a basipod, both with an undivided setiferous median edge. In Entomostraca, however, considered to embrace the Cephalocarida, Branchiopoda, and Maxillopoda, the maxillulary stem is subdivided into four endites, the proximal one representing the "proximal endite," which holds true for the ground plan of each of the three taxa (Walossek 1993:118). This again outdates any statements concerning a larger head for the ground plan of Arthropoda.

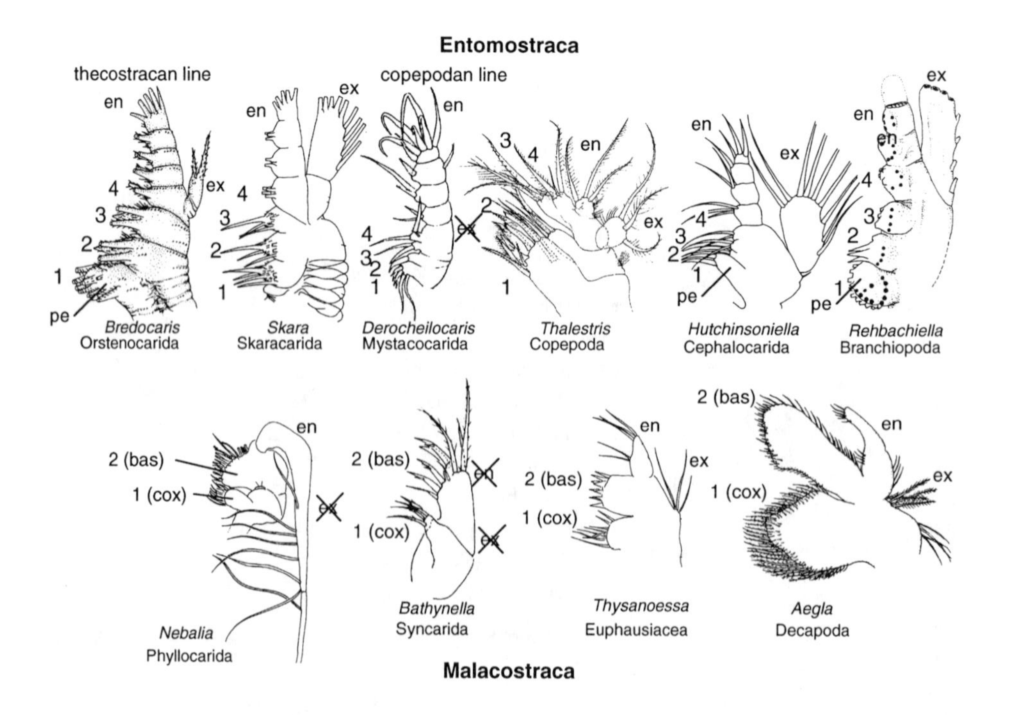


FIGURE 5.7
Examples of maxillules developed among Eucrustacea (from different sources, not to scale).

## *Rehbachiella kinnekullensis* Müller, 1983, and the Monophyly of Branchiopoda

### *Morphology*

Examination of about 150 specimens of this Upper Cambrian "Orsten" fossil, representing different developmental stages, permitted the reconstruction of a consecutive series of thirty ontogenetic stages (fig. 5.8). Increase in size and morphological development is very gradual. The series enabled not only monitoring morphogenetic changes in full detail, but also allowed the recognition of two different sets of larvae, which show minute but distinct differences particularly in the early phase of larval growth. The first stage, an orthonauplius of about 160 µm in length, has compound eyes, a head shield of moderate size, a large labrum and well-developed naupliar limbs of roughly the same size. The maxillules appear as a pair of spinelike setae at stage 2 and develop into buds at stage 3. At stage 4 the maxillae appear as buds, and from the fifth instar onward the trunk segments appear in two steps. At first they are only partially separated off in front of the rear portion, and at a second stage, they are

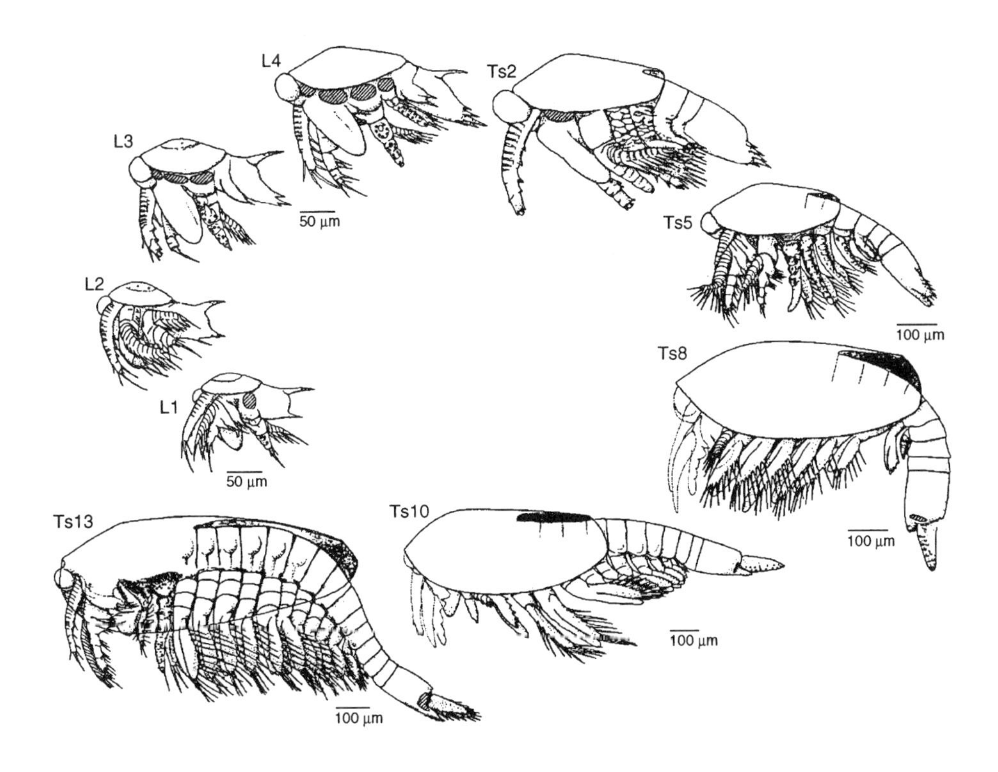


FIGURE 5.8
Life cycle of *Rehbachiella kinnekullensis* Müller, 1983, from nauplius to instar TS13 (thirtieth stage; combination of Walossek 1993, figs. 6, 7, 15).

fully demarcated. Hence it takes four "naupliar" stages before trunk development starts and twenty-six more stages until all thirteen trunk segments are developed.

The largest stage identified through well-preserved specimens is about 1.5–1.7 mm long. Of its thirteen trunk segments twelve bear limbs. The rear portion remains undivided and carries furcal paddles. The larval shape of the posterior trunk limbs, the undivided abdomen, the growth curves of different characters, and the presence of much larger, but unassignable fragments are evidence that even this thirtieth instar was not the final one.

The segments of the maxillae become progressively included into the naupliar head. This first affects the maxillulary segment, and at that stage the head shield reaches back to this segment (back to the mandibular segment in the nauplius). Much later, the maxillary segment is included when several trunk limbs have already appeared and the shield has extended considerably laterally and posteriorly. Later on the shield continuously grows out further backward and eventually covers much of the trunk. Yet it never fuses with any of the trunk segments, hence being still a simple head shield. It is noteworthy that the maxillules finally cease in growth and never develop more than four endites on their protopod, becoming heavily armed with setae and brushlike spines. The maxillae, on the other hand, never modify into special mouthparts but retain the shape of a trunk limb (see above).

*Development of Feeding Structures and Mode of Life*

In the first instars, all naupliar limbs serve as feeding and locomotory aids, as is known from free-swimming and feeding cirriped or copepod nauplii. With the progressive addition of post-mandibular limbs, a new feeding and locomotory apparatus gradually develops. This process includes the development of a deep thoracic sternal food groove and phyllopodial legs with more than two hundred setae each along their inner edge. While the posterior set bears double rows of closely spaced setules, most likely for filtering particles from incoming water, the anterior series of setae may have served for retaining and ejecting unsuitable particles. Comb spines occur in a median set for distracting particles from more posterior limbs and passing them over to the long anteriorly curved setae of the proximal endites, which point into the sternal groove and swept the particles toward the oral region.

The anterior naupliar apparatus remains functional until the thirtieth instar although modified in size, with reduction of the antennae and of the mandibular palp occurring progressively. Notably, in extant Anostraca the larval apparatus also remains functional until the latest phase of development, when all thoracopods already operate as filter limbs (see above).

Larger specimens are much less well preserved than the early larvae. It is possible that these later larvae above the millimeter level were capable of swimming above the bottom and, hence, could no longer be preserved in the same way as the smaller ones — either because of predation or because of their distance from the fossilization zone below the bottom layer.

*Systematic Status and Implications*

A filter apparatus of the specific type developed in *Rehbachiella* is present only in branchiopod Crustacea. It is an open system with currents entering the median path between the limb pairs from the ventral side. Food transport takes place within the depth of a sternal groove. Phyllocarid Malacostraca have developed quite a different filter apparatus, which is closed posteriorly (as with the basket feeding apparatus of Euphausiacea, etc.), has currents entering from the anterior, and in which food is being passed over fairly distally to the maxillae. There is no filter groove, and filtratory setation extends far up onto the endopod. Moreover, the armature with setae is quite different; in Branchiopoda and in *Rehbachiella* the filter setae are found at the posterior edge of the endites and retention setae are located anteriorly, while in Phyllocarida the reverse is the case. Lastly, the limbs of Phyllocarida have stems consisting of coxa and basis, and filtration is also a matter of the elongate, setiferous endopods.

Interestingly, the Cephalocarida provide us with a precursor state for the branchiopod apparatus. In this group, the limb structure with a soft basipod carrying lobate endites, including a small proximal endite, and their armature is basically similar to that of Branchiopoda, but there is no sternitic groove on the trunk, and no filtering setation is developed (scraper feeding). Again, the maxillules of Cephalocarida, inserting far laterally, have an elongate pusher endite (= proximal endite) to reach into the deep paragnath channel, in contrast to the brush maxillule of *Rehbachiella* (four endites on limb stem) and of extant Branchiopoda (only proximal endite retained). Cephalocarida have a five-segmented endopod, while that of Branchiopoda is maximally four-segmented *(Rehbachiella)*.

The assumption of the monophyly of Branchiopoda has never been validated. Moreover, Preuss (1951, 1957), who investigated the limb musculature of Anostraca and Phyllopoda based on a severe misunderstanding of limb morphology, proposed a distinction of these two groups, which led, at least in Germany, to the abandonment of the embracing taxon Branchiopoda. The more recent approach of Fryer (1987) returned to the starting point of any phylogenetic approach by placing each candidate taxon equally apart from one another rather than presenting any suggestion as to how the relationships between the taxa might be understood.

Figure 5.9 shows Walossek's (1993) hypothesis of the phylogeny of Branchiopoda. He identified the complex filtratory apparatus with all its morphological and functional requirements as an apomorphy of the Branchiopoda, thus characterizing the group as a monophylum (character 2 in fig. 5.9). Its identical morphology and presumably also its similar function enabled him to recognize *Rehbachiella* as a member of this crustacean taxon. All branchiopod taxa and *Rehbachiella* possess a "neck organ," present at least in the early larval stages (osmoregulatory organ). It remains uncertain, however, whether this could be another apomorphy of Branchiopoda or unites Branchiopoda and Maxillopoda, because a similar organ is present also in *Bredocaris,* a Cambrian representative of thecostracan Maxillopoda (character 1 in fig. 5.9; see also below). This question requires further study.

Assumptions as to the position of *Rehbachiella* within the Branchiopoda depend largely on assumptions of relationships within the Branchiopoda. There seem to exist two major lineages, which can be differentiated in particular by their eye morphology.

Euanostraca have displaced their compound eyes on to stalks, and there seems to be evidence that this is not the same type of stalked eyes as in the Malacostraca (character 3 in fig. 5.9). In the Phyllopoda, on the contrary, the compound eyes become internalized early in ontogeny and are displaced within a cuticular pocket toward the dorsal side of the head, a feature described by Claus as early as 1873 (character 4 in fig. 5.9). During the ontogeny of *Rehbachiella, t*he eyes are progressively displaced onto a frontal protrusion, which seems to indicate that this species was already a member of the anostracan lineage. This assumption would be in accordance

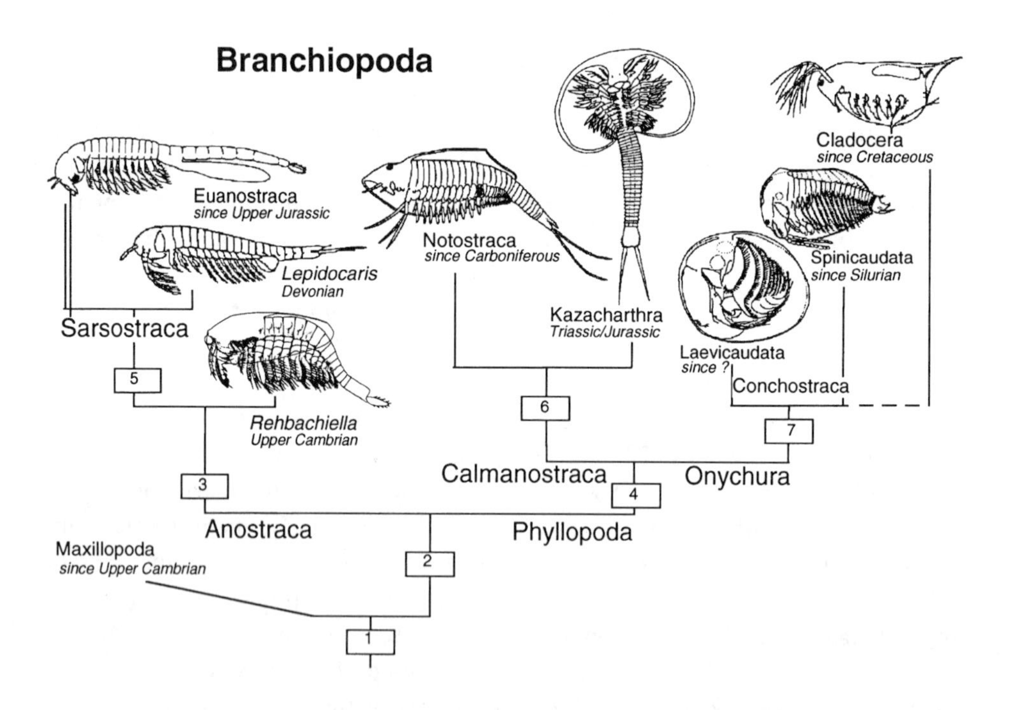


FIGURE 5.9

Presumed phyletic relationships of and within Branchiopoda; apomorphies: *(1)* (possibly uniting Branchiopoda and Maxillopoda) = osmoregulatory "neck organ," *(2)* (for Branchiopoda) = complex post-mandibular filter-feeding apparatus with sternal food groove, *(3)* (for Anostraca) = compound eyes raised at front/naupliar dorsal organ reduced in early larval development, *(4)* (for Phyllopoda) = compound eyes internalized and shifted dorsally, *(5)* (for Sarsostraca) = reduction of lateral expansion of head shield rims, *(6)* (for Calmanostraca) = bottom-dwelling life with modification of anterior trunk limbs for omnivorous feeding habits, polymetamery of trunk, and long, multiannulate furcal rami; *(7)* (for Onychura) = bivalved secondary shield behind larval head shield. Relationships within Onychura not resolved (slightly modified from Walossek 1995, fig. 6, not to scale).

with the early reduction of the neck organ during ontogeny in *Rehbachiella* and in extant Euanostraca, while this organ is retained up to the adult or becomes even enhanced in the Phyllopoda.

Accepting *Rehbachiella* as a member of the Anostraca assumes that the head shield, well-developed in this fossil, became progressively reduced along the anostracan lineage. Euanostraca possess it only in their hatching nauplius (Walossek 1993, fig. 53B), while in *Lepidocaris* from the Devonian the shield is retained up to later larval stages. As a matter of fact, the latter two taxa do not really lack the shield, but merely its lateral extensions are effaced, which gives the head of the adults a smoothly rounded, capsule-like appearance (character 5 for Sarsostraca = *Lepidocaris* and Euanostraca in fig. 5.9). By analogy, such effacement of the shield margins occurs also in the chelicerate lineage (the prosomal shield) — Xiphosura have a large shield, while that of Eurypterida and Scorpionida is only slightly pronounced, and in Araneae the prosoma is saclike.

The Phyllopoda comprise two taxa, the Calmanostraca and Onychura (= Diplostraca; shared apomorphy as mentioned above). Calmanostraca include the Notostraca and the fossil Kazacharthra. Notostraca have a fossil record from the Carboniferous and are subdivided into only two genera, *Triops* and *Lepidurus*. Apart from an obvious lack of changes in several morphological features, the morphology reveals a high degree of adaptive modification from the original branchiopod plan toward a bottom-dwelling, omnivorous to predatory mode of life. Again, segmentation of the trunk has obviously increased (more than thirty segments with more than seventy limbs). Yet their "conservative" construction has led many workers to treat Notostraca misleadingly as primitive among the Crustacea. These workers even reverted the taxon's structural changes — apomorphies of Notostraca (and Kazacharthra) — into plesiomorphies, that is, they interpreted them as characters inherited from a stem species they share with other crustacean groups.

Kazacharthra are a diverse group that existed only between the Upper Triassic and Lower Jurassic, i.e., they appeared later in the fossil record than the Notostraca, radiated rapidly into various taxa and became extinct rather soon. Kazacharthra share various derived features with Notostraca (character 6 for Calmanostraca in fig. 5.9), such as the large, flat head shield carrying an eye spot (compound eyes underneath) associated with the "neck organ," an ample, duplicature-like anterior rim on the ventral side of the shield narrowing and merging with the flat, plate-shaped labrum, limb morphology, polymetamery of the trunk, and modification of the furcal paddles to long, densely annulated rods.

Notostraca have modified the phyllopodial filter limb series of the ground plan of Branchiopoda in such a way that the anterior legs are no longer used for filtration but for bottom dwelling and capturing of prey. They are held far laterally and have an elongate, firm, but secondarily subdivided basipod carrying slender endites. Toward the posterior, the limbs become progressively more phyllopodium-like, and the limbs of the polypodous series behind the eleventh leg, where the gonopore is located, are very flat and attain more and more a shape similar to the larval limbs of the other Branchiopoda. New finds of Kazacharthra described by McKenzie et al. (1991) revealed eleven trunk limbs rather than six, as were described earlier. These limbs are almost identical to the anterior two trunk limbs of Notostraca and do not show a progressive transition toward a more phyllopodous shape (reconstruction in fig. 5.9). In contrast with Notostraca, the Kazacharthra show much more morphological plasticity. Some species are quite large and have rigid spinose head shields and telson plates. Remarkably however, the Kazacharthra had well-developed multiannulated antennules and biramous second antennae, according to McKenzie et al. (1991), while these are traditionally considered to be largely reduced in extant Branchiopoda.

As a matter of fact, well-developed second antennae also occur in the extant Onychura, though with symmetrical rami. With regard to what is known from *Rehbachiella,* the retention of both pairs of antennae in their plesiomorphic state in members of the two different lineages of Branchiopoda makes it more likely that reduction, or, at least modification, of these, had occurred convergently in either lineage (homoplasy).

Onychura (= Diplostraca), embracing the spinicaudate and laevicaudate conchostracans and the "cladocerans," can be characterized by the development of a unique secondary shield during larval development — at a stage called heilophora (character 7 in fig. 5.9). This secondary shield, which is not a "carapace," grows out from the tergite of the maxillary or possibly the first trunk segment and, later, may enclose the whole body, while the head shield remains present and becomes the sclerotic cover of the head, best seen in Laevicaudata or daphniid "cladocerans" (the validity of this group is uncertain, cf. Martin and Cash-Clark 1995). In addition to the locomotory second antennae, the Onychura retain the phyllopodous limb apparatus for filtering, which is a ground plan character of the Branchiopoda (deviations possible, particularly in the predatory forms among the "cladocerans").

## Contribution of Skaracarida, *Bredocaris,* and *Dala* to the Phylogeny of Maxillopoda

The taxon Maxillopoda was established by Dahl (1956) to comprise a number of small crustaceans but has not been universally accepted, mainly because these share no obvious apomorphies (see discussion in Boxshall 1992). Boxshall (1983) strongly criticized its validity, while Grygier (1983, 1984) was in favor of it. Grygier proposed ground plan characters for the Maxillopoda that enabled us to identify one of the "Orsten" crustaceans, *Bredocaris,* as a representative of a particular maxillopodan taxon, the Thecostraca (Müller and Walossek 1988b; Walossek and Müller 1992). Moreover, the systematic placing of *Bredocaris* led us also to reevaluate the affinities of Skaracarida, previously (Müller and Walossek 1985a) considered similar only to Mystacocarida and Copepoda.

Our thoughts on the phylogeny of Maxillopoda are based on the consideration of Recent taxa as well as on fossils from the Upper Cambrian "Orsten" of Sweden.

These fit, in our view, not only nicely into the taxon but also contribute significantly to the understanding of the ground pattern characters of Maxillopoda and evolutionary trends of and within the group. Full preservation of details and, in one case, of the larval cycle gives additional, useful insight into early life strategies of Maxillopoda and further development of these within the group. They also present apomorphic characters for Maxillopoda, validating its monophyly.

*Orstenocarida*

*Bredocaris admirabilis* is a very small form, about 0.85 mm long in the supposed adult state, which possibly lived in or at the interface between water and soft bottom (Müller and Walossek 1988b). Its larval cycle includes five "naupliar" stages, starting with a larva about 240 µm long and having lateral eyes, a head shield of moderate size, naupliar limbs of the feeding and locomotory type, and the buds of the maxillules. This indicates that this first instar is already a metanauplius and not the first stage or orthonauplius, as is present in *Rehbachiella* (for definition see Walossek 1993:82–95). The maxillule gains its final shape in the next stage, while the maxilla and four thoracopods appear progressively on the hind body but remain anlagen until the molt to the largest stage (fig. 5.10).

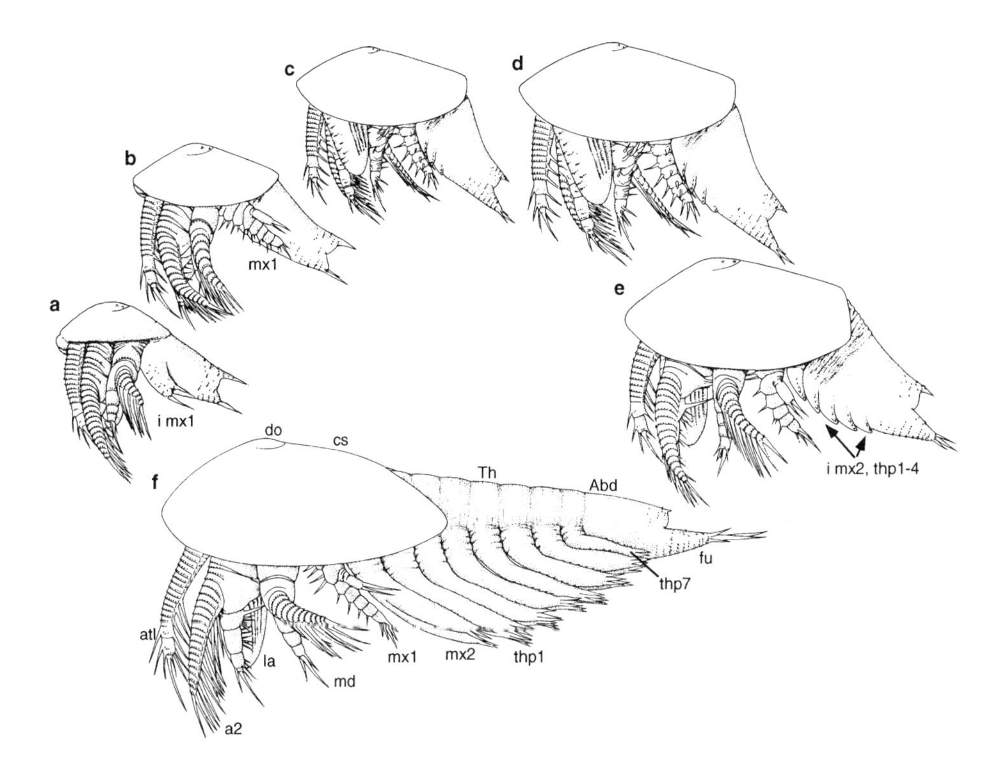


FIGURE 5.10
Larval cycle of *Bredocaris admirabilis* Müller, 1983. *(a–e)* First to fifth "naupliar" instars. *(f)* Presumed adult (modified from Müller and Walossek 1988b, fig. 4, stage N3 [c] added).

This stage, twice as large as the last "naupliar" stage, has a set of eight serial trunk limbs, the first of which represents the second maxilla, clearly recognizable by the distinct head boundary behind the corresponding segment (Müller and Walossek 1988b, pl. 3, fig. 8). This stage is interpreted as the final one, i.e., the adult, because of its complete development of the thoracic region. Its effacement of trunk, abdominal and limb segmentation, and the missing furcal articulation are seen as a special adaptation to meiofaunal life and an indication of paedomorphosis, features well-known from meiofaunal organisms (cf. Westheide 1987).

The second maxillae and trunk legs may have served mainly for locomotion, while their contribution to feeding was very likely only limited. Feeding was mainly or exclusively done by the pre-maxillary head limbs, i.e., the second antennae, mandibles, and the brush-shaped maxillules.

*Skaracarida*

The slender *Skara anulata* Müller, 1983 (fig. 5.11), and *S. minuta* Müller and Walossek, 1985, have the same gross morphology, but the former is 1.2 mm long and the latter only 0.7 mm long. The body comprises a head region with a small head shield having very slightly extended margins, and a slender trunk consisting of eleven segments and a small telson with articulated, two- to three-segmented furcal rami. The forehead region protrudes from the shield and bears a tubular or cone-shaped organ distally (Müller and Walossek 1985a, pls. 4, 5). Lateral eyes are missing, and the possibility that an internal naupliar eye was present can only be assumed.

The anterior three head limbs are similar to those of *Bredocaris* and *Rehbachiella*, with the exception that the mandibular coxae are rather small. The two pairs of maxillae and the first trunk limb are similar to each other. Together with the anterior limbs, they form a cephalothoracic locomotory and feeding apparatus capturing food particles between them in a post-oral feeding chamber (Müller and Walossek 1985a, pl. 11). The first trunk segment bears a small tergite, while all other segments are tubular. The seventh to ninth trunk segments carry three large spines ventro-

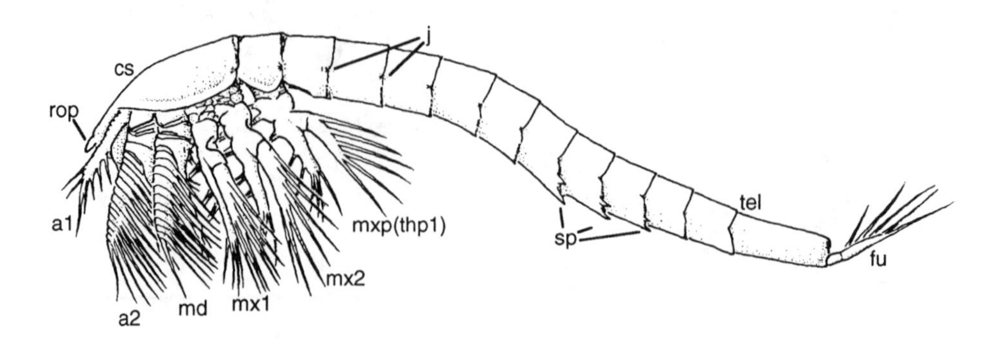


FIGURE 5.11
Corrected reconstruction of *Skara anulata* Müller, 1983 (from Walossek and Müller 1992, fig. 2; trunk in original reconstruction of Müller and Walossek 1985a, fig. 1 erroneously showed two additional segments).

posteriorly, and almost all segments have small marginal bristles to protect the intersegmental articulating membranes, as is the case with many small benthic crustaceans. Lateral pivot joints connecting the trunk segments permitted only a dorsoventral flexure of the trunk.

In addition to differences in size and some proportions, the feeding aids in particular serve to distinguish between the two species: the larger species has longer legs and fewer but stronger setae than the smaller one, while its mandibular coxa bears only the two setae, as in the early larval state of a mandible in various crustaceans (e.g., copepods; see Dahms 1990, fig.6).

Dala peilertae

This species was originally described from only a few fragments (Müller 1983). Walossek and Müller (1992) suggested that *Dala* might represent an additional fossil maxillopod, based on new evidence from material collected since. This includes the fragmentary record of mandibles and maxillules on a distorted head portion and an indication that the series of eight trunk limbs incorporates a trunk limb — shaped second maxilla — as in Cephalocarida, *Bredocaris,* and *Rehbachiella.* Accordingly, the thorax bears seven pairs of thoracopods, as in the ground plan of Maxillopoda (cf. Grygier 1983). The abdomen comprises four ring-shaped segments and a telson with paddle-shaped furcal rami, another ground plan feature of Maxillopoda (see below). *Dala* was most likely a good swimmer, as can be determined from the slender limbs (fig. 5.12). Yet locomotion and feeding were coupled, because the median

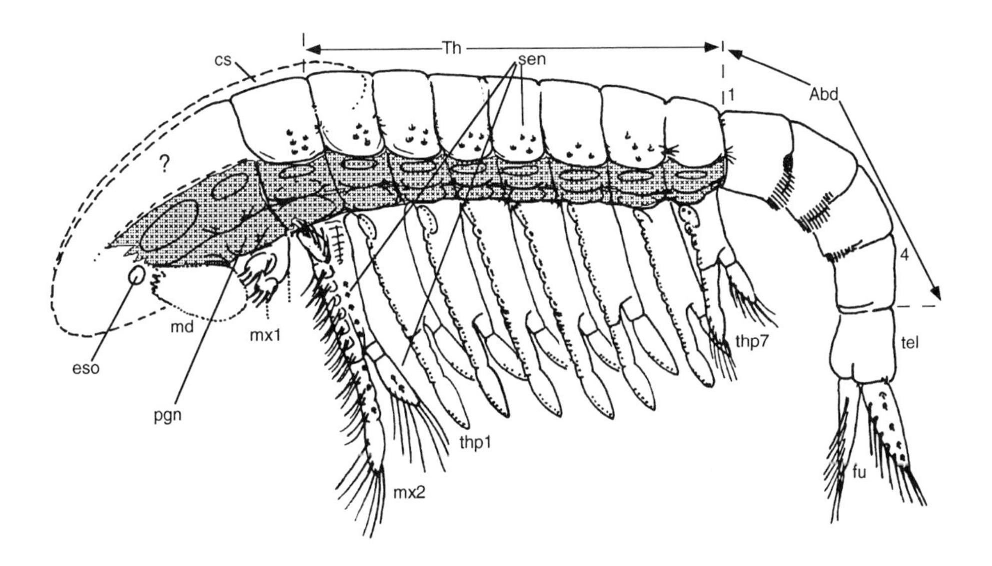


FIGURE 5.12

Reconstruction of *Dala peilertae* Müller, 1983, based on new material (total length approximately 2 mm, uncertain structures as stippled lines). Note the trunk limb–shape of the maxilla and existence of maxillules and mandibles anterior to them (left limbs omitted [ovals filled with lines]).

setal armature of the limb stem, including the proximal endite, is well developed and more elaborate than in *Bredocaris* and any extant Maxillopoda.

*Evolutionary Trends and Monophyly of Maxillopoda*

Taking *Dala, Bredocaris,* and Skaracarida into consideration as representatives of the Maxillopoda, several trends are recognized in the evolution of this taxon that may serve to characterize it as a monophylum (reconstructed phylogeny in fig. 5.13). These include the following:

1. the trunk tagmosis is reduced from a segmental stage developed in Cephalocarida and Branchiopoda to seven thoracomeres and four abdominal segments (7 + 4);
2. the use of thoracopods is exclusively for locomotion (the cirriped status with filter limbs is clearly derived);
3. feeding is principally a matter of the head appendages alone.

While *Dala* reveals the trunk tagmosis as given above, its coupling of locomotion and feeding in the trunk-limb series indicates that reduction of feeding aids occurred progressively in the stem lineage of Maxillopoda, placing *Dala* on the stem lineage (character 1 in fig. 5.13). Yet setal armature was already less developed in *Dala* than in the Cephalocarida and Branchiopoda.

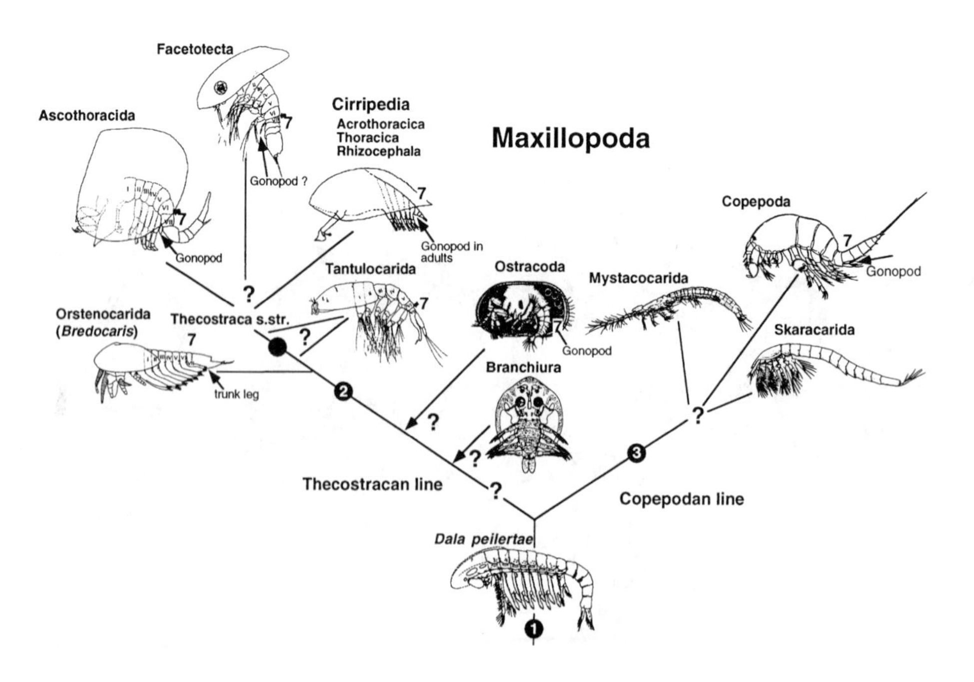


FIGURE 5.13

Assumed phylogeny of the Maxillopoda, including "Orsten" fossils. Question marks emphasize uncertainties. "7" indicates the seventh thoracomere bearing limbs or, at least, the gonopores (not to scale). Apomorphies supporting nodes 1–3 are discussed in text.

Newman (1983) suggested that paedomorphosis (neoteny) might have played a significant role in the evolution of Maxillopoda. This is apparent from two features. Larval development of Copepoda, which represents the plesiomorphic state among Maxillopoda, includes eleven successive stages, but the development is terminated rather than completed (Walossek 1993:87–88). The last stage of Copepoda, the adult, is even termed "copepodid 6" by copepodologists to emphasize its larval appearance. At this stage, the mandible still bears the basipod and the two rami, as do equivalent larval stages of other Eucrustacea. Accordingly, the mandibular shape cannot be considered as a plesiomorphy but is simply a larval feature.

Plesiomorphies in the ground plan of Maxillopoda *s.str.* are, for example, the moderate size of the head shield, the presence of compound eyes (members of the thecostracan lineage), the anameric development (Copepoda), the telson with articulated furcal paddles (members of the thecostracan lineage, such as the extant Ascothoracida and Cirripedia), and, last but not least, the second maxillae with the shape of a trunk limb *(Bredocaris, Dala).*

Maxillopoda presumably had a benthic swimming or meiofaunal mode of life originally, as is deduced from *Bredocaris,* Skaracarida, and *Dala,* and which is still recognizable today in various Copepoda, the Mystacocarida, and various Ostracoda. From this mode, the two major lineages (as yet unnamed), the copepod and the thecostracan lineages (fig. 5.13), evolved independently.

The thecostracan lineage retains in its ground plan the compound eyes, the well-developed head shield, feeding with the head limbs, the trunk segmentation (7 + 4), the serial design of the thoracopods (no maxillipeds) and the conical telson with articulate furcal paddles, best seen in the Ascothoracida. Also plesiomorphic is the shape and armature of feeding/locomotory nauplii, as present in extant Cirripedia.

While the earliest evolutionary pathway of the thecostracan lineage is difficult to reconstruct, there is a set of taxa clearly characterized by a synapomorphy in its distinctive larval cycle. In the first phase of this cycle, comprising maximally six instars, the second maxillae, which remain on the larval hind body, and the thoracopods do not differentiate into functional limbs. Trunk segmentation is completely suppressed until a metamorphic jump to the second phase, at which trunk segmentation (thorax and abdomen), thoracopods, and articulated furcal rami appear simultaneously (character 2 in fig. 5.13). Accordingly, the maximum number of six early instars is only superficially concordant with the number of "nauplii" of Copepoda. Again, the seventh instar of the Thecostraca *sensu* Grygier (1984), comprising the Ascothoracida, Facetotecta, and Cirripedia (named "cypris" in the Cirripedia), is not equivalent to the first copepodid larva of Copepoda, which is the seventh in the series of this group. Rather, the seventh thecostracan larva represents the completely segmented final developmental stage, hence it is approximately equivalent to a late copepodid or even the adult of Copepoda (Walossek 1993:87–88 and table 4).

*Bredocaris* shares this special life cycle of the core group of the Thecostraca but has maxillae in the shape of a trunk limb, indicating its stem-group position. In extant representatives of the Thecostraca, such as Ascothoracida and Cirripedia, the trunk segmentation appears in the second phase of the larval cycle (cypris and equivalent stages), together with articulated furcal rami. Again, the cirripede larval cycle starts with a true swimming and feeding orthonauplius, while the first larva of *Bredocaris* already shows the buds of the maxillule and a mandible with a small grinding plate. Correlated with the larval sequence of *Rehbachiella,* it already represents the third stage of an original anamorphic series. This loss of the earliest larval stage, the lack of segmentation of the thorax and abdomen in the presumed adult of *Bredocaris,* its reduction of trunk-limb segmentation, and lack of articulation of the furcal paddles are, hence, considered specialities of the fossil taxon, most likely in accordance with life at the sediment-water interface. Accordingly, the poorly developed subdivision of the trunk limbs of *Bredocaris* cannot be used as evidence for a developmental path of the "ur"-crustacean limb from an unsegmented "ur"-type, as claimed by Fryer (1992).

The Tantulocarida were formally established as a separate crustacean taxon by Boxshall and Lincoln (1983), not least because of their lack of head limbs, naupliar stages, and uncertainties about their trunk tagmosis. Huys (1991), however, demonstrated that this group had the same number of trunk segments as was present in the ground plan of Maxillopoda. Again, the recent description of the female gonopore on the first thoracomere suggests close relationships of the Tantulocarida with the Thecostraca (Huys et al. 1993).

All extant representatives of the thecostracan lineage, as given above, have evolved very distinctive lifestyles, departing from the presumed original benthic mode of life toward sessile or parasitic modes. This transition was apparently paralleled by the modification and reduction of head limbs and the mouth region and the development of special attachment devices. Yet this seems to have occurred independently in each group, since it affected the limbs and other head structures to different degrees. The exception is the sessile barnacles, which retain all their head appendages.

In contrast to *Bredocaris,* all extant members of the Thecostraca have changed the tagma boundary between the thorax and abdomen in such a way that the last, seventh thoracomere bearing the gonopore shifted to the abdominal tagma. Its limbs, well-developed in *Bredocaris,* have been modified to penis structures in the males or been lost in the females (cf. Grygier 1983). Yet the internal musculature reveals the original set of seven thoracomeres as known from facetotectan cypris larvae (Grygier 1987, fig. 4).

The copepod lineage comprises, in our view, three taxa, the Skaracarida, Mystacocarida, and the Copepoda, all of which share a meiobenthic mode of life, at least in their ground plan. These three taxa share several derived characters (character 3 in fig. 5.13), such as:

1. the shortening of the margins of the head shield;
2. the reduction of the furcal paddles to rod-shaped extensions (modified in the interstitial mystacocarids to clawlike structures);

3. the modification of the first thoracopod into a maxilliped; and, in parallel,
4. the achievement of a cephalothoracic feeding apparatus; and
5. the complete loss of the compound eyes.

In contrast to the ground plan of Maxillopoda and the thecostracan lineage, the feeding apparatus of the head has been extended in this group of taxa by one more pair of limbs, the maxillipeds. The remaining thoracopods are either retained for locomotion in Copepoda, reduced in Mystacocarida, or completely missing in the Skaracarida. A formal name for this taxon has not yet been proposed. The taxon Copepodoidea established by Beklemishev (1952) included Copepoda, Mystacocarida, Cirripedia, Ascothoracida, and probably Branchiura (*fide* Grygier 1983), which more or less parallels Dahl's (1956) Maxillopoda.

The interrelationships of the three taxa of the copepod lineage are still difficult to assess. Mystacocarida and Skaracarida may be linked by their reduction of trunk legs, though this is a reductive feature and may be the result of convergence. On the other hand, the Skaracarida and Copepoda share the frontal rostrum and the specific shape of the head shield. Again, in the Skaracarida, the first trunk segment bears a tergite that is closely attached to the posterior rim of the head shield, foreshadowing the situation in Copepoda where this segment is entirely incorporated within a cephalothorax (apomorphy of Copepoda). In the Mystacocarida, the maxilliped segment is still entirely free from the head. Specializations in ontogeny are difficult to assess: Copepoda possess the most primordial sequence among Maxillopoda, larvae of Skaracarida are unknown, and the ontogeny of Mystacocarida is largely modified, preventing further consideration.

In contrast to Mystacocarida and Skaracarida, the Copepoda have modified, convergently to the Thecostraca, their tagma boundary between the thorax and abdomen by having set off their seventh thoracomere, which bears the genital opening (limbs modified to copulatory supports only in males), from the thorax. Within the Copepoda, this boundary has even shifted one segment further anteriorly.

We consider the appearance of spines on the seventh trunk segment in the Skaracarida an indication that tagmosis had already shifted in this taxon from the 5 + 7 + 4 level, as developed in the ground plan of Maxillopoda, to the 5 + 6 + 5 level, which provides further support for their close affinities with the Copepoda. According to Huys (1991), Mystacocarida still reveal a set of seven thoracomeres indicated by remnants of the musculature of their limbs. Thus, their tagmosis is in accordance with the maxillopodan ground plan, and their external expression of five thoracopods (maxillipeds and four vestigial limbs) cannot serve to ally Mystacocarida with the Branchiura (which have four well-developed swimming thoracopods and no maxilliped).

A further apomorphy of the copepodan lineage may be seen in a specific subdivision of the limb stem of maxillules, maxillae, and maxillipeds (Walossek and Müller 1992): the four endites of the limb stem are retained (= entomostracan synapomorphy, according to Walossek 1993:118), but this still has to be investigated in more detail.

Plesiomorphies of the copepodan lineage include the benthic lifestyle and the retention of an anamorphic larval cycle, as expressed in the extant Copepoda.

Two taxa have been variously proposed as candidates to be included in the Maxillopoda, but these are difficult to assess: the parasitic Branchiura and the Ostracoda. The Ostracoda possibly were originally benthic, but developed a similar plasticity of lifestyles as the Copepoda. The Branchiura reveal many adaptive modifications toward ectoparasitism, while, for example, their large head shields and the possession of compound eyes are plesiomorphic features. Although the ontogeny of both taxa reveals various deviations from that of other Crustacea, it shows relicts of an anamorphic development, at least with regard to the development of the trunk limbs. Accordingly, Branchiura and Ostracoda should have branched off from the thecostracan lineage at least before the *Bredocaris* level. Likewise, they could have branched off from the copepodan lineage only at an early stage, i.e., before the reduction of the compound eyes and development of the cephalothoracic feeding apparatus including the first pair of thoracopods modified to maxillipeds. As a further alternative, both taxa may have originated separately from either lineage (arrows with question marks in fig. 5.13).

The hypothesis presented here is in general accordance with the cladogram of Boxshall and Huys (1989), who also recognized two different lineages within Maxillopoda. Deviations are mainly a matter of different assumptions on the characters of their data matrix. Examples (citing character numbers of Boxshall and Huys, with our interpretations in parentheses) are as follow:

- character 1: biramous a1 plesiomorphic for Crustacea (apomorphy of Malacostraca);
- characters 3, 6, 16: number of trunk somites — Boxshall and Huys consider the non-metameric telson as a trunk somite (product of the budding zone);
- character 9: presence or absence of a "carapace" (loss or merely reduction of lateral margins of the shield in copepodan line);
- character 29: absence of naupliar head shield as plesiomorphic to Maxillopoda (never absent in any Arthropoda);
- character 31: lateral gut caeca in shield as apomorphic within Maxillopoda (present even in earliest Euarthropoda = plesiomorphy of Crustacea *s.l.*);
- character 34: absence of post-maxillulary limb buds on "naupliar" trunk apomorphic for Thecostraca (buds in *Bredocaris* and the cirriped *Ibla quadrivalvis* [Anderson 1987]; mx2 buds also in other thecostracans).

Regrettably, erroneous counting of the number of trunk segments in Skaracarida and Tantulocarida by Boxshall and Huys (1989) prevented them from recognizing the trunk tagmosis as an apomorphy of Maxillopoda, which should comprise seven thoracomeres plus four limbless abdominal somites and a conical telson with furcal rami, as is supported by the reinterpretation of the trunk of *Dala*. These interpretative differences led Boxshall and Huys to assume a position of Ostracoda and Branchiura closer to the Thecostraca *s.str.*, which is at best unclear (see above). *Bredocaris* was not considered at all in their analysis.

In summary, much uncertainty still remains, but the Maxillopoda can be characterized as a monophylum in particular by specific changes in tagmosis as the consequence of a neotenic origin of the group. Considering the origin of Maxillopoda, the morphology of their maxillule is clearly of the entomostracan and not of the malacostracan type (fig. 5.7). The larval cycle, as revealed by the Copepoda, is plesiomorphic in its anamorphic design, but there are no details that correlate with that of any Malacostraca, in contrast to the view of Newman (1983, but see 1992). The presence of a neck organ on the head shield of *Bredocaris*, which is almost identical to that of *Rehbachiella*, may point in the direction of affinities between Maxillopoda and Branchiopoda (Walossek 1993). This has yet to be investigated in more detail, particularly since Cephalocarida may have lost this character on their way to miniaturism and a meiobenthic mode of life.

## Conclusions

Three-dimensional fossilization and preservation of larval development, as in the "Orsten" and similar localities, has proven a much more reliable source of information about the evolutionary path of structures and functional systems of Arthropoda than hypotheses derived exclusively from extant material or molecular data. However, our data base is still incomplete. Information is needed particularly on the early evolutionary phase, such as from the so-called "protarthropods" (Onychophora and Cambrian lobopodians, Tardigrada, Pentastomida) or already "arthropod-like" forms. Such steps have already been undertaken in various ways (e.g., Bergström and Hou, this volume). Likewise the new hypotheses of Dewel and Dewel (1996), which they derived from reinvestigations of segmentation and brain development of Tardigrada, are interesting. Dewel and Dewel depart from traditional views of the evolution of the arthropod head and may eventually lead to a better understanding of the homology of structures and even to the reconsideration of fossil taxa in their phylogenetic context. Application of the concept of phylogenetic systematics appears a necessity to us in that it enables us to test hypotheses and consequences of proposed evolutionary scenarios.

Molecular as well as computer-based cladistic analyses, particularly if applied data sets conflict with well-established character evidence, have not yet proven to be adequate methods. For example, Averof and Akam (1995) in their analysis, which appears modern, suggest that insects could have emerged from a crustacean-like ancestor independently from myriapods and after the major crustacean radiations. Confusion is complete (see above for ground patterns).

Any consequent reconstruction of phylogeny requires the consideration of ground patterns of taxa and evaluation of the consequences of decisions made. Schematic thinking in classificatory terms prejudices categories, which may turn out to be paraphyletic assemblages in which character sets of unrelated taxa are mixed in a single pool.

Fossils can play a major role in phylogeny reconstruction, in particular if synapomorphies are no longer expressed in the morphology of extant taxa. Understanding of early arthropod fossils as specialized species on the one hand and as examples of particular evolutionary plateaus on the other helps to reconstruct, as a first step, the major trends in arthropod development, such as tagmosis (e.g., head formation) and use of limbs, the major feeding and locomotory devices, and the development of reproductive strategies. In the past, fossils revealed only a very limited number of features due to deficiencies in preservation. The increasing number of discoveries of well-preserved fossils has overcome this obstacle now. Examples are the Early Cambrian Chengjiang fauna, which provides us with evidence from the oldest time horizon discovered so far, and those of the exceptional "Orsten" type of preservation. Both have already demonstrated that various, in some cases critical, aspects of the phylogeny of Arthropoda have in the past been misunderstood.

Information has progressively accumulated in the last decade regarding similar occurrences of the "Orsten" type in other regions and of different ages (for preservation and methodology, see, e.g., Müller 1985, 1990; summary of records in Walossek et al. 1993). Such occurrences are widely distributed in Sweden, Poland, Great Britain, Australia, Canada (Newfoundland), and Siberia. In the earlier stages of research it was mainly the arthropods that were studied, but the "Orsten" type of preservation is not restricted to this group. Examples of other soft-integumented organisms studied more recently are the problematic Cambrogeorginida (Müller and Hinz 1992), the vermiform Palaeoscolecida (Müller and Hinz-Schallreuter 1993), the parasitic Pentastomida (Walossek and Müller 1994, Walossek et al. 1994), and minute Tardigrada from the Middle Cambrian of Siberia (Müller et al. 1995; Walossek et al. in prep.). Adaptation of the "Orsten" methodology will surely uncover more such fossils in the future and very probably also more organisms other than Arthropoda.

Again, the research on the "Orsten" has touched upon life in an environment hitherto unconsidered in many ways, the benthic flocculent zone. This zone is inhabited by a small-sized fauna of organisms that play a significant role at lower levels of the trophic web, including many detritivores. Organic matter has continually sunk down to the sea floor since the earliest appearance of algae, and, hence, was available for the earliest destruents at the bottom. These may thus have preceded the origin of carnivorous organisms, i.e., members of higher levels of the food chain. The discovery of such minute forms down in the Cambrian might even have an impact on our understanding of early metazoan evolution in general, which may have started at a small scale long before that date.

ACKNOWLEDGMENTS

Thanks are due to Rainer Willmann, Göttingen, for his constructive comments on different versions of the manuscript and various fruitful discussions on arthropod phylogeny and phylogenetic systematics, and to Euan N. K. Clarkson, Edinburgh, who critically reviewed the manuscript and improved the language. Kluwer Adademic Press kindly gave permission to include two figures from volume 31 of *Hydrobiologia,* containing an article by D. W. Particular thanks are due to the Deutsche Forschungsgemeinschaft for their long-term support of the "Orsten" project and to Jan Bergström for funding travel by D. W. to the Swedish Museum of Natural History in Stockholm. Last, we are indebted to the editor, G. Edgecombe, Sydney, for his effort and patience in getting our manuscript finalized and the book published.

REFERENCES

Anderson, D. T. 1987. The larval musculature of the barnacle *Ibla quadrivalvis* Cuvier (Cirripedia, Lepadomorpha). *Proceedings of the Royal Society of London,* series B, 231:313–338.

Averof, M. and M. Akam. 1995. Insect-crustacean relationships: Insights from comparative developmental and molecular studies. *Philosophical Transactions of the Royal Society of London,* series B, 347:293–303.

Ax, P. 1985. Stem species and the stem lineage concept. *Cladistics* 1:279–287.

Barlow, D. I. and M. A. Sleigh. 1980. The propulsion and use of water currents for swimming and feeding in larval and adult *Artemia.* In G. Persoone, P. Sorgeloos, O. Roels and E. Jaspers, eds., *The Brine Shrimp Artemia, 1: Morphology, Genetics, Radiobiology, Toxicology,* pp. 61–73. Wetteren, Belgium: Universa Press.

Beklemishev, W. N. 1952. *Principles of Comparative Anatomy of Invertebrates.* 2d ed. Moscow: Nauka.

Bergström, J. 1992. The oldest arthropods and the origin of the Crustacea. *Acta Zoologica* 73:287–291.

Boxshall, G. A. 1983. A comparative functional analysis of the major maxillopodan groups. In F. R. Schram, ed., *Crustacean Issues.* Vol. 1, *Crustacean Phylogeny*, pp. 121–143. Rotterdam: Balkema.

Boxshall, G. A. 1992. Synopsis of group discussion on the Maxillopoda. *Acta Zoologica* 73:335–337.

Boxshall, G. A. and R. Huys. 1989. New tantulocarid, *Stygotantulus stocki,* with an analysis of the phylogenetic relationships within the Maxillopoda. *Journal of Crustacean Biology* 9:126–140.

Boxshall, G. A. and R. J. Lincoln. 1983. Tantulocarida: A new class of Crustacea ectoparasitic on other Crustaceans. *Journal of Crustacean Biology* 3:1–16.

Briggs, D. E. G. 1978. The morphology, mode of life, and affinities of *Canadaspis perfecta* (Crustacea, Phyllocarida), Middle Cambrian, Burgess Shale, British Columbia. *Philosophical Transactions of the Royal Society of London,* series B, 281:439–487.

Briggs, D. E. G. 1992. Phylogenetic significance of the Burgess Shale crustacean *Canadaspis. Acta Zoologica* 73:293–300.

Calman, W. T. 1909. Crustacea. In R. Lankester, ed., *A Treatise on Zoology, Part 7: Appendiculata,* 3:1–346. London: Adam & Charles Black.

Chen, J.-Y., J. Bergström, M. Lindström, and X.-G. Hou. 1991. The Chengjiang fauna: Oldest soft-bodied fauna on Earth. *National Geographic Research and Exploration* 7 (1): 8–19.

Cisne, J. L. 1975. Anatomy of *Triarthrus* and the relationships of the Trilobita. *Fossils and Strata* 4:45–63.

Cisne, J. L. 1982. Origin of the Crustacea. In L. G. Abele, ed., *The Biology of Crustacea.* Vol. 1, *Systematics, the Fossil Record, and Biogeography,* pp. 65–92. New York: Academic Press.

Claus, C. 1873. Zur Kenntnis des Baus und der Entwicklung von *Branchipus stagnalis* und *Apus cancriformis. Abhandlungen der königlichen Gesellschaft der Wissenschaften in Göttingen* 18:93–140.

Costlow, J. D. and C. G. Bookhout. 1957. Larval development of *Balanus eburneus* in the laboratory. *Biological Bulletin* 112:313–324.

Dahl, E. 1956. Some Crustacean Relationships. In K. G. Wingstrand, ed., *Bertil Hanström: Zoological Papers in Honour of his Sixty-fifth Birthday,* pp. 138–147. Lund: Zoological Institute.

Dahms, H.-U. 1990. Naupliar development of *Paraleptastacus brevicaudatus* Wilson, 1932 (Copepoda: Harpacticoida: Cylindropsyllidae). *Journal of Crustacean Biology* 10: 330–339.

Dewel, R. A. and W. C. Dewel. 1996. The brain of *Echiniscus viridissimus* Peterfi, 1956 (Heterotardigrada): A key to understanding the phylogenetic position of tardigrades and the evolution of the arthropod head. *Zoological Journal of the Linnean Society, London* 116:35–49.

Fisher, D. C. 1981. The role of functional analysis in phylogenetic inference: Examples from the history of the Xiphosura. *American Zoologist* 21:47–62.

Fryer, G. 1983. Functional ontogenetic changes in *Branchinecta ferox* (Milne-Edwards) (Crustacea, Anostraca). *Philosophical Transactions of the Royal Society of London,* series B, 303:229–343.

Fryer, G. 1987. A new classification of the branchiopod Crustacea. *Zoological Journal of the Linnean Society* 91:357–383.

Fryer, G. 1992. The origin of the Crustacea. *Acta Zoologica* 73:273–286.

Grygier, M. J. 1983. Ascothoracida and the unity of Maxillopoda. In F. R. Schram, ed., *Crustacean Issues.* Vol. 1, *Crustacean Phylogeny,* pp. 73–104. Rotterdam: Balkema.

Grygier, M. J. 1984. Comparative Morphology and Ontogeny of the Ascothoracida: A Step Toward a Phylogeny of the Maxillopoda. Ph.D. diss., University of California, San Diego.

Grygier, M. J. 1987. New records, external and internal anatomy, and systematic position of Hansen's Y-larvae (Crustacea; Maxillopoda; Facetotecta). *Sarsia* 72:261–278.

Hennig, W. 1966. *Phylogenetic Systematics.* Urbana: University of Illinois Press.

Hinz, I. 1987. The Lower Cambrian microfauna of Comley and Rushton, Shropshire/ England. *Palaeontographica* A 198:41–100.

Huys, R. 1991. Tantulocarida (Crustacea: Maxillopoda): A new taxon from the Temporary Meiobenthos. *P.S.Z.N.I.: Marine Ecology* 12:1–34.

Huys, R., G. A. Boxshall, and R. J. Lincoln 1993. The tantulocaridan life cycle: The circle closed? *Journal of Crustacean Biology* 13:432–442.

Kaestner, A. 1967. *Lehrbuch der Speziellen Zoologie.* Vol. 1, *Wirbellose;* Vol. 2, *Crustacea,* pp. 849–1242. Stuttgart: Fischer.

Kraus, O. and M. Kraus. 1994. Phylogenetic system of the Tracheata (Mandibulata): On "Myriapoda"-Insecta interrelationships, phylogenetic age, and primary ecological niches. *Verhandlungen des naturwissenschaftlichen Vereins in Hamburg (NF)* 34:5–31.

Lauterbach, K.-E. 1983. Zum Problem der Monophylie der Crustacea. *Abhandlungen des naturwissenschaftlichen Vereins in Hamburg (NF)* 26:293–320.

Lauterbach, K.-E. 1986. Zum Grundplan der Crustacea. *Abhandlungen des naturwissenschaftlichen Vereins in Hamburg (NF)* 28:27–63.

Lauterbach, K.-E. 1988. Zur Position angeblicher Crustacea aus dem Ober-Kambrium im Phylogenetischen System der Mandibulata (Arthropoda). *Abhandlungen des naturwissenschaftlichen Vereins in Hamburg (NF)* 30:409–467.

Lauterbach, K.-E. 1989. Das Pan-Monophylum: Ein Hilfsmittel für die Praxis der Phylogenetischen Systematik. *Zoologischer Anzeiger* 223:139–156.

Martin, J. W. and C. E. Cash-Clark. 1995. The external morphology of the onychopod "cladoceran" genus *Bythotrephes* (Crustacea, Branchiopoda, Onychopoda, Cercopagididae), with notes on the morphology and phylogeny of the order Onychopoda. *Zoologica Scripta* 24:61–90.

McKenzie, K. G., P.-J. Chen, and S. Majoran. 1991. *Almatium gusevi* (Chernyshev 1940): Redescription, shield shapes, and speculations on the reproductive mode (Branchiopoda, Kazacharthra). *Paläontologische Zeitschrift* 65:305–317.

Müller, K. J. 1979. Phosphatocopine ostracodes with preserved appendages from the Upper Cambrian of Sweden. *Lethaia* 12:1–27.

Müller, K. J. 1982. *Hesslandona unisulcata* sp. nov. with phosphatised appendages from Upper Cambrian "Orsten" of Sweden. In R. H. Bate, E. Robinson, and L. M. Sheppard, eds., *Fossil and Recent Ostracods,* pp. 276–304. Chichester: Ellis Horwood.

Müller, K. J. 1983. Crustacea with preserved soft parts from the Upper Cambrian of Sweden. *Lethaia* 16:93–109.

Müller, K. J. 1985. Exceptional preservation in calcareous nodules. *Philosophical Transactions of the Royal Society of London,* series B, 311:67–73.

Müller, K. J. 1990. Upper Cambrian "Orsten." In D. E. G. Briggs and P. R. Crowther, eds., *Palaeobiology: A Synthesis,* pp. 274–277. Oxford: Blackwell Scientific Publications.

Müller, K. J. and I. Hinz. 1992. Cambrogeorginidae fam. nov.: Soft-integumented problematica from the Middle Cambrian of Australia. *Alcheringa* 16:333–353.

Müller, K. J. and I. Hinz-Schallreuter. 1993. Palaeoscolecid worms from the Middle Cambrian of Australia. *Palaeontology* 36:549–592.

Müller, K. J. and D. Walossek. 1985a. Skaracarida: A new order of Crustacea from the Upper Cambrian of Västergötland, Sweden. *Fossils and Strata* 17:1–65.

Müller, K. J. and D. Walossek. 1985b. A remarkable arthropod fauna from the Upper Cambrian "Orsten" of Sweden. *Transactions of the Royal Society of Edinburgh, Earth Sciences* 76:161–172.

Müller, K. J. and D. Walossek. 1986a. *Martinssonia elongata* gen. et sp. n.: A crustacean-like euarthropod from the Upper Cambrian of Sweden. *Zoologica Scripta* 15:73–92.

Müller, K. J. and D. Walossek. 1986b. Arthropod larvae from the Upper Cambrian of Sweden. *Transactions of the Royal Society of Edinburgh, Earth Sciences* 77:157–179.

Müller, K. J. and D. Walossek. 1987. Morphology, ontogeny, and life-habit of *Agnostus pisiformis* from the Upper Cambrian of Sweden. *Fossils and Strata* 19:1–124.

Müller, K. J. and D. Walossek. 1988a. Eine parasitische Cheliceraten-Larve aus dem Kambrium. *Fossilien* 1:40–42.

Müller, K. J. and D. Walossek. 1988b. External morphology and larval development of the Upper Cambrian maxillopod *Bredocaris admirabilis. Fossils and Strata* 23:1–70.

Müller, K. J. and D. Walossek. 1991. Ein Blick durch das "Orsten"-Fenster in die Arthropodenwelt vor 500 Millionen Jahren. *Verhandlungen der deutschen zoologischen Gesellschaft* 84:281–294.

Müller, K. J., D. Walossek, and A. Zakharov. 1995. "Orsten" type phosphatised soft-integument preservation and a new record from the Middle Cambrian Kuonamka Formation in Siberia. *Neues Jahrbuch für Geologie und Paläontologie, Abhandlungen* 131:101–118.

Newman, W. A. 1983. Origin of the Maxillopoda: Urmalacostracan ontogeny and progenesis. In F. R. Schram, ed., *Crustacean Issues*. Vol. 1, *Crustacean Phylogeny,* pp. 105–120. Rotterdam: Balkema.

Newman, W. A. 1992. Origin of Maxillopoda. *Acta Zoologica* 73:319–322.

Preuss, G. 1951. Die Verwandtschaft der Anostraca und Phyllopoda. *Zoologischer Anzeiger* 147:49–64.

Preuss, G. 1957. Die Muskulatur der Gliedmaßen von Phyllopoden und Anostraken. *Mitteilungen des Zoologischen Museums Berlin* 33:221–256.

Sanders, H. L. 1963. The Cephalocarida: Functional morphology, larval development, comparative external anatomy. *Memoirs of the Connecticut Academy of Arts and Sciences* 15:1–80.

Schram, F. R. 1986. *Crustacea.* New York: Oxford University Press.

Schram, F. R. and M. J. Emerson. 1991. Arthropod pattern theory: A new approach to arthropod phylogeny. *Memoirs of the Queensland Museum* 31:1–18.

Scourfield, D. J. 1926. On a new type of crustacean from the Old Red Sandstone, *Lepidocaris rhyniensis. Philosophical Transactions of the Royal Society of London*, series B, 214: 153–187.

Siewing, R. (ed.) 1985. *Lehrbuch der Zoologie.* Vol. 2, *Systematik.* Stuttgart, New York: Fischer.

Storch, V. and U. Welsch. 1991. *Systematische Zoologie,* 4th ed. Stuttgart, New York: Fischer.

Størmer, L. 1939. Studies on trilobite morphology, 1: The thoracic appendages and their phylogenetic significance. *Norsk Geologisk Tidsskrift* 19:143–273.

Størmer, L. 1944. On the relationships and phylogeny of fossil and Recent Arachnomorpha: A comparative study on Arachnida, Xiphosura, Eurypterida, Trilobita, and other fossil Arthropoda. *Skrifter Utgitt av Det Norske Videnskaps-Akademi i Oslo, I. Matematisk-Naturvidenskapelig Klasse* 5:1–158.

Walossek, D. 1993. The Upper Cambrian *Rehbachiella a*nd the phylogeny of Branchiopoda and Crustacea. *Fossils and Strata* 32:1–202.

Walossek, D. 1995. The Upper Cambrian *Rehbachiella,* its larval development, morphology, and significance for the phylogeny of Branchiopoda and Crustacea. *Hydrobiologia* 31:1–13.

Walossek, D., I. Hinz-Schallreuter, J. H. Shergold and K. J. Müller. 1993. Three-dimensional preservation of arthropod integument from the Middle Cambrian of Australia. *Lethaia* 26:7–15.

Walossek, D. and K. J. Müller. 1990. Stem-lineage crustaceans from the Upper Cambrian of Sweden and their bearing upon the monophyletic origin of Crustacea and the position of *Agnostus*. *Lethaia* 23:409–427.

Walossek, D. and K. J. Müller. 1991. Lethaia Forum: *Henningsmoenicaris* n. gen. for *Henningsmoenia* Walossek and Müller, 1990: Correction of name. *Lethaia* 24:138.

Walossek, D. and K. J. Müller. 1992. The "alum shale window": Contribution of "Orsten" arthropods to the phylogeny of Crustacea. *Acta Zoologica* 73:305–312.

Walossek, D. and K. J. Müller. 1994. Pentastomid parasites from the Lower Paleozoic of Sweden. *Transactions of the Royal Society of Edinburgh, Earth Sciences* 85:1–37.

Walossek, D., J. E. Repetski, and K. J. Müller. 1994. An exceptionally preserved parasitic arthropod, *Heymonsicambria taylori* n. sp. (Arthropoda incertae sedis: Pentastomida), from Cambrian-Ordovician boundary beds of Newfoundland, Canada. *Canadian Journal of Earth Sciences* 31:1664–1671.

Walossek, D. and H. Szaniawski. 1991. *Cambrocaris baltica* n. gen. n. sp.: A possible stem-lineage crustacean from the Upper Cambrian of Poland. *Lethaia* 24:363–378.

Wehner, R. and W. Gehring. 1990. *Zoologie*. 22d ed. Stuttgart, New York: Thieme.

Westheide, W. 1987. Progenesis as a principle in meiofauna evolution. *Journal of Natural History* 21:843–854.

Whittington, H. B. and J. E. Almond. 1987. Appendages and habits of the Upper Ordovician trilobite *Triarthrus eatoni*. *Philosophical Transactions of the Royal Society of London*, series B, 317:1–46.

Willmann, R. 1989. Palaeontology and the systematization of natural taxa. In N. Schmidt-Kittler and R. Willmann, eds., *Phylogeny and the Classification of Fossil and Recent Organisms: Abhandlungen des naturwissenschaftlichen Vereins in Hamburg (NF)* 28: 267–291.

Wilson, G. D. F. 1992. Computerized analysis of crustacean relationships. *Acta Zoologica* 73:383–389.

# CHAPTER 6

## Fossils and the Interrelationships of Major Crustacean Groups

*Frederick R. Schram and Cees H. J. Hof*


## Abstract

We review new sources of information uncovered in the last ten years relevant to the phylogenetic relationships of the Crustacea and the most recent efforts to perform a cladistic analysis of the group. Previous databases exhibit flaws of various kinds, and we assemble a new database derived from these previous efforts that contains ninety binary and multistate morphological characters for sixty-four fossil and Recent crustacean taxa; polarization occurred using three uniramian outgroups. Analysis using PAUP 3.1.1 performed on the entire character set but deleting all the extinct taxa produced 165 equally parsimonious trees of 553 steps, from which a majority rule consensus tree reveals in series a clade of remipedes at the base, two paraphyletic maxillopodan clades, a distinct phyllopodan clade in the sense of Schram (1986), and a distinct malacostracan clade. Analysis of the entire data set including fossil and Recent taxa yielded 9700+ trees of 614 steps. A majority rule consensus of these revealed a great deal of structure to the tree, from which we can detect, in series, distinct arrays of Cambrian stem-group taxa, remipedes, a branchiopodan/cephalocaridan/*Lepidocaris* clade, two distinct maxillopodan clades, a phyllocarid clade, and a complex clade of "derived" malacostracans with some unsuspected arrangements, particularly in regard to syncarids and hoplocarids. From this latter analysis it appears that it is equally probable to conceive of most Cambrian forms either in a distinct clade of their own or as a stem-group transition series. However, deletion of the sperm and internal soft-anatomy characters favors a stem-group transition series for Cambrian forms. Therefore, the stem lineage hypothesis of Walossek and Müller receives some support from this analysis. Manipulation in MacClade of the consensus tree to force various traditional clades put forth by earlier workers can add anywhere from two to twelve extra steps to the most parsimonious arrangement. While we believe our venture offers some advantages over earlier efforts, we also accept that we have exhausted neither the possibilities of expanding to a more inclusive, total-evidence database nor the variety of ways in which we could explore this data set.

The last two decades have seen a great many discoveries that have significantly changed the way we look at the phylogeny and evolution of the Crustacea. Fossils have made great contributions in this regard, but discoveries based on new and phylogenetically interesting living forms and on the sequencing of nucleic acids have also added new levels of sophistication to our understanding. We shall of course focus our efforts here on the fossil record, but we cannot completely divorce our deliberations from knowledge gained from advances in these other fields.

Nevertheless, it seems appropriate to pause and take stock of the current situation and try to produce a synthesis. In this regard we have essentially come full circle from what began in 1982 with the publication of the *Biology of Crustacea*, volume 1 (Abele 1982). In that work three papers analyzed crustacean relationships from three distinct viewpoints. Cisne (1982) examined crustacean origins based on the study of unusually well-preserved trilobites from North America (Cisne 1981). He related his findings to earlier observations of several authors (especially Sanders 1963a; also Hessler and Newman 1975) who saw connections between the trilobites or trilobite-like forms and the cephalocarids as possible archetypal crustaceans. This viewpoint held that all arthropods with biramous limbs came from a single stock and stood in contrast to those arthropods that arose from a uniramous lineage. Schram (1982), at that time strongly influenced (Schram 1978) by the work of Anderson (1973) and Manton (1977), which interpreted developmental and functional morphological evidence as indicating a polyphyletic origin of arthropods, argued strongly against the trilobite/cephalocarid theory. Schram believed that, while some relationship existed between trilobites and cheliceriforms (chelicerates + pycnogonids), trilobitomorphs should remain as a separate lineage of arthropods. He also believed that progenetic paedomorphosis operated as an important mechanism for the origin of higher taxa of crustaceans. In this regard, he concluded that cephalocarids arose as specialized deposit feeders whose evolution occurred under an r-selection regime. Schram, however, did not offer any really coherent alternative to cephalocarids as a model for an "ur-crustacean." Hessler et al. (1982) examined issues in the evolution of crustaceans from the viewpoint of evidence derived from living forms. This essentially complemented the efforts of Cisne, and the two chapters together form a summary of the trilobite/cephalocarid theory.

All three of these contributions shared one thing: the authors framed their ideas with an almost complete lack of reference to phylogenetic systematics. Each of the contributions grew out of a long, well-established, evolutionary systematic tradition in carcinology. In that tradition, discussions of phylogeny often entailed exclusive focus on single points of morphology that one or the other authority thought significant. The consensus in that tradition, well summarized by Dahl (1963), viewed crustaceans as composed of distinct lineages arrayed in a kind of "phylogenetic lawn" and whose exact relationships to each other remained vague (except that cephalocarids probably occupied some kind of central position).

From that point in 1982 changes began to occur. More rigorous forms of analysis slowly began to be factored into discussions of the phylogeny of Crustacea. The edited volume *Crustacean Phylogeny* (Schram 1983a) contained several papers reflecting this. Briggs (1983) utilized principal components analysis to sort out Cambrian arthropods from the Burgess Shale based on the earlier work of Briggs and Whittington (1981) and then produced a phenogram based on polarization of head appendages, moving from few in number to larger numbers progressively incorporated into the arthropod cephalon. While Hessler (1983) and Watling (1983) produced trees that looked like cladograms, Sieg (1983) offered the first real cladistic analysis of any crustacean group in his review of the tanaidaceans, which he produced essentially by hand. Subsequently Schram (1984) presented the first computer-derived cladistic analysis in crustacean studies when he examined the phylogenetic relationships within the Eumalacostraca *s.str.* This last paper is also significant because it attempts to compare results with and without the input from the fossil record.

Schram (1986) developed this theme of linking fossils as integral elements further in the analysis of the phylogeny of all crustaceans. He performed the first cladistic analysis of all crustaceans, utilizing the computer program PHYSIS to generate his trees. Although his database admittedly contained many problems, he generated it by simply taking from the literature all those features of anatomy previous workers had thought important, and he then subjected that data to a single treatment. This effort focused on incorporating data from the newly discovered class of crustaceans, the Remipedia (Yager 1981). Thus in this analysis biramous trunk limbs and antennules were viewed as plesiomorphies. The most parsimonious arrangement to come from Schram's analysis indicated (fig. 6.1A) four major clades within the Crustacea, which he defined as classes: Remipedia, Malacostraca, Phyllopoda, and Maxillopoda. Alternative assumptions about character polarities, such as polyramous trunk limbs and uniramous antennules as plesiomorphies, produced longer trees, even while shifting the relative positions of remipedes and cephalocarids (fig. 6.1B). Thus, the remipedes suggested, in contrast to Schram's 1982 effort, a clear alternative to the trilobite/cephalocarid theory of crustacean origins (Schram 1983b).

The cladograms of Schram (1986) have received widespread attention, but not universal acceptance (e.g., see the critiques of Fryer 1988 and Dahl 1992). Nevertheless, they have assisted other workers in focusing their own efforts. Jamieson (1991) found considerable congruence with the scheme of Schram when analyzing the possible evolution of sperm in crustaceans (fig. 6.2). In addition, Grygier (1987), Boxshall and Huys (1989), and Høeg (1992), while pointing out some limitations of Schram's assessment of characters, essentially carried on with the methods of cladistic analysis while building more reliable databases in regard to maxillopodan crustaceans.

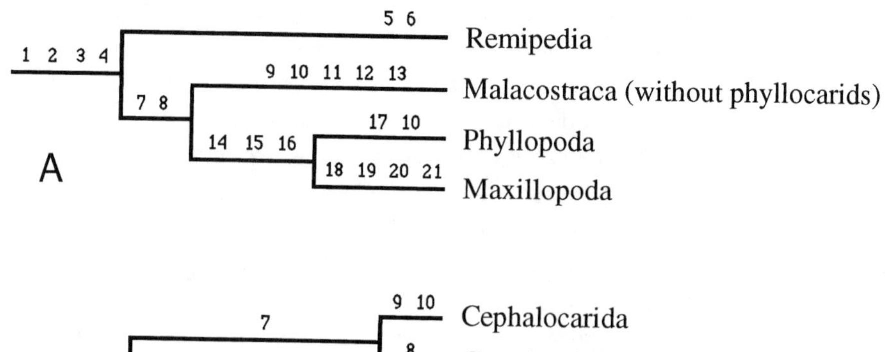


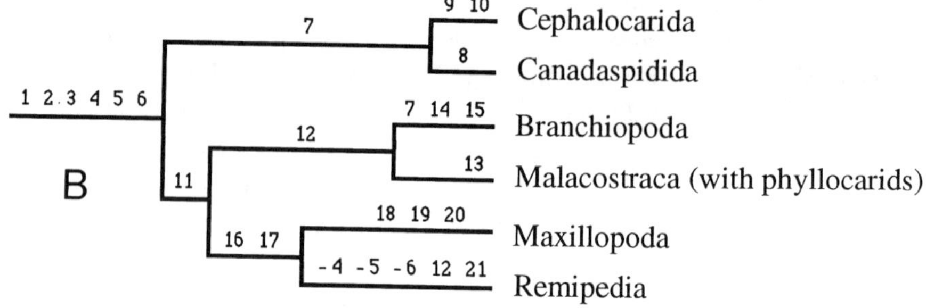


FIGURE 6.1
Simplified trees distilled from the analyses in Schram (1986), see text for discussion. *(A)* The most parsimonious tree that viewed biramous trunk limbs as the plesiomorphic state among crustaceans. The original total tree length was 109 steps with a consistency index (CI) of 0.615. *(B)* An alternative tree derived at the time of the original analysis in which certain character polarities were reversed, such as polyramous trunk limbs as the plesiomorphic state, resulted in a different arrangement of groups, a tree of 115 steps, and a CI of 0.583. For character states, see Schram (1986, fig. 43.1).

## Significant Sources of New Information

Since the late sixties, but most especially in the last decade, significant bodies of new information have emerged in the literature that have allowed the development of more complete and accurate databases. This information either clarifies areas of knowledge that formerly presented real problems or allows us to formulate new questions concerning the origin and evolution of the various crustacean groups. Interestingly, much of this advance has come from the fossil record.

### Burgess Shale

The restudy and redescriptions of the famous Burgess Shale fauna from the Middle Cambrian of British Columbia have captured the imagination of scientists and laypersons alike. The affinities of most of the Burgess arthropods lie outside the Crustacea, but some forms do seem relevant to our consideration. Among these latter, *Canadaspis perfecta* (see Briggs 1978) has attracted the most attention. Briggs believed that *Canadaspis* had affinities with the Phyllocarida, since it possessed eight pairs of thoracic limbs and a seven-segment, albeit limbless, abdomen. Briggs observed two sets of antennae with the antennules greatly reduced and the antennae somewhat less so. The mandible appeared to lack any palp. The maxillules had differentiated only slightly, and the maxillulary segment was perhaps only partially fused to the head, and the maxillae had differentiated hardly at all from the thoracic limbs, and the free maxillary segment seemed more thoracic than cephalic in nature.

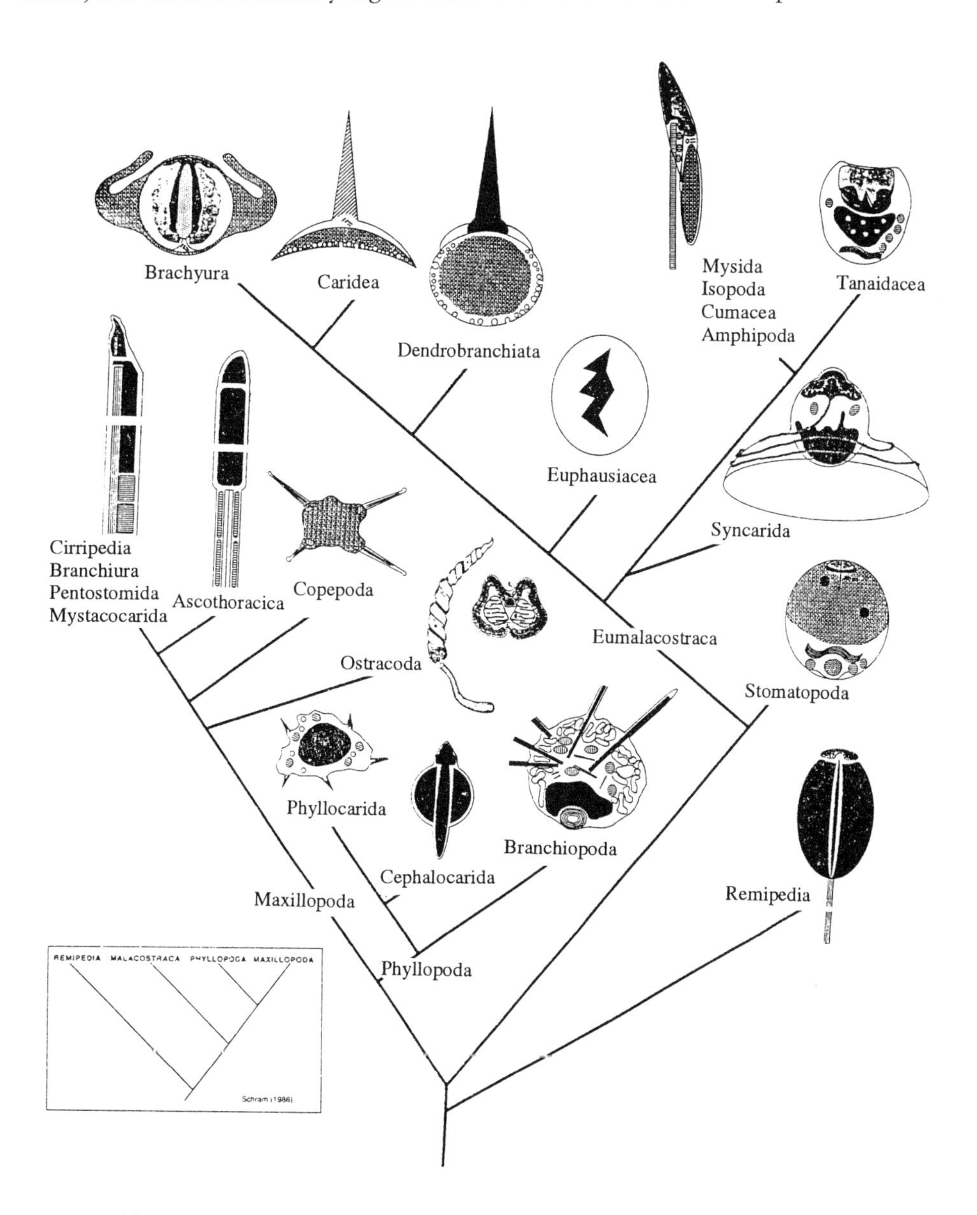


FIGURE 6.2

The evolution of crustacean sperm (modified from Jamieson 1991). Note the general correspondence with the major classes as arranged in figure 6.1A.

Other Burgess Shale species may have affinities with the Crustacea. Briggs himself allied the genus *Perspicaris* with *Canadaspis*. Whittington (1974) tentatively placed *Plenocaris* among the phyllocarids, but Briggs (1983) rejected *Plenocaris* as a phyllocarid and thought it unlikely to be a crustacean unless it was lying somewhere outside any recognized living class. Briggs (1983) leveled the same judgment against *Waptia* but conceded that *Odaraia* showed great similarity with branchiopods (albeit that those similarities largely appeared as plesiomorphies). Briggs thought it likely that *Priscansermarinus* indeed had affinities to the barnacles (Collins and Rudkin 1981).

Other productive localities for Cambrian arthropods have received attention in the literature. Briggs and Robison (1984) described exceptionally well-preserved arthropods from the Middle Cambrian of Utah, Robison (1984) found additional localities from Idaho, and Hou and Bergström (1991) extended the range of these evolutionarily interesting faunas to China.

The anatomical features of *Canadaspis* drew the attention of theoreticians. Dahl (1983a, but especially 1984) objected to Briggs's interpretation of phyllocarid affinities. Dahl reinterpreted several features of the anatomy of *Canadaspis* and pointed out that the nature of the thoracic limbs, as well as what might appear as a ten-segment thorax rather than a 2 + 8 (posterior head + thorax) pattern, combined to deny *Canadaspis* possible phyllocarid affinities. Dahl (1983a,b), however, held a staunch position against phylogenetic systematics as applied to Crustacea. He believed it sufficient to point out features that could argue against an affinity and, in the evolutionary systematic tradition, did not think it necessary to develop any alternative hypotheses as to those relationships. In response to Dahl, Briggs (1992) reaffirmed his own interpretations of *Canadaspis* while freely admitting that problem areas remained, such as the form of the thoracopods and the lack of pleopods, features Briggs perceived as possible apomorphies of *Canadaspis*. Briggs pointed out that many features of *Canadaspis* anatomy, such as the exact details concerning the head, remain ambiguous because of the nature of the fossils themselves. Nevertheless, Briggs believed that new views about the nature of the evolution of the crustacean head (see discussions of Walossek and Müller 1990 below) may explain away many of these anomalies. Briggs (1992) thus believed that *Canadaspis* should remain a phyllocarid (albeit possibly a derived one), but he conceded that one might also view it as a stem-group crustacean convergent to the phyllocarid form.

All of this basic taxonomic research on the Burgess Shale fauna has resulted in a series of papers analyzing cladistically the relationships of arthropods, with special attention to the creatures from the Middle Cambrian. Briggs and Fortey (1989) produced the first cladogram of Burgess arthropods. Their analysis, characterized as preliminary, focused on twenty-eight taxa, of which three came from among modern forms, six came from faunas other than the Burgess Shale, and nineteen came from the Burgess Shale itself. They excluded any uniramians from the analysis and utilized forty-six characters in PAUP, finding a consistency index of 0.38 (two slightly different concepts of uniramians exist: the first, in the sense of Manton, in which onychophorans, myriapods, and hexapods form a monophyletic group; and the second, which we prefer, in which only the myriapods and hexapods form a coherent taxon [this last is sometimes referred to as the Atelocerata or Tracheata]). The tree they obtained, which placed the Burgess genera *Marrella* and *Branchiocaris* at the base, argued for a highly derived condition for trilobites and suggested the possibility of the Crustacea as a paraphyletic taxon. In regard to this paraphyly, Briggs and Fortey remarked (1989:243) that "the characters used to diagnose living crustaceans may be primitive or convergently acquired," thus admonishing arthropod phylogenists, and crustacean workers in particular, to pay some attention to the basic definitions of higher taxa.

Briggs (1990) addressed the issue of arthropod polyphyly. He maintained that one can deal with the whole array of arthropods and the characters they share and obtain an imperfect picture of their relationships (a state of affairs that Dahl 1983a could not accept), or one can focus on differences and gaps in our knowledge of both fossil and living forms and of necessity accept a scheme of polyphyly (in the sense of Anderson 1973, Manton 1977, or Schram 1978). Briggs also raised two other very intriguing points. First, he suggested that in regard to cephalic tagmosis, arthropods may not have acquired and fixed these features early in their evolution. He suggested that crustaceans may have stabilized head segmentation later in their history by a sequential process, which one supposes might have reversed itself in individual cases. Second, he believed that trunk tagmosis in fact appears more stable than that of the head and thus more important in defining arthropod structural plans. Certainly, these observations by Briggs found some acceptance in the different approach to arthropod phylogeny put forth by Schram and Emerson (1991).

Briggs et al. (1992, 1993) expanded the database of Briggs and Fortey (1989), utilizing twenty-five Cambrian taxa in a total array of forty-six and scoring 134 features. Their efforts included not only cladistics but also phenetic procedures. They wanted to sort out the cladistic relationships as well as measure the morphological disparity (*sensu* Gould 1991) evident in their database. In this analysis, they included uniramians and used the Burgess Shale marine onychophoran *Aysheaia* as an outgroup. Although the consistency index in this analysis fell to 0.27, they did obtain a "crustaceomorph" clade, a lineage that includes obvious crustaceans and some "crustacean-like" creatures from the Cambrian (fig. 6.3). This lineage included six subclades: (1) the remipede *Speleonectes;* (2) *Skara* and the cephalocarid *Lightiella;* (3) a maxillopodan group (with the copepod *Calanus* linked to the mystacocarid *Derocheilocaris*, and the branchiuran *Argulus* linked to the Cambrian genera *Martinssonia* and *Bredocaris*); (4) *Plenocaris;* (5) another "maxillopodan" clade (that connected the lobster *Homarus* with the barnacle *Lepas* and the ostracode *Cypridina*); and (6) a branchiopod/phyllocarid lineage (which also included the Burgess genera *Odaraia* and *Waptia*).

It seems clear from all this that one must retain an open mind concerning changing past assumptions about defining major groups, about the characters thought important in the evolution of those groups, and about the actual taxa we may or may not be willing to include within any analysis.

## Cambrian "Orsten"

A most interesting array of incredibly preserved fossils has emerged from the Cambrian "Orsten" deposits of Västergötland, Sweden. Walossek and Müller cover this material in greater detail in chapter 5 of this volume, but we can here address several taxa and issues of relevance to crustacean relationships. Schram (1986), based on the limited information published up to that time (e.g., Müller 1983), lumped many of these taxa into a catchall category of larval forms. Since then, a beautiful series of detailed descriptions and analyses have appeared (e.g., see Müller and Walossek 1985, 1986a, 1988; Walossek and Müller 1990; Walossek and Szaniawski 1991; Walossek 1993) that make clear that many of these fossils represent actual crustacean adult or subadult stages of development as well as abundant and interesting real larval material (Müller and Walossek 1986b; Walossek 1993; Walossek et al. 1993).

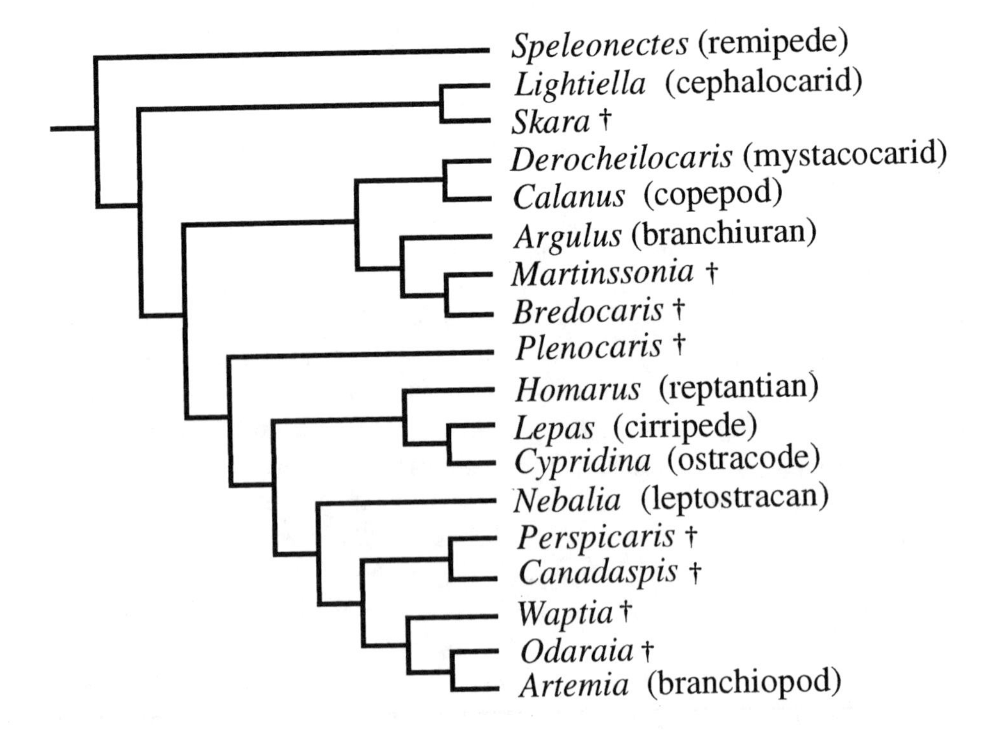


FIGURE 6.3
Part of the cladogram of Briggs et al. (1993) that concerns the clade of "crustaceomorphs."

The true larval forms bear many features of interest in their own right. These include nauplius larvae not attributable to other known adult forms in the fauna, the nauplius A and nauplius B of Müller and Walossek (1986b). Some of the larvae relate to specific adult forms, in particular for *Martinssonia elongata* (Müller and Walossek 1986a), *Bredocaris admirabilis* (Müller and Walossek 1988), and *Rehbachiella kinnekulensis* (Walossek 1993). Müller (1983) described this latter species on the basis of larval forms alone, and adults have still not been documented (Walossek 1993).

The most intriguing idea regarding crustaceans to emerge from the "Orsten" research centers on the recognition of two different kinds of species. First, a small group of taxa seems to fall within the borders of recognized living crustacean groups. For example, *Bredocaris* appears to lie within the Maxillopoda (Müller and Walossek 1988; Walossek and Müller 1992; Walossek 1993) and *Rehbachiella* seems attributable to the Branchiopoda (Walossek 1993). Suggestions that *Walossekia quinquespinosa* bears some relationship to *Rehbachiella,* that *Dala peilertae* may have affinities with cephalocarids (Müller 1983), or that the phosphatocopines resemble ostracodes (Müller 1979, 1982) must await further study (Walossek and Müller 1992; Walossek 1993).

Second, an intriguing array of fossils suggests crustacean forms but seems to variously "fall short of the mark" in many ways. Walossek and Müller (1990) perceive these as stem-group crustaceans, distinct from forms such as *Bredocaris* and *Rehbachiella,* which they believe lie within the "crown group" or true Crustacea. These former include, at least, *Henningsmoenicaris scultata, Cambropachycope clarksoni, Goticaris longispinosa* (Walossek and Müller 1990), *Martinssonia elongata* (Müller and Walossek 1986a), and *Cambrocaris baltica* (Walossek and Szaniawski 1991). The arguments here involve a fascinating redefinition of the crustaceans. Walossek and Müller take up the ideas of Ax (1985) on stem species and stem lineages. A series of stem species lead to a crown group. The stem species form a transition series that increasingly accumulates apomorphies until the core apomorphies by which one can define the crown group come to complete expression. In reference to crustaceans, Walossek and Müller (1990:411) recognized six principal apomorphies: (1) there should exist bipartite feeding, with a naupliar apparatus (built around the antennules, antennae, and mandibles) that functions early in development, and a post-naupliar apparatus (extending from the maxillule posteriad) that operates during the juvenile and adult phases; (2) the mouth region has a labrum roofing the atrium oris just outside the mouth proper and a set of paragnaths that arise from the mandibular sternum; (3) the posterior limbs on the trunk exhibit specializations for swimming or suspension feeding, with no true filtration achieved by any movements of the exopods; (4) the telson bears a terminal anus flanked by paddle-shaped caudal rami (probably effective in steering during swimming); (5) one can recognize an anamorphic ontogenetic sequence that starts with a nauplius (or at least a recognizable egg-nauplius stage); and (6) the feeding function of the naupliar limbs persists at least until the adult apparatus appears after a series of molts. What is immediately striking about this definition cuts to the core of our current understanding of prime crustacean apomorphies. Walossek and Müller did not include the presence of the sensory antennules and antennae, two sets of maxillae (which together with the antennular sets and the mandibles traditionally define a five-segment head), nor the presence of gills (which they believe [Walossek and Müller 1990:410] truly evolved only in hoplocaridans and eumalacostracans). However, Walossek and Müller (1992) do include the fourth limb developed as a maxillule in the derived crustacean ground plan.

We are confused concerning the relationship of the labrum and hypostome as homologous or nonhomologous structures. Walossek and Müller (1990:416) first indicate that no homology exists between these. They report that *Skara, Bredocaris, Rehbachiella, Walossekia,* and the phosphatocopines have a labrum. However, in *Martinssonia, Henningsmoenicaris,* and even *Agnostus,* they observed the mouth with a poorly defined hypostome as "an incipient step toward the development" of a true labrum/atrium oris (Walossek and Müller 1990:417). In effect, this would seem to imply that the hypostome acts as a homologous, phylogenetic precursor to the true labrum.

Where fossil larval material exists, it seems clear that the stem crustaceans and the arachnomorphs have larvae with initially four sets of functional limbs. Thus the crustacean nauplius with three sets of limbs would seem to clearly characterize Crustacea *s.str.*

The issue of leg segment homologies in Early Cambrian forms receives a most intriguing treatment (both in Walossek and Müller 1990, fig. 5; and Walossek 1993, fig. 54). Stem forms, such as *Cambrocaris, Goticaris, Henningsmoenicaris,* and *Martinssonia,* have subtriangular limb bases with medially directed spines and a separate proximal endite. *Agnostus* (a trilobite) has subrectangular limb bases with no proximal endite. Thus all stem crustaceans seem to have a separate endite on the proximal aspect of the coxa (fig. 6.4A). Walossek (1993) believes this endite is the homologous precursor of the true coxa in the crown crustaceans. Thus, the crustacean basis is the homologue of the trilobite coxa, the crustacean coxa being a unique structure. How this idea might be reconciled with the suggestions of Itô (1989) on the origin of the copepod basis, a theory involving remipedes and cephalocarids (fig. 6.4B), and how this impinges on possible segment homologies with uniramians, must await further study and consideration.

Finally, Walossek and Müller's interpretations raise issues with regard to segment numbers in trunk limb endopods. *Agnostus* and trilobites have seven segments in the endopod (or telopod of some authors) while crustaceans supposedly have reduced this to five segments. We can find this number further reduced in "more advanced" crown crustaceans, such as in living remipedes where we find four segments in the endopod. *Tesnusocaris,* a fossil remipede, however, clearly has nine endopodal segments and thus seems to stand outside the pale of consideration here (fig. 6.4C). It would seem that use of segment numbers in the thoracic endopods might prove

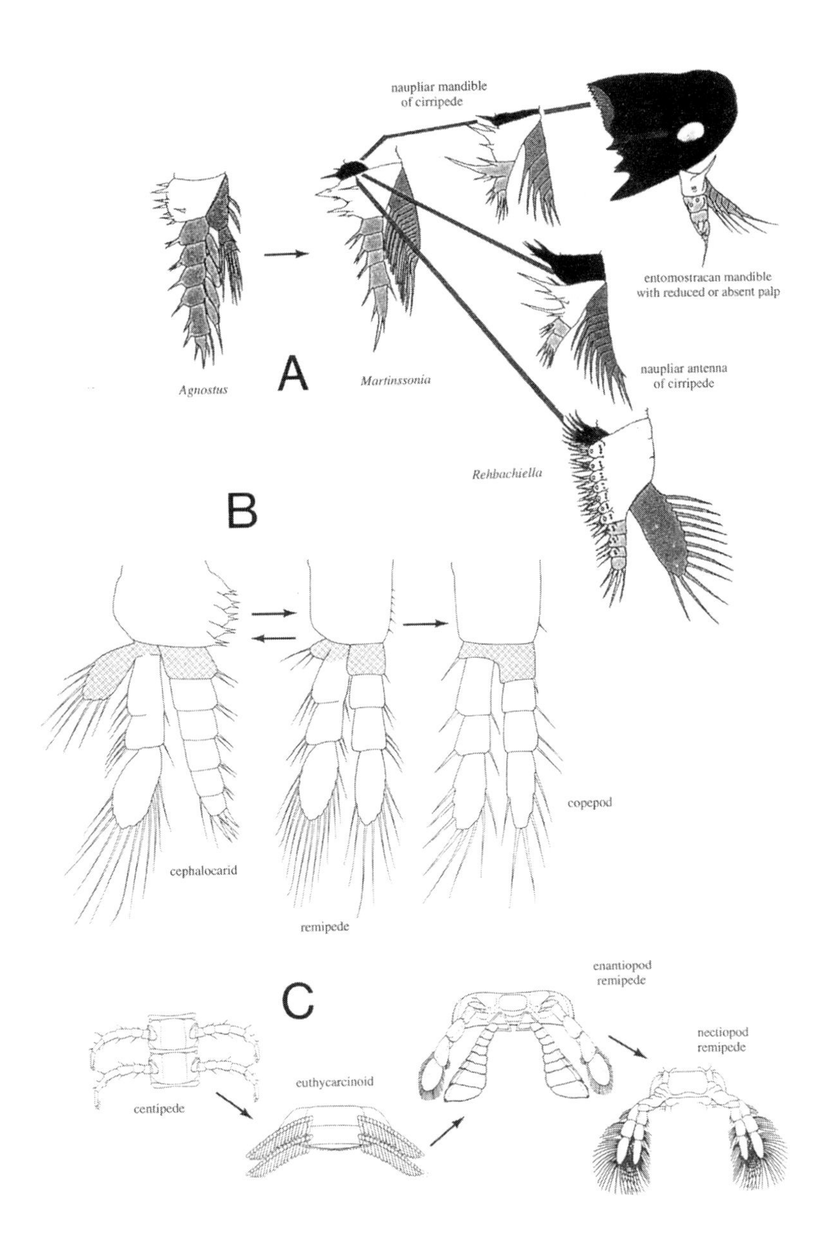


FIGURE 6.4

Various hypotheses for the evolution of crustacean limbs. *(A)* The ideas of Walossek and Müller (1990) and Walossek (1993) concerning the origin of the crustacean coxa from a free endite that first evolved in stem-group crustaceans. *(B)* The idea of Itô (1989) that the basis in copepods formed from the fusion of segments or parts of segments of the exopod and endopod of the remipede. (Itô believed that the remipede limb could in turn be derived from that of the cephalocarid, but Schram and Emerson [1991] maintained that the opposite could be true.) *(C)* The idea of Emerson and Schram (1990) that the biramous limb (e.g., nectiopod remipedes) formed from the proximal fusion of uniramous limbs (e.g., centipedes), proceeding first through a diplopodous stage (with separate sternites and fused tergites, as in the euthycarcinoids) and then through a duplopodous stage (with fused sternites and tergites but no fusion of limb parts, as in enantiopod remipedes).

useful but that we must judiciously employ this feature rather than build too much on it.

## Other Interesting Forms

In addition to Burgess Shale and Swedish "Orsten" fossils, other interesting forms have appeared in the literature, and we might review some of these here. One of these groups (Jones and McKenzie 1980) is the Cambrian Bradoriida, typically placed in the Ostracoda. Although until recently little of the anatomy inside the shell of most bradoriids *s.str.* survived preservation (Huo and Shu 1983), some workers often combined this enigmatic group with the Phosphatocopina (e.g., see McKenzie et al. 1983). Although conceding the basic oligomeric (or reduced and fused) nature of the ostracode bodyplan (McKenzie 1983), McKenzie (1991) nonetheless gave bradoriids a central role as postulated ancestors for all crustaceans. Part of this seems tied in with the past understanding of phosphatocopines as ostracodes, however, Walossek and Müller (1992) and Walossek (1993) suggest that new studies of phosphatocopines will remove these "Orsten" forms from the ostracodes. How this will affect the standing of bradoriids is only now beginning to emerge (the ostracode status and phosphatocopine affinity of bradoriids are rejected by Hou et al. 1996). Certainly McKenzie's separation of bradoriids into Lipabdomina and Abdomina most likely should eventually disappear, because it splits the category on the basis of presence or absence of a primitive feature, an abdomen.

One must rank the Thylacocephala as among the most problematic of fossil arthropod groups. The literature on these animals for the most part focuses on description of species and elucidation of their peculiar form (see Rolfe 1992 for a short summary and introduction to the literature). Their huge compound eyes, raptorial limbs, and the fact that their bodies are completely enclosed within their carapaces mark these as among the most distinctive and peculiar predators of the Paleozoic. Pinna et al. (1985) interpreted Thylacocephala as possibly related to cirripedes (a kind of barnacle precursor); Secretan (1985) suggested possible malacostracan affinities, and Schram (1990) thought the parallels to remipedes very striking. This last grew in part from a cladistic analysis based on the characters used in 1986 (Schram 1986) but which he did not publish (Schram 1990) because without the definitive crustacean characters verified for thylacocephalans such an analysis would prove interesting but not particularly convincing. Indeed, workers on these fossils have yet to verify the definitive crustacean characters on these creatures and so the case for them as crustaceans must remain "not yet proven" (Rolfe 1992).

Another fossil problematicum exists in the Cycloidea. This peculiar group exhibits a variety of forms ranging from tiny, caplike nodules to large, flattened plates. All of them possess radiating ridges on the body that mark segment boundaries, often preserve geniculate grasping limbs anteriorly, and (when preserved) have rather robust walking limbs. Various workers since 1835 have placed the cycloids in almost every major group of arthropods at one time or another, including trilobites, limuline merostomes, branchiuran crustaceans, or within the copepods. Schram et al. (1997), on the basis of a cladistic analysis using only maxillopodan types and of information garnered from the study of Mazon Creek cycloids, conclude that these fossils occurring from the Carboniferous through the Triassic have close affinities to the Copepoda (a position that does not seem to hold here when the entire array of crustaceans is analyzed together). However, whatever their taxonomic affinities, consideration of their general habitus suggests that cycloids may have filled the niche of crabs in the Late Paleozoic and Early Mesozoic. Cycloids illustrate nicely, however, how patient study of a group by a succession of workers over many decades (in this case more than a century) can slowly reveal information that can allow an increasingly clear understanding of a group's relationships and also clarify the extent of adaptive radiations that led to present-day forms.

One final fossil of considerable phylogenetic interest occurs in the Devonian, *Lepidocaris rhyniensis* Scourfield (1926, 1940). Scourfield allied his Lipostraca to the Anostraca in part due to his belief that *Lepidocaris* possessed reduced maxillae, and Sanders (1963b) agreed. Schram (1986) reinterpreted the reduced maxilla as merely the raised pore of the maxillary gland, and this then made possible the recognition of the aberrant first thoracopod of Scourfield as the true maxilla. In light of that, Schram believed that the strongest affinities of Lipostraca were with cephalocarids, and indeed cephalocarids and lipostracans appeared as sister groups in his cladistic analysis. Walossek (1993) agreed with Schram's interpretation of *Lepidocaris* anatomy but argued that the lipostracans' closest affinities remained with anostracans and that *Lepidocaris* acted as an intermediate step between a primitive branchiopod form, like *Rehbachiella* from the Cambrian "Orsten," and the Euanostraca.

## Contending Hypotheses About Crustacean Origins

Three groups of living crustaceans have achieved considerable notoriety as possible ancestral types: Branchiopoda, Cephalocarida (= Brachypoda), and Remipedia. Testing which of these might have stood closest to an ancestor should be a focus of any cladistic analysis of crustaceans.

### Branchiopoda

The Branchiopoda have long had a position as models for the ancestral crustaceans, since the time of Borradaile (1917, 1926), who maintained that the leaflike, multiramous limbs of this group more closely resembled that of an ancestral crustacean than those of any other group. Although this idea held sway for some time, the discovery of the first cephalocarid, *Hutchinsoniella,* subsumed the idea of the primitive nature of the multiramous limb into a new cephalocarid theory of crustacean origins (see below). Fryer (1992), however, has made new arguments to retrieve the idea that branchiopods lie closest to the most primitive type of Crustacea. He believed that the largely unsegmented trunk limbs of branchiopods represent a state prior to the development of articulating joints in the crustacean limb. This idea is at odds with what we know about the limbs of all the potential crustacean outgroups within the arthropods, all of whom have true jointed limbs. Fryer's hypothesis, however, clearly continues in the tradition of evolutionary systematics, and the phylogenetic tree he puts forth has few concrete links between groups (Fryer 1992:285, fig. 6).

Fryer's idea of the lack of segmentation in the branchiopod limb attracted the attention of Walossek (1993) when the latter discussed the place of *Rehbachiella* in the scheme of crustacean and specifically branchiopod evolution. However, the analysis of Walossek focuses on the potential functional significance of this arrangement. He suggested that it reflects a need to more easily alter the pressure around the surfaces of a vibrating limb that may operate under conditions of low Reynolds number. In short, one could interpret the effective lack of limb segmentation as an apomorphy of the Branchiopoda *s.str.*

## Cephalocarids

The cephalocarid theory of crustacean origins began with the description (Sanders 1957) of *Hutchinsoniella macracantha*. This largely derived from the strong resemblance of the maxillae of that animal to the thoracopods and on a long anamorphic series in the course of development. Individual cephalocarid limb types, both larval and adult, lend themselves to interpretation as precursors to limb types found in other groups (Sanders 1963a,b). Certainly the similarities to other groups have their striking aspects. Sanders himself (1963a) pointed out, however, that these similarities derived more from the primitive nature of the structures themselves rather than from any derived conditions. Dahl (1956), in fact, had quickly united the cephalocarids and branchiopods into a group he called the Gnathostraca, a position from which he may have retreated (Dahl 1963) too quickly. McKenzie et al. (1983) also placed cephalocarids within the Branchiopoda.

Schram (1982) called into question the generally accepted viewpoint of the then assumed primitive position of the cephalocarids. Although certain features of cephalocarids appear primitive, such as the "thoracic" nature of the maxillae, Schram believed that many aspects of the bodyplan and biology of cephalocarids revealed them to be highly derived and specialized forms that progenetic paedomorphosis had enabled to adapt to a distinctive habitat in the upper flocculent zones of muddy bottoms. This interpretation found little support at first. Subsequent cladistic analysis (Schram 1986) seemed to indicate that *Hutchinsoniella* and *Lepidocaris* shared several apomorphies and that these two formed a clade that sorted as a sister group to the branchiopods *s.str.* Fryer (1988) rejected Schram's hypothesis, repeating a theme that has emerged many times in his papers, namely that "cladograms should not be allowed to obscure the fundamental biological attributes of animals" (Fryer 1992:284). McKenzie (1991), however, continued to maintain cephalocarids as a sister group to branchiopods.

Walossek (1993) pointed out that the supposed primitive phylogenetic position of cephalocarids has never been confirmed by any cladistic analysis and that cephalocarids do not appear as primitive forms at all with respect to the morphology and ontogeny of "Orsten" forms. Walossek turned his attention to considering the nature of the developmental sequence of cephalocarids as well as other crustaceans. In this regard, he found clear indications that the anamorphic sequences of the Cambrian form *Rehbachiella* and anostracans possess what he considered primitive features, namely, long and slow, gradual additions of both segments and legs through successive molts. Cephalocarids, on the other hand, have a distinctly different pattern, one that suggests possible independent control of segment addition and limb development since these sequentially accrue at different rates during the course of ontogeny. Indeed, the analysis of developmental sequences by Walossek may emerge as one of the most important contributions to come from the study of the "Orsten" material. Walossek, however, advances a detailed analysis of different forms of phyllopodous limbs among crustaceans and outlines various aspects of the functioning of filter-feeding systems evident in distinct groups. As a result, he rejected the cladistic analysis of Schram (1986) that placed all thoracic filter feeders together into a single class Phyllopoda.

Abundant evidence now seems available to indicate that cephalocarids possess some highly derived features. How this affects their phylogenetic position one cannot say at this point. Most workers opposed to the analysis of Schram (1986) focus their arguments on emphasizing the supposed primitive features of the cephalocarids. Only Walossek seconds Schram's observation concerning the many derived aspects of cephalocarid anatomy and biology. Whether cephalocarids should remain as a primitive and separate class (albeit with many unique and derived features of their own) or find sister group status with some other crustacean group can find some kind of resolution only within a cladistic analysis.

## Remipedes

Most recently, a relatively new class, the Remipedia (Yager 1981), has received attention as a possible primitive type. Schram (1983b) believed that the unregionalized nature of the remipede trunk and the possession of a long series of biramous paddles as trunk limbs suggested an alternative for crustacean origins to that advocated in the cephalocarid theory. The biramous nature of the trunk limbs in his remipede theory harkened back to the suggestions of Cannon and Manton (1927) concerning primitive crustacean limb types. However, a problem with the remipedes as a supposed primitive type surfaced almost immediately for many people (e.g., see Hessler 1992) since the derived features of remipedes (mainly in regard to their specialized mouth parts) appear even more obvious than those of cephalocarids. Yet, certain features of remipede internal anatomy especially in regard to the nervous system, digestive system, serial cephalic glands (Schram and Lewis 1989), and possibly the reproductive system (Itô and Schram 1988; Yager 1989) also appear to support some special phylogenetic status for remipedes.

The supposed primitive position of remipedes obtained heightened significance with the identification of fossil members of the class from the Mississippian of Texas and the Pennsylvanian of Illinois. *Tesnusocaris goldichi,* with its distinctive antennular and antennal form, fanglike maxillules, prehensile maxillae and maxillipedes, and fusion of the first trunk segment to the head has clear affinities to living remipedes (Emerson and Schram 1991). The material, in addition, displays unusual features that suggest: (1) a rather different interpretation of the origin of the biramous limb from the fusion of two separate uniramous limbs (Emerson and Schram 1990); and (2) the possibility of a significant resynthesis of ideas concerning arthropod segment evolution as a whole (Schram and Emerson 1991). Even Emerson and Schram found these conclusions rather startling and difficult to accept, and while some of the evidence for their position is indeed rather intriguing, its effectiveness as a paradigm awaits further analysis (see Emerson and Schram 1997).

In regard to remipedes, some molecular data (various references in Boxshall et al. 1992; Felgenhauer et al. 1992) seem to indicate that this group may have affinities with the maxillopodans, although apparently some problems remain concerning the interpretation of these data (Abele et al. 1990). In light of McKenzie's (1991) views on the possible central position of Maxillopoda in crustacean evolution, perhaps this should not concern us (although McKenzie himself places remipedes surprisingly among the hoplocaridans, eumalacostracans, and phyllocaridans). At this point, one can only conclude the same thing about remipedes as about cephalocarids—while they apparently have many primitive features, they also possess some striking derived features, and their exact phylogenetic position must remain an open issue. Clearly, the resolution of the conflict over the position of the remipedes vis-à-vis the stem-group crustaceans and the possible affinities of remipedes with maxillopodans could possibly come from a cladistic analysis of Crustacea.

## The Role of Molecules

We cannot leave the issue of contending hypotheses without some comment on the important new molecular techniques. Early efforts at DNA determination (Bachmann and Rheinsmith 1973) focused largely on cytological aspects while indicating some phylogenetic possibilities to the method. This line of research never developed further. Several papers, however, have appeared recently that address various issues in crustacean phylogeny using sequencing techniques. These include works by Abele et al. (1989) concerning pentastomids, Kim and Abele (1990) and Abele (1991) involving decapods, Spears et al. (1992) examining brachyurans, and Abele et al.

(1992) dealing with maxillopodans. The results have proven interesting. Comparisons of molecular and morphological trees for decapods as a whole and brachyurans in particular revealed complementary features (Abele 1991; Spears et al. 1992). However, such comparisons for maxillopodans suggest that molecular data cannot support a monophyletic status for this class. The 18S rDNA technique has found the most use in this regard, but it may have limited application. For example, Abele (pers. comm.) believes that while 18S rDNA has something to tell us at the level of class in maxillopodan relationships, parts of the mitochondrial DNA would probably function better to elucidate relationships within orders, for example in Thoracica (barnacles). As useful as these techniques seem, however, one must recognize the limits of these procedures. Holmquist (1983) and Roderigo et al. (1994) demonstrated the limited capacity of sequencing to resolve so-called deep branches in evolutionary trees. This is due to the fact that in a random set of changes in base pairs along a nucleic acid sequence repeated transitions and transversions at previously altered sites can effectively mask the earlier substitutions with the result that the phylogenetic signal theoretically implicit in base pair changes becomes scrambled and disappears. Sequencing information, therefore, does have limits, and one should remain cautious concerning the interpretation of trees derived only from sequence analysis.

On the other hand, Wheeler et al. (1993) assessed side by side molecular and traditional morphological character databases in regard to phylogenetic relationships among all arthropods. The use of molecular techniques alone often produces strange combinations of taxa. The morphological data seem to contain insufficient information to completely resolve all clades. However, when the data are combined into a single database, a tree with a high degree of resolution emerges, but it basically resembles the tree derived from morphology alone. In this instance, the molecular data, while problematic alone, served to complement the morphology and further resolve the polytomies resulting from that analysis, a useful thing indeed.

Molecular data have a role to play in understanding phylogeny. However, that role must remain that of an additional tool available to scientists, not that of replacement for "old" techniques. Not only must molecular sequences receive careful interpretation and morphologists develop tolerance for this new source of information, but "outside spectators" must not force researchers into pigeonholes by questioning whether a morphologist can really add molecules to his or her research program or whether a sequencer only distracts from a research protocol by addressing morphological information.

## What is a Crustacean?

It would appear that the progress made in understanding fossil forms in the last decade, especially the fossils from the Cambrian faunas (see above), requires adjustments to the commonly accepted definition of Crustacea. Certain features of this definition remain fixed in the secondary literature, such as the five-segment head (two pairs of antennae, mandibles, and two pairs of maxillae), presence of a carapace, biramous appendages, and the presence of a nauplius larva or egg-nauplius stage in development, as evidenced most recently by the introductory review of Barnes and Harrison (1992). Precisely which aspects of this classic definition might be more applicable than others, given the comments above, could be tested in an unordered and unweighted analysis of characters designed to preclude a priori judgments about character assessments.

Probably the central feature of this definition includes the five-segment head, which we can perceive as actually a whole suite of characters, as noted by all the major exegetes from Calman (1909) to Schram (1986). However, as noted above, many authors have begun in different ways to question this supposedly most crustacean of features. The head conditions in arthropods as a whole reveal wide variations, as first noted by Briggs (1983). Furthermore, Walossek and Müller (1990) and Walossek (1993) cautioned us to apply care when addressing this issue for the crustaceans proper. Indeed, if one puts aside preconceptions about what "theoretically" constitutes a head, a different pattern emerges. Table 6.1 lists some of the major crustacean groups and notes the adult condition of cephalization. The number of head segments reflects how many segments fuse to form a "cephalic unit." Although the number five most commonly occurs, the range actually extends from none (for rhizocephalans who have effectively "lost" their head), through four (in some stem-group crustaceans), up to eight to ten (in selected malacostracans).

Walossek and Müller (1990) rightly separate the issue of head segment number from head appendage number. Even more variation occurs when one considers head appendages, since fusion of segments and specialization of limbs appear in fact to operate independently. Because of the observed variety, one might have difficulty polarizing these characters and, unlike in the case of the head segment count, five head appendages is not the most common state. Forms like cephalocarids, *Lepidocaris, Rehbachiella,* and *Bredocaris* have "maxillae" similar in form to those of the thoracopods, and an extreme development along these lines occurs in *Skara* where all the post-mandibular limbs have a "thoracic" form. One could score branchiopods as having four limbs, too, since the "maxillae" often mature as vestigial or greatly reduced structures. A common theme among many crustaceans seems to be the incorporation of a "maxillipede" into the head array as occurs in remipedes, all the peracarid eumalacostracans (of which Isopoda in table 6.1 serves only as an example), copepods, cycloids, and mystacocarids (noteworthy in that they appear to have a maxillipede without having incorporated the segment into the head at all). Many groups add additional limbs to the head array. Ostracodes effectively carry on as "encapsulated heads," the rest of their body having undergone extreme reduction (often termed oligomerization); some cycloids add additional limbs into the maxillipedal array, and many malacostracans have elaborate maxillipedal batteries such as stomatopods (five characteristically), decapods (at least three, and big-clawed forms like Astacidea actually could be said to have four since that claw does not function as a walking leg), and tanaidaceans (which also have an anteriorly directed

TABLE 6.1
Selected crustacean groups and the number of segments and appendages in the "cephalon" based purely on the form of the limbs and the shape of the "head field," as well as the superficial state of the head armature

| TAXON | HEAD SEGMENTS | HEAD APPENDAGES | HEAD SHIELD | CARAPACE |
|---|---|---|---|---|
| ***selected modern types*** | | | | |
| Remipedia | 6 | 6 | + | - |
| cephalocarids | 5 | 4 | + | - |
| *Lepidocaris*† | ?5 | 4 | + | - |
| Anostraca | 5 | 5* | - | - |
| Notostraca | 5 | 5* | + | + |
| Conchostraca | 5 | 5* | -√ | + |
| Cladocera | 5 | 4 | + | + |
| Ostracoda | ?7 | 7 | -√ | + |
| Branchiura | 6 | 3# | + | + |
| Pentastomida | ? | 2 | - | - |
| Ascothoracida | 5 | 4 | + | + |
| Rhizocephala | 0 | 0 | - | - |
| Cirripedia | ?5 | 3° | + | + |
| Mystacocarida | 5 | 6 | + | - |
| Skaracarida† | 5 | 6 | + | - |
| Copepoda | 6(7) | 6 | + | - |
| Cycloidea† | ?6 | 6–7 | + | + |
| Tantulocarida | ?5 | 0 | + | - |
| *Bredocaris*† | 5 | 4 | + | - |
| Stomatopoda | 10 | 10 | + | + |
| Anaspidacea | 6 | 5 | + | - |
| Bathynellacea | 5 | 5 | + | - |
| Tanaidacea | 7 | 7 | + | + |
| Isopoda | 6 | 6 | + | - |
| Euphausiacea | 5 | 5 | + | + |
| Decapoda | 8 | 8–9 | + | + |
| ***stem-group crustaceans*†** | | | | |
| *Henningsmoenicaris* | 5 | 5 | + | - |
| *Cambropachycope* | 4 | 4 | + | - |
| *Goticaris* | 4 | 4 | + | - |
| *Martinssonia* | 5 | 5 | + | - |
| *Cambrocaris* | ? | ? | ? | - |

* = maxillae vestigial, √ = anterior head free from carapace, † = extinct, # = maxillules and maxillae as unique attachment organs, ° = antennules and antennae lost or modified.

claw on the second thoracopod). A few groups lose all head appendages, such as rhizocephalans and tantulocarids, and some lose the antennae (the cirripedes proper). One can spend much time fretting over whether the maxillae that resemble trunk limbs are primitive or secondarily derived (an issue we can resolve with an unordered analysis of characters), but we believe that, at the very least, one should begin to view the subject of head form and development as much more variable than it has been considered in the past.

The question of biramous limbs now seems irrelevant as a defining feature of Crustacea. The cladistic analyses of Briggs et al. (1992, 1993) clearly group all the non-uniramous arthropods in a single clade. This confirms the arguments of Hessler and Newman (1975) for recognition of the "schizoramous" arthropods in opposition to the uniramous forms, as derived from the work of Manton (1977). Schram and Emerson (1991) also believed that multiramous limbs serve to characterize crustacean, arachnomorph, and various other Burgess-type arthropods. Biramous limbs serve as an apomorphy for all non-uniramous arthropods rather than as a defining feature of any particular group of arthropods.

Another issue of relevance here with regard to limb structure concerns the nature of the crustacean coxa. Traditional views have tended to homologize the basal-most segment of all arthropod limbs as a coxa, although variations occurred when discussions arose over a supposed "pre-coxal" segment more proximal to the body wall than the coxa proper (e.g., see Hansen 1925). However, as mentioned above, Walossek and Müller (1990) and Walossek (1993) identify and place great significance on a distinct endite found at the limb base of stem-group and some primitive crown-group crustaceans. These workers postulate, as we have seen above, that this endite develops into the crustacean coxa and conclude that the crustacean basis is the homologue of the coxa of other arthropods. Walossek has produced a most intriguing hypothesis, of which the most inviting aspect remains its potential ability to provide further unity to the Crustacea. How this hypothesis relates to the proximal enditic gnathobases on the limbs of the strange larva D of Müller and Walossek (1986b) awaits clarification; might these indicate another separate evolution of a coxa in cheliceriform arthropods as well?

The coxal endite theory (fig. 6.4) is opposed by two other equally intriguing hypotheses about the origin and evolution of crustacean limbs. Itô (1989), from his study of remipede, cephalocarid, and copepod external setal anatomy and internal muscle arrangements, postulated that the copepod basis resulted from the fusion of all or parts of remipede proximal ramal segments. The basis in that case arises as a new structure unique to copepods (possibly by extension to other crustaceans as well). This idea of lateral and medial elements fusing to form new limb segments finds a central role in the hypothesis of Emerson and Schram (1990) concerning the formation of a biramous limb from two adjacent uniramous and diplopodous ones. Perhaps we might find elements of truth in all these theories, or possibly only one will prevail. That we have multiple contending hypotheses will insure a healthy degree of debate and research on this issue.

## Earlier Cladistic Analyses

Only Schram (1986) and Wilson (1992) have attempted across-the-board cladistic analyses of all Crustacea with single databases. Schram focused on analyzing the information available in the literature at that time and incorporating as much information as possible from fossil forms into the analysis. Performed in the early 1980s, his approach was synthetic in that it took only what anyone could find in the standard literature, and it probably could not help but have some serious flaws in terms of assessing apomorphies. Schram's analysis employed a Wagner option within the program PHYSIS. The result, distilled into a series of cladograms for individual presentation and discussion, indicated four major classes of Crustacea (fig. 6.1): a group with unregionalized bodies and highly carnivorous forms, the Remipedia; a group of taxa with highly regionalized bodies, the Malacostraca (but not containing the phyllocarids); and two groups lacking to some degree posterior appendages (the old Entomostraca *s.l.*); a cluster of orders sharing leaflike, multiramous limbs, the Phyllopoda (containing the Branchiopoda, Cephalocarida, and Phyllocarida); and the highly reduced and/or modified class, the Maxillopoda. The principal constraints in the polarized character matrix revolved around the assumptions that a lack of tagma and biramous trunk limbs represented more primitive conditions, tagma and multiramous limbs more derived ones. Schram did not favor this system for any other reason than its apparent parsimony. In his database, the scheme with four classes (fig. 6.1A) contained 109 steps, with a consistency index of 0.615. When he reversed the polarities concerning tagma and trunk limb rami, the resulting tree (fig. 6.1B) contained 115 steps with a consistency index of 0.583.

This scheme of crustacean classes, in addition to recognizing four more or less coherent groups, also seemed to coincide with a coherent pattern of crustacean feeding types (Schram 1986). The remipedes appeared to represent a rather specialized form of carnivory; malacostracans *s.str.* displayed a variety of variations on the theme of particle grappling by various combinations of mouthparts; the Phyllopoda all shared a kind of filter feeding that employed multiramous, foliaceous, trunk limbs, and the Maxillopoda utilized within their reduced bodyplans a variety of feeding types with parasitism as a major theme. As already mentioned above, this arrangement also seemed to coincide with major sperm morphotypes (Jamieson 1991), with remipedes possessing a "primitive" aqua sperm, malacostracans displaying an acrosome (possibly evolved from endoplasmic reticulum rather than from the Golgi apparatus), phyllopodans with amoeboid sperm, and maxillopodans generally with a flagellum (with the exception of the ascothoracids) originating from the anterior portion of the nucleus (fig. 6.2).

Wilson (1992) has advanced the only other cladistic analysis of all crustaceans. He employed a data list of fifty-nine characters developed by a committee of the researchers in attendance at the workshop on crustacean phylogenetics held at Kristeneberg, Sweden, in 1990. A single tree of 242 steps (fig. 6.5A) resulted from the PAUP analysis of these data, with a consistency index (CI) of 0.702, homoplasy index (HI) of 0.533, retention index (RI) of 0.493, and rescaled consistency index (RC) of 0.346. Wilson ran the data unordered and unweighted.

Several strange combinations appear in this tree, such as remipedes as a sister group to the branchiurans and thecostracans, and cephalocarids as a sister group to these (thus both cephalocarids and remipedes appear embedded in the maxillopodans), and *Canadaspis* sorts as a sister group to ostracodes. However, the "phyllocarids" rejoin the eumalacostracans in a malacostracan clade that itself acts as a sister group to the branchiopods and *Lepidocaris*. How we might explain these strange combinations emerges from the Wilson data matrix itself, which on close examination reveals several problems. To point out only a few examples, according to the matrix, remipedes lack a naupliar process on the antennae (but no one has yet collected a remipede larva), several other remipede features are mis-scored, all eumalacostracans have biramous antennules (ignoring the slightly variable condition [Brusca and Wilson 1991] for groups like isopods), and *Canadaspis,* while lacking rami on the mandibles, nonetheless has an exopod (?) with one segment.

In an attempt to understand what these data could tell us, we rescored the Wilson data set, correcting errors, taking care of character inconsistencies, and scoring multistate features more extensively (one can obtain, if interested, the details of that rescoring from the authors since out of space considerations we present only the results here). The new analysis with PAUP 3.1.1 produced not one tree, but nine equally parsimonious trees with a length of 283, CI of 0.784, HI of 0.625, RI of 0.531, and RC of 0.416. The 50% majority rule consensus tree we obtained (fig. 6.5B) is different from that of Wilson (1992) but still contains some strange links. Remipedes appear as a sister group to eumalacostracans (recalling the suggestion of McKenzie 1991); these are linked to a clade containing barnacles, argulid fish lice, and tantulocarids, and cephalocarids appear as a sister group to most crustaceans except for a miscellaneous array of maxillopodans. *Canadaspis,* however, merges again with the phyllocarids.

Clearly the analysis of Wilson (1992), with its emphasis in the character database on information from mouthparts, is greatly at odds with that of Schram (1986), with its emphasis on body tagma features. Any expanded analysis of crustaceans should attempt to reconcile and merge these differences.

## A New Analysis

We decided to develop a new data matrix in which many of the mouthpart features that figure strongly in Wilson's (1992) analysis are combined with information on body tagma. We also include more data about sperm and soft anatomy of the nervous system as well as a few features derived from consideration of other fossils. We finally arrived at ninety characters, about half of them multistate. Often the variety present within an order or group of orders was sufficient to require multiple states in the matrix for that taxon, and during the actual analyses we instructed the program to treat these as polymorphisms. We scored characters for the most part merely as to presence (or absence) of a particular state, since we intended to run the analyses unordered and unweighted. Subsequently we did experiment with some limited weighting and selective ordering of some features.

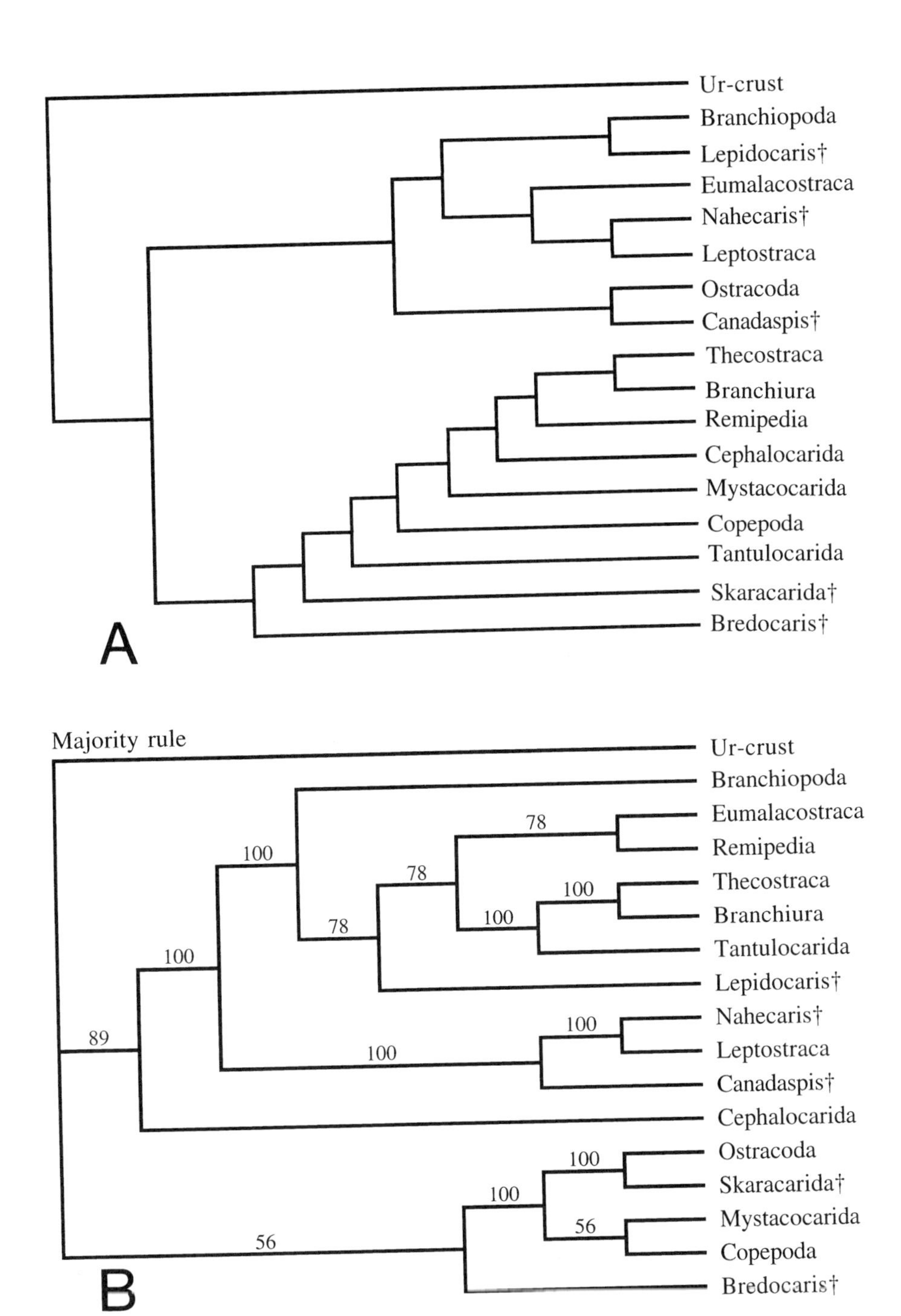


FIGURE 6.5

*(A)* The single tree that resulted from the analysis of Wilson (1992), with a length of 242, CI of 0.702, HI of 0.533, RI of 0.493, and RC of 0.346. *(B)* 50% majority rule consensus of nine trees that emerged from the analysis after our corrections of the original Wilson (1992) data matrix, with a length of 283, CI of 0.784, HI of 0.625, RI of 0.531, and RC of 0.416.

## Characters Used

We made a concerted effort to produce a master list of characters that reconciled what most authors have suggested in the recent past. We offer that list here with comments on the individual features.

1. Antennules: *(0) uniramous; (1) biramous; (2) triramous.*

Arthropod workers generally agree that the antennules, or first antennae, of crustaceans are homologues of the antennae of other arthropod groups. Schram (1986) believed that the biramous nature of these helped define crustaceans. Certainly the presence of biramous antennules finds counterparts in few other taxa among the arthropods, where the antennae are uniformly uniramous, as indeed they are among many crustacean groups as well.

One could argue that either crustacean antennules arose as biramous structures that secondarily lost a ramus to return in many instances to a uniramous state, or that biramous antennules arose from primitively uniramous antennules convergently among different groups of crustaceans. There are merits to both views. The antennules and eyes associate with the most anterior, or acronal, portions of the head, which according to Schram and Emerson (1991) find affinities with the first monosegment of the arthropod body. However, their concept of duplomeres and the origin of biramous limbs derived from Arthropod Pattern Theory (Emerson and Schram 1990; Schram and Emerson 1991) does not help resolve this issue, since the antennules/antennae of all arthropods seem associated with the unfused anteriormost monosegments. In terms of outgroup analysis, all other arthropods that possess an antenna have a uniramous limb in this location. This implies that the biramous and triramous conditions noted among crustaceans are apomorphies.

2. Antennule development: *(0) well-developed; (1) reduced.*

Characters 1 and 2 are not redundant. Character 1 deals with the form of the antennules, character 2 with the degree of development of that form whatever it may be. The antennules seem well expressed in most arthropod groups that possess them. Consequently, reduced or absent conditions register as apomorphies.

3. Antennule condition: *(0) all rami flagellate; (1) "exopod" scalelike; (2) ramus of a few or only one segment.*

A multisegmented, flagellate condition most often typifies this limb. It is difficult if not impossible to characterize the rami of this limb in terms of the traditional definitions of exopod and endopod. The limb rami are typically expressed as well-developed flagella. Deviations from this state appear to be apomorphies.

4. Second pair of head appendage as antennae in adults: *(0) absent; (1) present.*

The presence of two sets of sensory antennae in the adult seems to characterize almost all true crustaceans (except those groups where one or more of the appendages disappears or becomes modified for other purposes). One cannot necessarily take this as a diagnostic of the group, however, since some of the Burgess Shale arthropods also develop a second antenniform limb. We might debate whether these creatures are stem-group crustaceans or something else, but we should really only resolve this within the context of a cladistic analysis of all arthropods.

The biramous nature of the limb that serves as the crustacean antennae, however, appears as a plesiomorphic feature, and so Schram (1986) erred in using this as a defining apomorphic feature of Crustacea.

5. Second pair of head appendages function: *(0) as a feeding limb; (1) as a sensory limb; (2) absent or otherwise reduced.*

We must distinguish between the function of this appendage and its structure (character 4). In this case, the second limb may or may not serve as a sensory limb, and its function can also change during the course of the life cycle. The condition here refers to the one that occurs in the adult state. Walossek (e.g., 1993) makes an important point, namely, that we should distinguish the function of this limb in the larvae from what we find in the adults, and that in the stem-group crustaceans from that in the crown group.

One can easily derive the biramous sensory antennae of adult crustaceans from the biramous, leglike, feeding structures found in crustacean larvae and those noted in stem-group crustaceans. That these limbs lose their feeding function at the close of the larval part of the life cycle and take up a sensory role in the adult stage, however, clearly does seem apomorphic.

6. Antenna form: *(0) biramous; (1) uniramous; (2) absent or otherwise modified.*

The appearance of secondarily uniramous antennae in several crustacean groups would correspond to similar events that often occur in the evolution of thoracic limbs from a biramous to a uniramous form. Thus the presence of uniramous antennae would seem to function as a further apomorphic state when evaluating individual crustacean groups.

7. Antenna expression: *(0) well-developed; (1) absent or otherwise modified.*

As with characters 1 and 2, we believe it is important to distinguish between structural form and the degree of development of the limb. Of course, a problem arises when the limbs are so poorly developed or even absent that we cannot adequately assess function, form, and degree of expression. Nevertheless, we wish to retain the information implicit in making judgments about these factors.

8. Antennal naupliar process: *(0) absent; (1) present.*

This refers to the structure on the developing antenna used by larvae to facilitate feeding. Walossek (1993) pointed out the importance of considering naupliar feeding in questions of crustacean phylogeny. Not all feeding antennae need necessarily have a well-developed process, but its presence almost certainly denotes a limb so used in feeding. The presence of such a process thus appears as an apomorphy associated with what we might view as a more plesiomorphic, purely locomotory function of the second limb.

9. Antennal exopod: *(0) flagellate; (1) single scale; (2) scale with a separate basal segment; (3) absent or otherwise reduced.*

No matter what the function of this limb, either in feeding or sensation, the most primitive expression of the exopod is as a flagellum (e.g., see Walossek and Müller 1990). Deviations from this condition are interpreted as apomorphies.

10. Antennal protopod of nauplius: *(0) < 1/2 total length of limb; (1) > 1/2 total length of limb.*

By outgroup and ingroup comparison, the generalized and most common state seems to be that of modestly developed protopods. A deviation of very large size from this norm denotes probable apomorphy, possibly related to special dynamic effects around the mouth in feeding larvae.

11. Form of compound eyes: *(0) sessile; (1) stalked; (2) lobed; (3) absent.*

By outgroup comparison to arthropods not included within either the stem-group or the crown-group crustaceans, the sessile condition for compound eyes would appear to be plesiomorphic. One should not construe, however, the other character states for this feature as a transition series, but merely as alternative apomorphies.

12. Compound eye condition: *(0) free; (1) encapsulated.*

This apomorphy characterizes a restricted set of orders within the branchiopod crustaceans.

13. Naupliar eye: *(0) absent; (1) present, inverse; (2) present, everse.*

Although the term *naupliar eye* refers to a specific organ in crustaceans, especially in larvae, medially located, single-ocellus optic organs are not unique to the heads of crustaceans. Similar structures, to which the naupliar eyes of crustaceans have been homologized, are noted in hexapods. Many papers of Elofsson (1963, 1965, 1966) have investigated these and the associated "frontal organ" structures in a phylogenetic context (Paulus 1979). Consequently, much use has been made of these features in the past in sorting out crustacean relationships. We could hardly ignore them but, of course, encountered problems when we tried to score fossil organisms for which we largely lack this information. We believe, nevertheless, that any complete database must include these features if for no other reason than that it can serve to sort out relationships among living forms and thus effectively relieve the load on other features that apply to fossils as well.

14. Dorsal frontal organ: *(0) absent; (1) present.*

15. Ventral frontal organ: *(0) absent; (1) present.*

16. Posterior median frontal organ: *(0) absent; (1) present.*

17. Tapetal cells: *(0) absent; (1) present.*

This feature exists only in the eyes of various maxillopodan crustaceans.

18. Bec oculair: *(0) absent; (1) present.*

This "capsule" in the front part of malacostracan heads contains the various elements of the naupliar eye/frontal organ complex.

19. Distal frontal organ: *(0) absent; (1) present.*

20. Antennal gland: *(0) absent; (1) present.*

Schram and Lewis (1989) have suggested that the series of glands in the head expresses a phylogenetically ancient condition of serial secretory glands in all crustacean body segments. These apparently have been lost for the most part or specialized for specific uses. Similar glands have been noted on other limbs, e.g., on the mandibles of remipedes (Schram et al. 1986) and some branchiopods (Fryer 1988), but we know little about the function and expression of those glands across the spectrum of crustaceans. Hessler (pers. comm.) has noted serially arranged coelomic cavities in *Hutchinsoniella.* This remains a potentially fertile area for future investigation.

21. Maxillulary gland: *(0) absent; (1) present.*

Although theoretically possible for any number of crustaceans, only nectiopod remipedes possess these, apparently as an autapomorphy.

22. Maxillary gland: *(0) absent; (1) present.*

The expression of this and character 20 in our tally came only from scoring presence or absence in adult forms. Larval or juvenile expression of these glands remains another area that might lead to further phylogenetic information in the future.

23. Labrum: *(0) absent; (1) present.*

Walossek in several papers has pointed out the importance of careful consideration of the anatomy around the mouth region. In his understanding, the labrum becomes a crucial feature in characterizing the crown crustacean condition. See the text above for discussion of homology problems between the labrum and hypostome.

24. Labrum expression in larva: *(0) moderate; (1) enlarged to extend well posterior of the mouth field.*

See comments on character 25.

25. Labrum expression in adults: *(0) moderate; (1) enlarged to extend well posterior of the mouth field.*

Characters 24 and 25 help to distinguish taxa within certain "phyllopodous" crustaceans. We defined enlarged expression as one in which the labrum extended far enough back to completely encompass the mandibles and at least part of the maxillules. Moderate expression, most commonly found among crustaceans, finds the labrum extending only to or partly over the mandibles. These features deal with aspects of the biology of the crustaceans independent of whether a labrum is present or not, i.e., character 23. Again, the functional significance of this development is not clear at this time, but it may have relevance regarding the channeling of fluid (possibly at a Reynold's number scale that indicates viscous flow) in and around the mouth during feeding.

26. Mandible endopod: *(0) > three segments; (1) two segments; (2) one segment; (3) absent.*

This feature, as well as character 27 and those related ones on other mouthparts, derives from the database of Wilson (1992), which in turn was influenced by the findings of Walossek and Müller in the Cambrian "Orsten" arthropods.

27. Mandible exopod: *(0) 13–11 segments; (1) 11–6 segments; (2) 5–3 segments; (3) 1–2 segments; (4) absent.*

The ranges for the segments in this feature are somewhat arbitrary, but they do avoid the use of thirteen character states. In part they reflect somewhat natural groupings of character expression. In the past, this feature has typically not been considered an important one. Its use, however, derives from consideration of Cambrian "Orsten" forms, species that otherwise lack information on many important features.

28. Base of mandible: *(0) developed with endites; (1) with distinct molar and incisor processes, and a variable tooth row; (2) as in condition 1, but with a well-developed lacinia mobilis; (3) grinding gnathobase with solitary molar process; (4) serrate blade; (5) toothed; (6) absent or otherwise reduced.*

This feature reflects functional processes in feeding. We believe, however, that if the distal rami of the mandible contain phylogenetically useful information, then the consideration of mandibular protopods must also figure in the scheme of things. These features, however, do not form any kind of transition series.

29. First post-mandibular limb (maxillula): *(0) as a trunklike limb; (1) as a mouthpart (but relatively unmodified, i.e., with distinct rami and/or endites); (2) as a fang.*

This expression of the feature tries to avoid making assumptions about whether this limb in fact serves as a mouthpart, since in many "stem" forms this seems not to be the case. The question of the fate of the first and second limbs posterior to the mandibles is an important one. As outlined above, for example, the question has direct relevance to assessing the phyletic position of an animal such as *Canadaspis*. This feature forms part of the complex of characters that would define a crustacean five-segment head, and as such its appearance in any unordered and unweighted cladistic analysis in tandem with other features could go a long way toward laying out a definition of what constitutes a crustacean.

30. First post-mandibular limb (maxillula) endites: *(0) 8–6 in number; (1) 5–4 in number; (2) 3–1 in number; (3) absent.*

Feature 30 (as well as 31 below), as was also noted for feature 27, has somewhat arbitrary subdivisions of feature expression, again to avoid needlessly long character lists and to combine what appear to be naturally occurring groupings. *Skara* and *Canadaspis* exhibit the "0" state with enditic lobes occurring on the protopod and the endopod segments.

31. First post-mandibular limb (maxillula) endopod: *(0) > four segments; (1) three segments; (2) two segments; (3) one segment; (4) absent; (5) otherwise modified.*

32. First post-mandibular limb (maxillula) exopod: *(0) flagelliform (multiarticulate); (1) one segment; (2) absent.*

Wilson (1992) has a longer list of character states, but we found it difficult to identify every state he used or to determine just what was meant (e.g., "rudimentary"). Consequently, we have adopted this shorter and simplified list.

33. Second post-mandibular limb (maxilla): *(0) as a trunklike limb; (1) as a mouthpart; (2) absent or vestigial.*

As with feature 29, we believe a neutral characterization of this limb handled in an unordered and unweighted analysis can go a long way to moving away from an a priori, theory-laden definition of crustaceans. In addition, we must point out that this is not necessarily a transition series.

Scoring some animals has proven difficult. For example, *Bredocaris* has a limb in this position that looks for all intents and purposes like the trunk limbs that follow it, except for a slightly more elaborate proximal endite. We follow Müller and Walossek (1988) and score this as a mouthpart. *Skara* proves even more difficult. Müller and Walossek (1985) recognize a maxillule, a maxilla, and a maxillipede. All three limbs look alike. The first two occur on segments fused to the head. The third limb, however, occurs on a separate, free, trunk segment and is the only trunk limb. So, do we have in *Skara* a series of trunklike limbs, some of which occur on segments fused to the head, or do we have a series of poorly differentiated mouthparts extending into the anterior trunk? We follow Müller and Walossek, aware that we may be using a theory-laden limb identification.

34. Second post-mandibular limb (maxilla) epipodites: *(0) no epipodal development; (1) scaphognathite; (2) small lobe.*

35. Second post-mandibular limb (maxilla) endites: *(0) 8–6 in number; (1) 5–4 in number; (2) 3–1 in number; (3) absent.*

See comments concerning characters 27 and 30.

36. Second post-mandibular limb (maxilla) endopod: *(0) > 6 segments; (1) 5–4 segments; (2) 3–1 segments; (3) vestigial or absent.*

See comments concerning characters 27 and 30. *Hutchinsoniella* clearly has six segments. The situation with *Odaraia* and *Branchiocaris* remains confused because of the nature of the preservation, but we believe the evidence is reasonable for the existence of a large number of segments in these animals (Briggs 1976, 1981). The condition in *Martinssonia* is difficult to assess due to preservation problems, but Müller and Walossek (1986a, fig. 9H) seem to indicate at least six segments.

37. Second post-mandibular limb (maxilla) exopod: *(0) flagelliform (multiarticulate); (1) two segments; (2) unsegmented; (3) vestigial or absent.*

The expression of this feature generally takes two forms: either the multiarticulate/flagellar structure seen on many but not all the stem forms, or a single, paddlelike, or vestigial/reduced segment. Only *Hutchinsoniella* has two clearly demarcated segments.

38. Second post-mandibular segment: *(0) free; (1) fused to head.*

We believe that we must be careful to distinguish what happens to specializations of the limbs from what happens to the segments themselves. This of course also extends to "maxillipedal" segments (see character 46). However, the significance of this feature lies in its status in defining Crustacea based on a five-segment head. Appearance in a tree with other related features, as stated above several times, could determine an effective characterization of the Crustacea.

39. Anterior trunk limb(s) as maxillipede(s): *(0) none, i.e., as a trunk limb; (1) one; (2) two; (3) three; (4) four; (5) five.*

This and the following seven characters proved difficult to score. The obvious alternative, given the variety of expression in individual limb structures, was to score the first five trunk limbs and segments individually for eight characters, i.e., a total of forty characters for the front part of the trunk alone. This would seem to place too much weight on this region. The arrangement we employed here seems to lose information by merging a range of structural modifications in the suite of maxillipedes with variable expressions (see the data matrix). Time limits did not allow us to explore the effects of scoring forty separate characters relevant to maxillipedes, and that remains an option someone probably should take up someday. (See also our comment concerning *Skara* under character 37).

40. Anterior trunk limb(s) endites: *(0) absent; (1) lamellate; (2) gnathobasic; (3) lobate.*

41. Anterior trunk limb(s) epipodites: *(0) absent; (1) simple flap; (2) stalked and distally lobate; (3) as a gill.*

42. Anterior trunk limb(s) protopod: *(0) simple; (1) lamellate.*

43. Anterior trunk limb(s) form: *(0) no maxillipedes; (1) short limb, developed as a manipulatory mouthpart; (2) flagellate distal elements; (3) prehensile or subchelate.*

44. Anterior trunk limb(s) exopod: *(0) > three segments; (1) two segments; (2) unsegmented; (3) absent.*

45. Anterior trunk limb(s) endopod: *(0) > seven segments; (1) six segments; (2) five segments; (3) four segments; (4) three segments; (5) two segments; (6) unsegmented.*

46. Anterior trunk limb(s) segment(s): *(0) free; (1) one fused to head; (2) two fused to head; (3) three fused to head; (4) four fused to head; (5) five fused to head.*

47. Posterior trunk limbs: *(0) all uniform; (1) thoracic and abdominal separation.*

This feature deals with regionalization of the trunk, an independent phenomenon from what may happen at the front end of the trunk in terms of maxillipede differentiation. To some extent there is some overlap with characters 65 and 66, although this feature deals with the general differentiation of limbs into types while the latter features focus on segment patterning. A problem also exists in scoring this feature when the actual manifestation of the "abdominal limbs" is to have none at all. We dealt with this here as a "1" feature and contend with the specific variations that can occur in this regard in character 54. Although one might argue equally that "?" would have been more appropriate here when the limbs are entirely absent (i.e., "character not applicable"), we preferred to use the "?" to deal with those animals (fossil or Recent) where the abdominal limb situation was simply not known (e.g., Hymenostraca) or with those living forms where this is really an inappropriate feature to include (e.g., Rhizocephala).

48. Posterior thoracic limbs form: *(0) uniramous; (1) biramous; (2) polyramous.*

Some overlap exists between this character and character 90. This feature deals with the form of posterior thoracic limbs, while we use character 90 as an attempt to deal with Arthropod Pattern Theory issues (Schram and Emerson 1991). See comments on characters 89 and 90.

49. Posterior thoracic limbs endites: *(0) no endites; (1) present.*

50. Posterior thoracic limbs epipodites: *(0) no epipodites; (1) flaps; (2) true gills; (3) oöstegites.*

51. Posterior thoracic limbs exopod: *(0) > three segments; (1) two segments; (2) one segment; (3) flagellate; (4) absent.*

We chose in this case to make a distinction between a limb that has three or more articles to the branch, each operating more or less independently (e.g., in nectiopodans), as opposed to a truly flagellate condition (e.g., in many eumalacostracan orders).

52. Posterior thoracic limbs endopod: *(0) > six segments; (1) five segments; (2) four segments; (3) three segments; (4) two segments; (5) one segment; (6) absent.*

In this, we follow the arguments of Emerson and Schram (1991) and Walossek (1993) on large numbers of articles in limb branches being more primitive. As noted above, Fryer (1988, 1992) disagrees, but we believe outgroup comparison indicates that lack of articulation in any of the limb parts is better viewed as an apomorphy especially in Branchiopoda.

53. Posterior thoracic limbs protopod: *(0) single segment; (1) two segments; (2) three segments.*

This is a vexing character, and books have been written about the nature of the protopod, i.e., whether it was originally a single-article structure. In truth, there is no clear consensus on just how the various leg segment patterns came to be. Emerson and Schram (1991) and Itô (1989) argued in different ways for a consideration of fusion of medial and lateral elements, in the former case with whole limbs involved and in the latter with parts of limb rami. Walossek (1993) made a case for the coxa having evolved from a separate endite in stem crustaceans. Many other authors (e.g., Hansen 1925; Brooks 1962) viewed a subdivision of the originally one-segment protopod as possible. How the leg protopod form might be explained does not concern us here, and indeed different processes may have prevailed in different groups.

54. Abdominal limbs: *(0) all present; (1) posterior limbs reduced or absent; (2) all limbs absent.*

See comments on character 47.

55. Abdominal muscles: *(0) simple; (1) caridoid musculature.*

This is a feature that has relevance only to various eumalacostracan orders.

56. Abdominal limb epipodites: *(0) absent; (1) present.*

Although potentially relevant to any group, this feature has special significance in uniting hoplocaridan orders.

57. Anterior abdominal limbs of male: *(0) plain, unmodified; (1) as a penis (petasma or other such structure).*

Although we could have scored various possibilities and patterns here, we preferred for the time being to simply note the presence or absence of male copulatory structures in this limb position. To make a priori decisions as to whether certain structures in this location in different groups may or may not be homologous we believe would be to act on the basis of too few comparative studies in this field.

58. Anterior abdominal limbs of female: *(0) plain, unmodified; (1) as a brood pouch.*

59. Male gonopore location (post-maxillary trunk segment numbers): *(0) posterior end of body; (1) segments 13–15; (2) segments 6–8; (3) segment 4; (4) segment 11.*

See discussion of character 60.

60. Female gonopore location (post-maxillary trunk segment numbers): *(0) posterior end of body; (1) segments 6–8; (2) segment 1; (3) segment 4; (4) segment 11; (5) segments 13–15.*

Characters 59 and 60 are another set of theory-laden features. Originally, the locations of crustacean gonopores were thought to be extremely variable and not subject to patterns outside of specific groups. Schram and Emerson (1991) advanced a theoretical framework wherein the gonopore location was indeed pattern specific. Without commenting on the status of their theory, we simply note here gonopore locations, with the outgroup uniramians simply bearing the gonopores at or near the posterior body terminus.

61. Body terminus: *(0) anal somite; (1) wide telson without furca; (2) wide telson with furca; (3) pointed telson without furca; (4) pointed telson with furca.*

Schram (1986), picking up on the suggestions of many others, tried to move toward a treatment of the terminus of the crustaceans as a distinct region in its own right. Walossek (1993) also makes reference to the structural and functional importance of this region, especially in regard to *Rehbachiella.* It is possible here that we have simplified and combined too much and that this character could be treated as two or three separate ones.

62. Caudal rami: *(0) simple in form; (1) absent; (2) developed as abreptors.*

This character continues in the vein of Schram (1986), who opted for treating the posterior end of the body as a distinct region with specific terms to connote the anatomical variety noted there. That work should be consulted for details. The nature of caudal rami, however, obtained increased significance with some of the observations of Walossek (1993) when he placed this feature among those that could serve to differentiate stem-group from crown-group crustaceans.

63. Uropodal rami: *(0) none; (1) one or two segments; (2) multisegmented.*

This, of course, is a feature that helps to link malacostracan groups.

64. Uropod numbers: *(0) none; (1) one set; (2) three sets.*

This character concerns only anatomical conditions within Malacostraca and specifically defining in part the Amphipoda.

65. Thorax: *(0) no distinct thorax (undifferentiated trunk); (1) 11 segments; (2) eight segments; (3) seven segments; (4) four segments; (5) < three segments.*

We recognize the thorax as the part of a regionalized trunk between the head, i.e., the maxillary segment, and the abdomen. The abdomen typically begins on the eighth or ninth post-maxillary segment behind the posterior gonopore (with some notable exceptions).

66. Abdomen: *(0) no distinct abdomen (undifferentiated trunk); (1) > seven segments; (2) six segments; (3) five segments; (4) four segments; (5) two segments; (6) one segment; (7) absent.*

The abdomen is the posterior portion of the regionalized trunk, which may or may not bear limbs. It is typically that region posterior to the posterior-most gonopore. Its limbs may be structurally different from those of the thorax, but they are often reduced or altogether absent (see character 54).

67. Carapace (posterior growth of maxillary segment): *(0) absent; (1) well-developed, covering anterior part of body; (2) short, covering only anteriormost thoracic segments; (3) bivalved/hinged, often enveloping most or all of the body.*

Schram (1986) made much of the different expressions of the carapace. Walossek (1993) believed the issue overdrawn, and we agree. The presence of a carapace seems to be one characteristic of the crustacean lineage, but the appearance and expression of the carapace are variable. For now we opt to note its presence or absence with some few major variants.

The carapace itself has diverse manifestations: some short (hemicaridean eumalacostracans), some fused to the thorax (eucarid malacostracans), some all-encompassing (barnacles, phyllocarids, and ostracodes), some leaving the head free (conchostracans and cladocerans), some wrapping around the side of the head (tantulocarids, mictaceans, eumalacostracans), some as shieldlike structures (branchiurans), some so short as to make it questionable whether a real carapace exists or not (remipedes, mictaceans). Combine these variations with the scattered occurrences of carapaces in various Cambrian Burgess Shale arthropods, which may or may not be crustaceans (like *Branchiocaris, Burgessia, Waptia, Odaraia, Plenocaris*), and the situation becomes more fluid. On the other hand, if these other Cambrian forms with carapaces sort out as stem-group crustaceans, then the issue of carapaces and the definition of crustaceans could become relevant again. This would imply that, while a carapace's presence or absence is not definitive, a tendency exists in the lineage that induces "crustaceomorphs" to produce carapaces. We probably should view the carapace as a character that is "evocative" of a crustacean condition but not truly definitive.

Finally, Dahl (1983a, 1991) presented a rather different viewpoint as to the origins of the carapace, at least in regard to Malacostraca. In so far as these issues impinged upon our evaluation of character states, we deferred to the rebuttal of Newman and Knight (1984) of this issue.

68. Head shield: *(0) present; (1) absent.*

We view the head shield as a typical arthropod structure, wherein the tergites of the "head segment" essentially fuse into a single unit, with some occasional development laterally. In this respect, an extreme development of the head shield posteriorly (i.e., from the level of the posterior portion of the maxillary segment) is a carapace, a thought reflecting in part, but not completely, the discussions of Secretan (1964) and Casanova (1993). Almost all crustaceans have such a shield, as do the members of the stem group. The lack of a head shield, as occurs among some of the branchiopods, could be viewed as a defining apomorphy for some of those groups.

69. Rostrum: *(0) absent; (1) present and fused to head shield/carapace; (2) present and hinged to head shield/carapace.*

Like the carapace, this feature often acts as a characteristic feature of crustaceans. Not all crustaceans have a rostrum, although when an animal has one, we almost automatically view its possession as quintessentially crustacean. In truth, if the rostrum can be said to characterize any group, it can only be said to do so for the malacostracans.

70. Gills (as distinct filamentory or platelike structures from the epipodites): *(0) none; (1) on thorax; (2) on abdomen.*

This is a feature that Walossek and Müller (1990) believed should have rather strict construction. In that respect, they indicated that only Malacostraca have true gills.

71. Male cones: *(0) none; (1) present.*

These are structures so far noted only on some peracarid eumalacostracans.

72. Head form: *(0) solid; (1) cephalic kinesis.*

This feature has long figured in discussions of malacostracan comparative anatomy. Schram (1978) thought it crucial in determining the distinct position of hoplocaridans. One can find, however, kinetic cephalons in groups outside crustaceans, specifically as a distinct possibility in euthycarcinoids.

73. Orthonauplius: *(0) none; (1) egg nauplius only; (2) present with fronto-lateral horns; (3) present without fronto-lateral horns.*

This character refers specifically to the larva that has only three sets of limbs: antennules, antennae, and mandibles. Anything more than these three is a type of metanauplius. The diagnostic significance for the larva with three sets of limbs is quite crucial (e.g., see Walossek and Müller 1990).

When trying to determine crustacean affinities, if no free nauplius larva exists in the life cycle, then one can usually detect a clear naupliar equivalent stage in the egg, where anlagen for the three naupliar limbs appear on the germinal disc. Although the three-limbed nauplius has a long established status in defining crustaceans, this has recently become more significant with the identification of fossilized larvae from other arthropod groups in the Cambrian "Orsten." The importance of four-segmented or four-limbed stages in trilobites and merostomes served to tie these latter groups together. Protaspid larvae have four segments (e.g., see Whittington 1959) in their earliest stages, and these seem to correspond to the four segments that Cisne (1981) confirmed in the head of adult trilobites (antennae and three leg-bearing units). Furthermore, Müller and Walossek (1986b) found four sets of limbs in their larva D (a larva with supposed cheliceriform, possibly even pycnogonid, affinities). This pattern of four segments in the head (antennae + three) prevails in many, but not all, of the non-trilobite and non-chelicerate Burgess Shale trilobitoids (e.g., see Bergström 1992). So the presence of a nauplius, or egg-nauplius stage, is a truly distinctive and defining apomorphy of Crustacea — and only convergent with the three limb pairs of the protonymphon larvae of extant pycnogonids.

74. Zoea larva: *(0) absent; (1) present.*

This is a defining larva of the various groups that are contained in the eucarid malacostracans.

75. Cypris larva: *(0) absent; (1) present.*

This is a distinctive crustacean larval type that defines a whole group of orders within the Cirripedia, as opposed to some others that are autapomorphic (e.g., the tantulus larva).

76. Larval organ (cuticular dorsal organ or neck organ): *(0) absent; (1) present.*

The significance of this structure is not yet clear. It is only just beginning to be studied (e.g., see Martin and Laverack 1992; Walossek 1993). It is included here, even though its condition in many groups is not known, under the assumption that its importance as a phylogenetically informative structure will increase in the future as more becomes known about it.

77. Organ of Bellonci: *(0) absent; (1) present.*

As in the case of character 76, little is known about these organs although much has been made of them in the past as potentially phylogenetically interesting. In many papers dealing with various maxillopodans Grygier (e.g., see Grygier 1987) often alludes to the potential importance of this structure. Its significance is not yet totally realized in this regard since so little is known about it.

78. Heart shape: *(0) long; (1) bulbous.*

Little has been done with internal anatomy in sorting out phylogenetic relationships, in part because surprisingly little is known about crustacean internal anatomy. This feature was used by Schram (1986) in helping to define Maxillopoda. In this regard here, we score thoracicans with a bulbous heart, extending the observations of Burnett (1987) that the rostral sinus acts as a functional heart. Watling (1983) attempted to utilize arterial patterns phylogenetically as well. Much more could be done in this regard, and future databases probably should give some attention to this matter.

79. Egg brooding: *(0) none; (1) in a thoracic pouch; (2) in a burrow with female; (3) in a special sac; (4) on pleopods; (5) under the carapace.*

This is the only behavioral feature present in the data set. Burkenroad (e.g., 1981) made much of this feature in assessing relationships among eucarids. More could probably be done with this character in this regard, but this feature may be better treated as a source for testing phylogenetic relationships developed from purely morphological or molecular approaches. These character states do not form a transition series.

80. Lattice organs: *(0) absent; (1) present.*

This feature has only recently entered the literature and the concerns of comparative morphologists. An apparently sensory structure on the anterior portion of the head shield/carapace, it has been found in all groups of thecostracan maxillopodans (Jensen et al. 1993). It serves as a good apomorphy for these groups, and although searches for this feature in other groups of crustaceans have been limited up to now, there are a number of groups, especially within the maxillopodans, in which it has been looked for and not been found (Jensen 1993; Høeg, pers. comm.).

81. Sperm flagellum: *(0) basal insertion; (1) apical insertion; (2) absent.*

This and the following sperm features are included since they have been shown repeatedly to bear interesting phylogenetic information. This information certainly helps to sort out living groups, but like characters 13–19 these sperm features contribute nothing toward sorting fossils. Our information came from several review references, in particular Felgenhauer and Abele (1991), Jamieson (1987, 1991), and Wingstrand (1972, 1978, 1988).

82. Sperm acrosome: *(0) central apical; (1) lateral; (2) absent.*

83. Sperm filamentous arms: *(0) none; (1) present.*

84. Sperm perforatorium: *(0) none; (1) present.*

85. Sperm nuclear membrane: *(0) present; (1) absent (chromatin diffuse).*

86. Sperm mitochondria: *(0) present; (1) absent.*

87. Spermatophore: *(0) none; (1) present.*

88. Sperm centriole: *(0) present; (1) doublet; (2) centriolar root homologue (cross-striated pseudoflagellum); (3) absent.*

89. Segment composition: *(0) monosegments; (1) diplosegments; (2) duplosegments.*

90. Appendage composition: *(0) primary uniramy; (1) biramy; (2) secondary uniramy.*

Characters 89 and 90 were included in this analysis in part to deal with the issues raised by Arthropod Pattern Theory (Schram and Emerson 1991). These issues, however, required a separate and complete treatment in their own right, as undertaken by Emerson and Schram (1997).

In terms of taxa employed, we decided to score each order of living crustaceans separately except for copepods and ostracodes, which we scored as single groups. We believe it better not to make a priori decisions and lump taxa into larger groups, as did Wilson (1992), e.g., having a single taxon Syncarida instead of three distinct groups: Bathynellacea, Anaspidacea, and Palaeocaridacea. The analyses confirmed the wisdom of this decision since sometimes individual components of supposedly more inclusive taxa sorted quite independently in the analysis. When we expanded the database with fossils that have figured prominently in the literature, we used individual taxa in most cases, e.g., the Cambrian Burgess Shale form *Canadaspis perfecta* or the Devonian *Lepidocaris rhyniensis*. In other cases, we scored whole orders, e.g., the Late Paleozoic malacostracan Pygocephalomorpha and Aeschronectida. In addition, when dealing with fossils, we included not only species that have appeared in the literature as "undoubted" crustaceans, such as *Skara annulata,* and species interpreted as stem-group crustaceans, e.g., *Cambropachycope clarksoni* or *Martinssonia elongata,* but we also decided to score creatures that have not been typically thought of as crustaceans although they have appeared from time to time in previous analyses of "crustaceomorphs," forms such as *Marrella splendens, Odaraia alata,* or *Branchiocaris pretiosa.*

In retrospect, we considered that we might have scored the entire matrix using individual species as did Briggs et al. (1993), i.e., using a specific species such as *Anaspides tasmaniae* rather than the entire order Anaspidacea. This would have reduced the number of polymorphic character states and made fossil and living taxa in our matrix more equivalent as well as equivalent with what is done in molecular sequence analyses. If we had done that, however, familiar groups in phylogenetic analyses, e.g., Thermosbaenacea, would have disappeared in favor of less familiar names, such as *Monodella stygicola.* In addition, we would then have had to decide whether we wanted to score more plesiomorphic or more apomorphic taxa (or something in between) and then defend our selections (e.g., answer questions such as whether *Theosbaena cambodjiana* is more or less derived than *Monodella stygicola* among the total array of thermosbaenaceans).

Generally, we scored the adult condition for the characters except when the character itself dealt with larval forms. However, earlier runs with our matrix following this rule exactly resulted in Rhizocephala and Tantulocarida appearing among the Cambrian stem forms. This was undoubtedly due to the large number of "inapplicable" scorings for many of the characters, which PAUP treats as missing data. To minimize this, we scored several "adult" features, especially among the rhizocephalans, with information garnered from the larvae. This resulted in a shift of these taxa to positions among the maxillopodans and further adjusted the sequence of stem-group taxa.

We rooted the analyses to multiple uniramian forms. A centipede and apterygote sufficed from among the living taxa and a euthycarcinoid from among the fossils. The latter forms a peculiar fossil group that figured in the hypotheses of Emerson and Schram (1990) and Schram and Emerson (1991). We favored uniramians as the outgroup in our analyses since these seemed indicated by both Arthropod Pattern Theory (Schram and Emerson 1991) and the recently published analyses concerning arthropod relationships focusing on Burgess Shale fossils and some other selected fossil and living forms (Briggs et al. 1993).

We used PAUP 3.1.1 for our analyses. Heuristic searches were necessary given the size and complexity of the matrix. In the first run of an analysis, we selected random stepwise addition with a deactivation of MULPARS. After that search, we saved the resulting trees. The second run used those trees in memory with an activated MULPARS option that specified saving only trees equal to or less than the length in memory. Sometimes the number of trees that resulted was small, e.g., nine trees from the rescored data matrix of Wilson (1992). At other times the total number of possible trees resulting was in the thousands, exceeding the memory limits in the computers available to us. We generally ran the analyses until we obtained a memory overflow. In those instances, we obtained a large enough number of trees so that the resulting consensus trees derived from them in repeated runs of the data were generally the same.

We favored 50% majority rule consensus trees, believing that strict consensus puts too heavy a constraint on the conclusions drawn from the data. It is our position that one does not strive for ultimate truth with these kinds of analyses, one merely organizes the information available at any one point in time. We freely admit that any database right now is imperfect, which would only be further confirmed with a strict consensus. Nevertheless, we believe that seeing what the most likely patterns are, given a particular database, has in fact heuristic value in focusing on the current characters and guiding future research to gather more fundamental data.

## Results

Our analyses ran the matrix unordered and unweighted, with uninformative characters deleted (table 6.2), and multistate characters indicating polymorphism (table 6.3). The initial results appear in figures 6.6 and 6.7. We excluded all fossils in the first set to see what pattern we could derive from the characters we used. We obtained 165 trees of length 553 (CI = 0.541, HI = 0.673, RI = 0.657, and RC = 0.355). The tree metrics, while low, remain reasonable for such a large matrix. The 50% majority rule consensus in figure 6.6 yielded a tree that, while mildly surprising in regard to a few points, appeared quite reasonable on the whole. Even the strict consensus of the data (fig. 6.7) contained a lot of structure but highlighted some problem areas. What is immediately obvious from these consensus trees is the easy recognition of several traditional groups. Malacostraca clearly emerges, even if some of the postulated relationships within it may appear strange. For example, the stomatopods occur as a sister group of the eucarids, syncarids appear paraphyletically, and while peracarids are monophyletic, many will still be bothered by the persistent linkage of isopods and amphipods. Among the "entomostracans" the leptostracans and cephalocarids continue to associate with the branchiopods (*sensu* Schram 1986), and the maxillopodans emerge as a paraphyletic group near the base of the tree, just above nectiopod remipedes. This arrangement is evocative of several themes that have appeared in other analyses: that remipedes are allied with maxillopodans has been suggested, as mentioned above, by some recent rDNA sequencing (e.g., see Boxshall et al. 1992), the basal position of the maxillopodans was advocated by McKenzie (1991), and the resurrection out of the last century of the old

TABLE 6.2

The two uninformative features, as well as the soft anatomy features deleted from the analysis in the cladogram of figure 6.10

| Characters Uninformative in Complete Analysis | Characters Excluded from Soft Anatomy |
|---|---|
| 21, 58 | 13–19, 55, 77–79, 81–88 |

TABLE 6.3
Data matrix for the cladistic analysis discussed in the text, with 90 characters for each taxon

| | 1 | 2 | 3 | 4 | 5 | 6 | 7 | 8 | 9 | 10 |
|---|---|---|---|---|---|---|---|---|---|---|
| centipede | 0 | 0 | 0 | 0 | - | - | - | - | - | - |
| | 0 | 0 | 0 | 0 | 0 | 0 | 0 | 0 | 0 | - |
| | - | - | - | - | - | - | - | 3 | 1 | 3 |
| | - | - | 1 | 0 | 3 | 2 | - | 1 | 1 | 0 |
| | 0 | 0 | 1 | - | 0 | 1 | 0 | 0 | - | - |
| | - | 0 | 0 | - | - | - | - | - | 0 | 0 |
| | 0 | 1 | 0 | - | 0 | 0 | 0 | 0 | 0 | 0 |
| | 0 | 0 | 0 | - | - | ? | ? | 0 | 0 | ? |
| | 0 | 0 | 0 | ? | 0 | 0 | 0 | 0 | 0 | 0 |
| euthycarcinoid | 0 | 0 | 0 | 0 | - | - | - | - | - | - |
| | 1 | 0 | ? | ? | ? | ? | ? | ? | ? | ? |
| | ? | ? | ? | ? | ? | ? | ? | ? | 0 | ? |
| | ? | ? | 0 | 0 | ? | ? | ? | ? | 0 | 0 |
| | 0 | 0 | 0 | ? | 0 | 0 | 0 | 0 | 0 | ? |
| | ? | 0 | 0 | 2 | ? | 0 | ? | ? | ? | ? |
| | 3 | 1 | 0 | 0 | 1 | 3 | 0 | 1 | 0 | ? |
| | ? | 1 | ? | ? | ? | ? | ? | ? | ? | ? |
| | ? | ? | ? | ? | ? | ? | ? | ? | 1 | 0 |
| apterygote | 0 | 0 | 0 | 0 | - | - | - | - | - | - |
| | 0 | 0 | 1 | 1 | 0 | 0 | 0 | 0 | 0 | - |
| | - | - | - | - | - | - | - | 3 | 1 | 3 |
| | - | - | 1 | 0 | 3 | 1 | - | 1 | 0 | 0 |
| | 0 | 0 | 0 | - | 0 | 0 | 1 | 0 | - | - |
| | - | 1 | 0 | 1 | - | - | - | - | 0 | 0 |
| | 0 | 3 | 0 | - | 5 | 1 | 0 | 0 | 0 | 0 |
| | 0 | 0 | 0 | - | - | ? | ? | 0 | 0 | ? |
| | 0 | 0 | 0 | 1 | 0 | 0 | 0/1 | 0 | 0 | 0 |
| Nectiopoda | 1 | 0 | 0 | 1 | 1 | 0 | 0 | ? | 1 | ? |
| | 3 | - | ? | ? | ? | ? | ? | ? | ? | 0 |
| | 1 | 1 | 1 | ? | 1 | 3 | 4 | 3 | 2 | 1 |
| | 0 | 2 | 1 | 0 | 2 | 0 | 3 | 1 | 1 | 0 |
| | 0 | 0 | 3 | 0 | 0 | 1 | 0 | 1 | 0 | 0 |
| | 0 | 2 | 0 | ? | ? | ? | ? | ? | 1 | 1 |
| | 0 | 0 | 0 | 0 | 0 | 0 | 0 | 0 | 0 | 0 |
| | 0 | 0 | ? | ? | ? | 0 | ? | 0 | ? | ? |
| | 0 | 0 | 0 | 1 | 0 | 0 | 1 | 3 | 2 | 1 |
| Aeschronectida | 2 | 0 | 0 | 1 | 1 | 0 | 0 | ? | ? | ? |
| | 1 | 0 | ? | ? | ? | ? | ? | ? | ? | ? |
| | ? | ? | ? | ? | ? | ? | ? | ? | 1 | ? |
| | 1 | ? | 0/1 | ? | ? | ? | 3 | 1 | 0 | ? |
| | ? | ? | 0 | ? | 3 | 0 | 1 | 1 | 0 | 0 |
| | 2 | 2 | 2 | 0 | ? | 1 | 0 | 0 | ? | ? |

TABLE 6.3 *(continued)*

| | 1 | 2 | 3 | 4 | 5 | 6 | 7 | 8 | 9 | 10 |
|---|---|---|---|---|---|---|---|---|---|---|
| | 1 | 1 | 1 | 1 | 2 | 2 | 1 | 0 | 2 | 2 |
| | 0 | ? | ? | ? | ? | ? | ? | ? | ? | ? |
| | ? | ? | ? | ? | ? | ? | ? | ? | 2 | 1 |
| Palaeostomatopoda | 2 | 0 | 0 | 1 | 1 | 0 | 0 | ? | ? | ? |
| | 1 | 0 | ? | ? | ? | ? | ? | ? | ? | ? |
| | ? | ? | ? | ? | ? | ? | ? | ? | ? | ? |
| | ? | ? | ? | ? | ? | ? | ? | ? | 5 | ? |
| | ? | ? | 3 | 3 | 3 | ? | 1 | ? | ? | ? |
| | ? | ? | ? | 0 | ? | 1 | 0 | 0 | ? | ? |
| | 2 | 1 | 1 | 1 | 2 | 2 | 1 | 0 | 2 | 2 |
| | 0 | ? | ? | ? | ? | ? | ? | ? | ? | ? |
| | ? | ? | ? | ? | ? | ? | ? | ? | 2 | 1 |
| Stomatopoda | 2 | 0 | 0 | 1 | 1 | 0 | 0 | ? | 2 | - |
| | 1 | 0 | 2 | 1 | 1 | 0 | 0 | 1 | 0 | 1 |
| | 0 | 1 | 1 | 0 | 0 | 0 | 4 | 1 | 1 | 2 |
| | 3 | 2 | 1 | 0 | 1 | 3 | 3 | 1 | 5 | 0 |
| | 1 | 0 | 3 | 3 | 3 | 5 | 1 | 1 | 0 | 0 |
| | 2 | 4 | 2 | 0 | 0 | 1 | 0 | 0 | 2 | 1 |
| | 1 | 1 | 1 | 1 | 2 | 2 | 1 | 0 | 2 | 2 |
| | 0 | 1 | 1 | 0 | 0 | 1 | 1 | 0 | 2 | ? |
| | 2 | 0 | 0 | 1 | 1 | 0 | 0 | 1 | 2 | 1 |
| Anaspidacea | 1 | 0 | 0 | 1 | 1 | 0 | 0 | 0 | 1 | - |
| | 1 | 0 | 2 | 1 | 0 | 0 | 0 | 1 | 0 | 0 |
| | 0 | 1 | 1 | - | 0 | 0 | 4 | 1 | 1 | 2 |
| | 3 | 2 | 1 | 0 | 2 | 2 | 3 | 1 | 1 | 2 |
| | 0 | 0 | 0 | 2 | 2 | 1 | 1 | 1 | 0 | 1 |
| | 3 | 1 | 1 | 0 | 1 | 0 | 1 | 0 | 2 | 1 |
| | 1 | 1 | 1 | 1 | 2 | 2 | 0 | 0 | 0 | 0 |
| | 0 | 0 | 1 | 0 | 0 | 1 | 1 | 1 | 0 | ? |
| | 2 | 0 | 0 | 1 | 0 | ? | 0 | ? | 2 | 1 |
| Palaeocaridacea | 1 | 0 | 0 | 1 | 1 | 0 | 0 | ? | 1 | ? |
| | 1 | 0 | ? | ? | ? | ? | ? | ? | ? | 1 |
| | 0 | ? | 1 | ? | 0 | 1 | 4 | 1 | 1 | ? |
| | 1 | 2 | 1 | ? | ? | 2 | 3 | 1 | 1 | ? |
| | ? | 0 | 1 | ? | 2 | 0 | 1 | 1 | 0 | 1 |
| | 1/3 | 1 | 1 | 0 | ? | 0 | 0 | 0 | ? | ? |
| | 1 | 1 | 1 | 1 | 2 | 2 | 0 | 0 | 0 | 0 |
| | 0 | 0 | ? | ? | ? | ? | ? | ? | ? | ? |
| | ? | ? | ? | ? | ? | ? | ? | ? | 2 | 1 |
| Bathynellacea | 1 | 0 | 1 | 1 | 1 | 0 | 0 | 0 | 1 | - |
| | 3 | - | ? | ? | ? | ? | ? | ? | ? | 0 |
| | 0 | 1 | 1 | - | 0 | 0 | 4 | 5 | 1 | 2 |
| | 4 | 2 | 1 | 0 | 2 | 2 | 3 | 1 | 0 | 0 |
| | 2 | 0 | 0 | 2 | 4 | 0 | 0 | 1 | 0 | 1 |
| | 2 | 3 | 0 | 2 | ? | - | - | - | 2 | 1 |

TABLE 6.3 *(continued)*

| | 1 | 2 | 3 | 4 | 5 | 6 | 7 | 8 | 9 | 10 |
|---|---|---|---|---|---|---|---|---|---|---|
| | 0 | 0 | 1 | 1 | 2 | 3 | 0 | 0 | 0 | 0 |
| | 0 | 0 | 1 | 0 | 0 | ? | ? | 1 | 0 | ? |
| | ? | ? | ? | ? | ? | ? | ? | ? | 2 | 1 |
| Belotelsonidea | 1 | 0 | 0 | 1 | 1 | 0 | 0 | ? | 1 | ? |
| | 1 | 0 | ? | ? | ? | ? | ? | ? | ? | ? |
| | ? | ? | 1 | ? | 0 | 3 | 4 | 5 | 1 | 2 |
| | ? | ? | 1 | ? | 2 | ? | ? | 1 | 0 | 0 |
| | ? | 0 | 0 | 3 | 2 | 0 | 1 | 0 | 0 | ? |
| | 4 | 1 | 0 | 0 | ? | 0 | 0 | 0 | ? | ? |
| | 4 | 1 | 1 | 1 | 2 | 2 | 1 | 0 | 1 | ? |
| | 0 | 0 | ? | ? | ? | 1 | ? | ? | ? | ? |
| | ? | ? | ? | ? | ? | ? | ? | ? | 2 | 2 |
| Waterstonellidea | 1 | 0 | 0 | 1 | 1 | 0 | 0 | ? | 1 | ? |
| | 1 | 0 | ? | ? | ? | ? | ? | ? | ? | ? |
| | ? | ? | ? | ? | ? | ? | ? | ? | 0 | ? |
| | ? | ? | 0 | ? | ? | ? | ? | ? | 0 | ? |
| | ? | ? | 0 | ? | ? | ? | 1 | 1 | ? | ? |
| | ? | ? | ? | 0 | ? | ? | 0 | 0 | ? | ? |
| | 1 | 1 | 1 | 1 | 2 | 2 | 1 | 0 | ? | ? |
| | ? | ? | ? | ? | ? | ? | ? | ? | ? | ? |
| | ? | ? | ? | ? | ? | ? | ? | ? | 2 | 1 |
| Lophogastrida | 1 | 0 | 0 | 1 | 1 | 0 | 0 | 0 | 1 | - |
| | 1 | 0 | 0 | 0 | 1 | 0 | 0 | 1 | 0 | 1 |
| | 0 | 1 | 1 | - | 0 | 0 | 4 | 2 | 1 | 2 |
| | 2 | 2 | 1 | 0 | 2 | 2 | 2 | 1 | 1 | 1 |
| | 1 | 0 | 1 | 2 | 2 | 1 | 1 | 1 | 0 | 2/3 |
| | 3 | 1 | 1 | 0 | 1 | 0 | 0 | 0 | 2 | 1 |
| | 1 | 1 | 1 | 1 | 2 | 2 | 1 | 0 | 0/1 | 1 |
| | 0 | 0 | 1 | 0 | 0 | 1 | 1 | 0 | 1 | ? |
| | ? | ? | ? | ? | ? | ? | ? | ? | 2 | 1 |
| Mysida | 1 | 0 | 0 | 1 | 1 | 0 | 0 | 0 | 1 | - |
| | 1 | 0 | 0 | 0 | 1 | 0 | 0 | 1 | 0 | 1 |
| | 0 | 0 | 1 | - | 0 | 0 | 4 | 2 | 1 | 2 |
| | 2/4 | 2 | 1 | 0 | 2 | 2 | 2 | 1 | 1 | 1 |
| | 1 | 0 | 1 | 2 | 2 | 0 | 1 | 1 | 0 | 3 |
| | 3 | 1 | 1 | 0 | 1 | 0 | 0 | 0 | 2 | 1 |
| | 1 | 1 | 1 | 1 | 2 | 2 | 1 | 0 | 0/1 | 0 |
| | 0 | 0 | 1 | 0 | 0 | 1 | 1 | 1 | 1 | ? |
| | 2 | 0 | 0 | 1 | 0 | 0 | 0 | 2 | 2 | 1 |
| Pygocephalomorpha | 1 | 0 | 0 | 1 | 1 | 0 | 0 | ? | 1 | ? |
| | 1 | 0 | ? | ? | ? | ? | ? | ? | ? | ? |
| | ? | ? | 1 | ? | 0 | ? | ? | ? | 1 | 2 |
| | ? | 2 | 1 | ? | 2 | ? | 3 | 1 | 2 | 1 |
| | ? | 1 | 1 | 3 | 3 | ? | 1 | 1 | 0 | 3 |
| | 3 | 2 | 1 | 0 | ? | ? | 0 | 0 | 2 | ? |

TABLE 6.3 *(continued)*

| | 1 | 2 | 3 | 4 | 5 | 6 | 7 | 8 | 9 | 10 |
|---|---|---|---|---|---|---|---|---|---|---|
| | 2 | 1 | 1 | 1 | 2 | 2 | 1 | 0 | 1 | ? |
| | 1 | 0 | ? | ? | ? | ? | ? | ? | 1 | ? |
| | ? | ? | ? | ? | ? | ? | ? | ? | 2 | 1 |
| Mictacea | 1 | 0 | 0 | 1 | 1 | 0 | 0 | ? | 1 | ? |
| | 2 | 0 | ? | ? | ? | ? | ? | ? | ? | ? |
| | ? | ? | 1 | ? | 0 | 0 | 4 | 2 | 1 | 2 |
| | 4 | 2 | 1 | 0 | 2 | 3 | 3 | 1 | 1 | 1 |
| | 0 | 0 | 1 | 3 | 2 | 1 | 1 | 0/1 | 0 | 3 |
| | 1/4 | 1 | 1 | 1 | ? | 0 | 0 | 0 | 2 | 1 |
| | 1 | 1 | 1 | 1 | 2 | 2 | 0 | 0 | 1 | 0 |
| | 0 | 0 | ? | ? | ? | ? | ? | ? | 1 | ? |
| | ? | ? | ? | ? | ? | ? | ? | ? | 2 | 1 |
| Isopoda | 0 | 0/1 | 0 | 1 | 1 | 1 | 0/1 | 0 | 3 | - |
| | 0 | 0 | 2 | 1 | 0 | 0 | 0 | 0 | 0 | 0/1 |
| | 0 | 1 | 1 | - | 0 | 0 | 4 | 2 | 1 | 2 |
| | 4 | 2 | 1 | 0 | 2 | 3 | 3 | 1 | 1 | 1 |
| | 1 | 0 | 1 | 3 | 2 | 2/3/4 | 1 | 0 | 0 | 3 |
| | 4 | 1 | 1 | 0 | 0 | 0/1 | 1 | 0 | 2 | 1 |
| | 1 | 1 | 1 | 1 | 2 | 3/4 | 0 | 0 | 0 | 2 |
| | 0 | 0 | 1 | 0 | 0 | 1 | 0/1 | 0 | 1 | ? |
| | 2 | 0 | 0 | 1 | 0 | 0 | 0 | 2 | 2 | 2 |
| Amphipoda | 0/1 | 0 | 0 | 1 | 1 | 1 | 0 | 0 | 3 | - |
| | 0 | 0 | 0 | 0 | 0 | 0 | 0 | 0 | 0 | 1 |
| | 0 | 0 | 1 | - | 0 | 0/1/2/3 | 4 | 2 | 1 | 2 |
| | 2/3/4 | 2 | 1 | 0 | 2 | 3 | 3 | 1 | 1 | 1 |
| | 0 | 0 | 1 | 3 | 2 | 1 | 1 | 0 | 0 | 2/3 |
| | 4 | 1 | 1 | 0 | 0 | 0 | 0 | 0 | 2 | 1 |
| | 1 | 1 | 1 | 2 | 2 | 2 | 0 | 0 | 0 | 1 |
| | 0 | 0 | 1 | 0 | 0 | 0 | 1 | 1 | 1 | ? |
| | 2 | 0 | 0 | 1 | 0 | 0 | 0 | 2 | 2 | 2 |
| Cumacea | 1 | 1 | 0 | 1 | 1 | 1 | 0/1 | 0 | 3 | - |
| | 2 | 0 | 0 | 0 | 0 | 0 | 0 | 0 | 0 | 0 |
| | 0 | 1 | 1 | - | 0 | 3 | 4 | 2 | 1 | 2 |
| | 3 | 2 | 1 | 0 | 2 | 3 | 3 | 1 | 3 | 1 |
| | 2 | 0 | 1 | 0/3 | 2 | 3 | 1 | 1 | 0 | 3 |
| | 0 | 1 | 1 | 0/1/2 | ? | 0 | 0 | 0 | 2 | 1 |
| | 1 | 1 | 1 | 1 | 2 | 2/3 | 2 | 0 | 0 | 0 |
| | 0 | 0 | 1 | 0 | 0 | 0 | ? | 0 | 1 | ? |
| | 2 | 0 | 0 | 1 | 0 | 0 | 0 | 2 | 2 | 1 |
| Tanaidacea | 1 | 0 | 0 | 1 | 1 | 0 | 0 | 0 | 1 | - |
| | 2 | 0 | 0 | 0 | 0 | 0 | 0 | 0 | 0 | 1 |
| | 0 | 1 | 1 | - | 0 | 0/3 | 4 | 2 | 1 | 2 |
| | 2/4 | 2 | 1 | 0 | 2 | 3 | 3 | 1 | 2 | 0 |
| | 2 | 0 | 1/3 | 1/3 | 2 | 2 | 1 | 0/1 | 0 | 3 |
| | 1/4 | 1 | 1 | 0 | 0 | 0 | 0 | 0 | 2 | 1 |

TABLE 6.3 *(continued)*

| | 1 | 2 | 3 | 4 | 5 | 6 | 7 | 8 | 9 | 10 |
|---|---|---|---|---|---|---|---|---|---|---|
| | 0/1 | 1 | 1/2 | 1 | 2 | 2/3 | 2 | 0 | 0 | 0 |
| | 1 | 0 | 1 | 0 | 0 | 0 | ? | 0 | 1 | ? |
| | 2 | 0 | 0 | 0 | 0 | 0 | 0 | ? | 2 | 1 |
| Spelaeogriphacea | 1 | 0 | 0 | 1 | 1 | 0 | 0 | 0 | 1 | ? |
| | 2 | 0 | ? | ? | ? | ? | ? | ? | ? | ? |
| | ? | ? | 1 | ? | 0 | 2 | 4 | 2 | 1 | 2 |
| | 3 | 2 | 1 | 0 | 2 | 3 | 3 | 1 | 1 | 1 |
| | 2 | 0 | 1 | 3 | 2 | 1 | 1 | 1 | 0 | 3 |
| | 2 | 1 | 1 | 1 | ? | 0 | 0 | 0 | 2 | 1 |
| | 1 | 1 | 1 | 1 | 2 | 2 | 2 | 0 | 0 | 0 |
| | 0 | 0 | ? | 0 | 0 | ? | ? | ? | 1 | ? |
| | ? | ? | ? | ? | ? | ? | ? | ? | 2 | 1 |
| Thermosbaenacea | 1 | 0 | 0 | 1 | 1 | 1 | 0 | 0 | 3 | - |
| | 3 | ? | ? | ? | ? | ? | ? | ? | ? | ? |
| | ? | ? | 1 | - | 0 | 0 | 4 | 2 | 1 | 2 |
| | 2 | 2 | 1 | 0 | 2 | 2 | 2/3 | 1 | 2 | 1 |
| | 1 | 1 | 1 | 3 | 6 | 2 | 1 | 1 | 0 | 3 |
| | 1 | 1 | 1 | 1 | 0 | 0 | 0 | 0 | 2 | 1 |
| | 0 | 1 | 1 | 1 | 2 | 3 | 2 | 0 | 0 | 0 |
| | 0 | 0 | ? | 0 | 0 | ? | ? | 1 | 5 | ? |
| | ? | ? | ? | ? | ? | ? | ? | ? | 2 | 1 |
| Euphausiacea | 1 | 0 | 0 | 1 | 1 | 0 | 0 | 0 | 1 | ? |
| | 1 | 0 | 2 | 0 | 1 | 0 | 0 | 1 | 0 | 1 |
| | 0 | 1 | 1 | 0 | 0 | 0 | 4 | 1 | 1 | 2 |
| | 3 | 2 | 1 | 0 | 2 | 2 | 2 | 1 | 0 | 0 |
| | 1 | 0 | 0 | 2 | 2 | 0 | 1 | 1 | 0 | 2 |
| | 2 | 1 | 1 | 0 | 1 | 0 | 1 | 0 | 2 | 1 |
| | 4 | 1 | 1 | 1 | 2 | 2 | 1 | 0 | 0/1 | 1 |
| | 0 | 1 | 3 | 1 | 0 | 1 | ? | 1 | 0 | ? |
| | 2 | 2 | 0 | 0 | 1 | ? | 1 | ? | 2 | 1 |
| Amphionidacea | 1 | 0 | 0 | 1 | 1 | 0 | 0 | 0 | 1 | - |
| | 1 | 0 | ? | ? | ? | ? | ? | ? | ? | ? |
| | ? | ? | 1 | - | 0 | 3 | 4 | 3 | 1 | 3 |
| | 4 | 2 | 1 | 1 | 2 | 2 | 3 | 1 | 1 | 0 |
| | 3 | 0 | 1 | 0 | 2 | 1 | 1 | 1 | 0 | 2 |
| | 0/4 | 1/3 | 1 | 0 | ? | 0 | 0 | 1 | 2 | 1 |
| | 3 | 1 | 1 | 1 | 2 | 2 | 1 | 0 | 1 | 1 |
| | 0 | 0 | 0 | 1 | 0 | 1 | ? | 1 | 1 | ? |
| | ? | ? | ? | ? | ? | ? | ? | ? | 2 | 1/2 |
| Dendrobranchiata | 1 | 0 | 0 | 1 | 1 | 0 | 0 | 0 | 1 | 0 |
| | 1 | 0 | 2 | 1 | 1 | 0 | 0 | 1 | 0 | 1 |
| | 0 | 0 | 1 | 0 | 0 | 1 | 4 | 1 | 1 | 2 |
| | 2 | 2 | 1 | 1 | 2 | 2 | 3 | 1 | 3 | 0/1 |
| | 3 | 0/1 | 1/2 | 0/3 | 2 | 3 | 1 | 1 | 0 | 2 |
| | 2/3/4 | 1 | 1 | 0 | 1 | 0 | 1 | 0 | 2 | 1 |

TABLE 6.3 *(continued)*

| | 1 | 2 | 3 | 4 | 5 | 6 | 7 | 8 | 9 | 10 |
|---|---|---|---|---|---|---|---|---|---|---|
| | 3 | 1 | 1 | 1 | 2 | 2 | 1 | 0 | 1 | 1 |
| | 0 | 0 | 3 | 1 | 0 | 1 | ? | 1 | 0 | 0 |
| | 2 | 0 | 0 | 0 | 0/1 | ? | 1 | 3 | 2 | 1 |
| Caridea | 1 | 0 | 0 | 1 | 1 | 0 | 0 | 0 | 1 | - |
| | 1 | 0 | 2 | 1 | 1 | 0 | 0 | 1 | 0 | 1 |
| | 0 | 0 | 1 | 0 | 0 | 0/3 | 4 | 1/3/5 | 1 | 2 |
| | 3 | 2 | 1 | 1 | 2 | 2 | 3 | 1 | 3 | 0/1 |
| | 3 | 0/1 | 1 | 0/2 | 2/6 | 3 | 1 | 1 | 0 | 2 |
| | 2/3 | 1 | 1 | 0 | 1 | 0 | 1 | 0 | 2 | 1 |
| | 1/3 | 1 | 1 | 1 | 2 | 2 | 1 | 0 | 1 | 1 |
| | 0 | 0 | 1 | 1 | 0 | 1 | 1 | 1 | 4 | 0 |
| | 2 | 2 | 0 | 1 | 0 | 0 | 1 | 0 | 2 | 1/2 |
| Euzygida | 1 | 0 | 0 | 1 | 1 | 0 | 0 | 0 | 1 | - |
| | 1 | 0 | 2 | 1 | 1 | 0 | 0 | 1 | 0 | 1 |
| | 0 | 0 | 1 | 0 | 0 | 0 | 4 | 1 | 1 | 2 |
| | 3 | 2 | 1 | 1 | 2 | 2 | 3 | 1 | 3 | 1 |
| | 3 | 0/1 | 1 | 2/3 | 2/4 | 3 | 1 | 0 | 0 | 2 |
| | 4 | 1 | 1 | 0 | 1 | 0 | 0 | 0 | 2 | 1 |
| | 1 | 1 | 1 | 1 | 2 | 2 | 1 | 0 | 1 | 1 |
| | 0 | 0 | 1 | 1 | 0 | ? | ? | ? | 4 | 0 |
| | 2 | 2 | 0 | 0 | 0 | ? | ? | ? | 2 | 2 |
| Reptantia | 1 | 0 | 0 | 1 | 1 | 0 | 0 | 0 | 1 | - |
| | 1 | 0 | 2 | 0/1 | 0/1 | 0 | 0 | 1 | 0 | 1 |
| | 0 | 0 | 1 | 0 | 0 | 0 | 4 | 1 | 1 | 2 |
| | 3 | 2 | 1 | 1 | 2 | 2 | 3 | 1 | 3 | 0/1 |
| | 0/3 | 0/1 | 1 | 0/2 | 2/6 | 3 | 1 | 0 | 0 | 2 |
| | 4 | 2 | 1 | 0 | 1 | 0 | 0/1 | 0 | 2 | 1 |
| | 1 | 1 | 1 | 1 | 2 | 2 | 1 | 0 | 1 | 1 |
| | 0 | 0 | 1 | 1 | 0 | 1 | 0/1 | 1 | 4 | 0 |
| | 2 | 0 | 0/1 | 1 | 1 | 0 | 1 | 0/2 | 2 | 2 |
| Leptostraca | 1 | 0 | 1 | 1 | 1 | 1 | 0 | 0 | 3 | - |
| | 1 | 0 | 0 | 0 | 0 | 0 | 0 | 0 | 0 | 1 |
| | 0 | 1 | 1 | - | 0 | 0 | 4 | 1 | 1 | 2 |
| | 3 | 1 | 1 | 0 | 2 | 2/3 | 2 | 1 | 0 | 0 |
| | 1 | 1 | 0 | 2 | 2/3 | 0 | 1 | 2 | 0 | 1 |
| | 2 | 1/2/3 | 0 | 1 | 0 | 0 | 0 | 0 | 2 | 1 |
| | 0 | 0 | 0 | 0 | 2 | 1 | 3 | 0 | 2 | 0 |
| | 0 | 1 | 1 | 0 | 0 | 1 | 1 | 0 | 5 | ? |
| | 2 | 2 | 1 | 0 | 0 | 0 | 0 | 0 | 2 | 1 |
| Hymenostraca | ? | ? | ? | ? | ? | ? | ? | ? | ? | ? |
| | ? | ? | ? | ? | ? | ? | ? | ? | ? | ? |
| | ? | ? | ? | ? | ? | ? | ? | ? | ? | ? |
| | ? | ? | ? | ? | ? | ? | ? | ? | ? | ? |
| | ? | ? | ? | ? | ? | ? | ? | ? | ? | ? |
| | ? | ? | ? | ? | ? | ? | ? | ? | ? | ? |

TABLE 6.3 *(continued)*

| | 1 | 2 | 3 | 4 | 5 | 6 | 7 | 8 | 9 | 10 |
|---|---|---|---|---|---|---|---|---|---|---|
| | 3 | 0 | 0 | 0 | ? | 1 | 3 | ? | ? | ? |
| | ? | ? | ? | ? | ? | ? | ? | ? | ? | ? |
| | ? | ? | ? | ? | ? | ? | ? | ? | 2 | 1 |
| Archaeostraca | 1 | 0 | 0 | 1 | 1 | 0 | 0 | ? | 0 | ? |
| | 1 | 0 | ? | ? | ? | ? | ? | ? | ? | ? |
| | ? | ? | ? | ? | ? | ? | ? | 5 | ? | ? |
| | ? | ? | ? | ? | ? | ? | ? | ? | ? | ? |
| | ? | ? | ? | ? | ? | ? | 1 | ? | ? | ? |
| | ? | ? | ? | 1 | ? | ? | ? | ? | ? | ? |
| | 3 | 0 | 0 | 0 | ? | ? | 3 | 0 | 2 | ? |
| | ? | ? | ? | ? | ? | ? | ? | ? | ? | ? |
| | ? | ? | ? | ? | ? | ? | ? | ? | 2 | 1 |
| Hoplostraca | ? | ? | ? | 1 | 0 | ? | 0 | ? | ? | ? |
| | 1 | 0 | ? | ? | ? | ? | ? | ? | ? | ? |
| | ? | ? | ? | ? | ? | ? | ? | 5 | ? | ? |
| | ? | ? | ? | ? | ? | ? | ? | ? | ? | ? |
| | ? | ? | ? | ? | ? | ? | 1 | ? | ? | ? |
| | ? | ? | ? | 1 | ? | ? | ? | ? | ? | ? |
| | 3 | 0 | 0 | 0 | ? | 1 | 3 | 0 | 0 | ? |
| | ? | ? | ? | ? | ? | ? | ? | ? | ? | ? |
| | ? | ? | ? | ? | ? | ? | ? | ? | 2 | 1 |
| *Canadaspis* | 0 | 1 | 2 | 1 | 1 | 1 | 0 | ? | 3 | ? |
| | 1 | 0 | ? | ? | ? | ? | ? | ? | ? | ? |
| | ? | ? | 1 | ? | 0 | 3 | 4 | 3 | 0 | 0 |
| | 0 | 1 | 0 | 0 | 3 | 0 | 2 | 0 | 0 | 0 |
| | 0 | 0 | 0 | ? | ? | 0 | 0 | 1 | 0 | 0 |
| | 2 | 0 | 0 | 2 | ? | - | - | - | ? | ? |
| | 0 | 1 | 0 | 0 | 2 | 1 | 3 | 0 | 0 | 0 |
| | 0 | ? | ? | ? | ? | ? | ? | ? | ? | ? |
| | ? | ? | ? | ? | ? | ? | ? | ? | 2 | 1 |
| *Nahecaris* | 1 | 0 | 0 | 1 | 1 | 0 | 0 | ? | 0 | ? |
| | 1 | 0 | ? | ? | ? | ? | ? | ? | ? | ? |
| | ? | ? | 1 | ? | 0 | ? | ? | 5 | 1 | 2 |
| | ? | ? | 1 | ? | 2 | ? | ? | ? | 0 | 0 |
| | 0 | 0 | 0 | ? | 2 | 0 | 1 | 1 | 0 | 0 |
| | ? | 1 | ? | 1 | ? | 0 | ? | ? | ? | ? |
| | 3 | 0 | 0 | 0 | 2 | 1 | 3 | 0 | 2 | 0 |
| | ? | ? | ? | ? | ? | ? | ? | ? | ? | ? |
| | ? | ? | ? | ? | ? | ? | ? | ? | 2 | 1 |
| *Lepidocaris* | 0 | 0 | 2 | 1 | 1 | 0 | 0 | 1 | 3 | 1 |
| | 3 | - | ? | ? | ? | ? | ? | ? | ? | 0 |
| | 0 | 1 | 1 | 1 | 1 | 3 | 4 | 3 | 1 | 2 |
| | 1/4 | 2 | 1 | 0 | 1 | 2 | 2 | 1 | 0 | 3 |
| | 0 | 1 | 0 | 2 | 6 | 0 | 0 | 1 | 1 | 0 |
| | 2 | 5 | 0 | 2 | ? | - | - | - | ? | ? |

TABLE 6.3 *(continued)*

| | 1 | 2 | 3 | 4 | 5 | 6 | 7 | 8 | 9 | 10 |
|---|---|---|---|---|---|---|---|---|---|---|
| | 0 | 0 | 0 | 0 | 1 | 2 | 0 | 0 | 0 | 0 |
| | 0 | 0 | ? | 0 | 0 | 0 | ? | ? | 3 | ? |
| | ? | ? | ? | ? | ? | ? | ? | ? | ? | 1 |
| Cephalocarida | 0 | 0 | 0 | 1 | 1 | 0 | 0 | 1 | 0 | 1 |
| | 3 | - | 0 | 0 | 0 | 0 | 0 | 0 | 0 | 0 |
| | 0 | 1 | 1 | 1 | 0 | 3 | 4 | 1 | 1 | 2 |
| | 0 | 1 | 0 | 2 | 0 | 0 | 1 | 1 | 0 | 3 |
| | 1 | 1 | 0 | 1 | 1 | 0 | 0 | 2 | 1 | 1 |
| | 1 | 0 | 0 | 2 | 0 | - | - | - | 2 | 1 |
| | 0 | 0 | 0 | 0 | 2 | 1 | 0 | 0 | 0 | 0 |
| | 0 | 0 | 1 | 0 | 0 | ? | 0 | ? | 3 | ? |
| | 2 | 0 | 0 | 1 | 0 | 1 | 0 | 3 | ? | 1 |
| Notostraca | 0 | 1 | 2 | 1 | 1 | 1 | 1 | 1 | 3 | 1 |
| | 0 | 1 | 1 | 0 | 0 | 1 | 0 | 0 | 1 | 0 |
| | 0 | 1 | 1 | 0 | 0 | 3 | 4 | 3 | 1 | 2 |
| | 4 | 2 | 2 | 0 | 3 | 3 | 3 | 1 | 0 | 3 |
| | 1 | 1 | 2 | 2 | 0 | 0 | 0 | 2 | 1 | 1 |
| | 2 | 5 | 0 | 2 | 0 | - | - | - | 4 | 4 |
| | 0/1 | 3 | 0 | 0 | 1 | 1 | 3 | 0 | 0 | 0 |
| | 0 | 0 | ? | 0 | 0 | 1 | ? | 0 | 3 | 0 |
| | 0 | 2 | 1 | 0 | 0 | 0 | 0 | 0 | 2 | 1 |
| Kazacharthra | ? | ? | ? | ? | ? | ? | ? | ? | ? | ? |
| | 3 | ? | ? | ? | ? | ? | ? | ? | ? | ? |
| | ? | ? | ? | ? | ? | ? | ? | ? | ? | ? |
| | ? | ? | ? | ? | ? | ? | ? | ? | ? | ? |
| | ? | ? | ? | ? | ? | ? | 0 | 2 | 1 | 0 |
| | 2 | 5 | 0 | 2 | ? | - | - | - | ? | ? |
| | 1 | 0/1 | 0 | 0 | ? | 1 | 3 | 0 | 0 | ? |
| | ? | ? | ? | ? | ? | ? | ? | ? | ? | ? |
| | ? | ? | ? | ? | ? | ? | ? | ? | 2 | 1 |
| Anostraca | 0 | 1 | 2 | 1 | 1 | 1 | 0 | 1 | 3 | 1 |
| | 1 | 0 | 2 | 0 | 1 | 0 | 0 | 0 | 0 | 1 |
| | 0 | 1 | 1 | 1 | 1 | 2/3 | 4 | 3 | 1 | 2 |
| | 4 | 2 | 2 | ? | 3 | 3 | 3 | 1 | 0 | 3 |
| | 1 | 1 | 0 | 2 | 6 | 0 | 0 | 2 | 1 | 1 |
| | 2 | 5 | 0 | 2 | 0 | - | - | - | 4 | 5 |
| | 1 | 0 | 0 | 0 | 1 | 1 | 0 | 1 | 0 | 0 |
| | 0 | 0 | 1/3 | 0 | 0 | 1 | 1 | 0 | 3 | 0 |
| | 2 | 2 | 1 | 0 | 0 | 0 | 0 | 0 | 2 | 1 |
| *Rehbachiella* | 0 | 0 | 0 | 1 | 0 | 0 | 0 | 1 | 0 | 0 |
| | 0 | 0 | ? | ? | ? | ? | ? | ? | ? | ? |
| | ? | ? | 1 | 1 | 0 | ? | ? | 0 | 1 | 1 |
| | 0 | 1 | 1 | 0 | 1 | 1 | 2 | 1 | 0 | 3 |
| | 0 | 0 | 0 | 2 | 2 | 0 | 1 | 1 | 1 | 0 |
| | 2 | 1/2 | 0 | 0 | ? | 0 | ? | ? | ? | ? |

TABLE 6.3 *(continued)*

| | 1 | 2 | 3 | 4 | 5 | 6 | 7 | 8 | 9 | 10 |
|---|---|---|---|---|---|---|---|---|---|---|
| | 0 | 0 | 0 | 0 | 1 | 5 | 3 | 0 | 0 | 0 |
| | 0 | 0 | 3 | 0 | 0 | 1 | ? | ? | ? | ? |
| | ? | ? | ? | ? | ? | ? | ? | ? | 2 | 1 |
| Conchostraca | 0 | 1 | 2 | 1 | 1 | 0 | 0 | 1 | 0 | 1 |
| | 0 | 1 | 1 | 0 | 0 | 1 | 0 | 0 | 1 | 0 |
| | 0 | 1 | 1 | 1 | 1 | 3 | 4 | 3 | 1 | 2 |
| | 4 | 2 | 2 | 0 | 3 | 3 | 3 | 1 | 0 | 3 |
| | 1 | 1 | 0 | 2 | 6 | 0 | 0 | 2 | 1 | 1 |
| | 2 | 5 | 0 | ? | ? | ? | ? | ? | 4 | 4 |
| | 0 | 2 | 0 | 0 | 0 | 0 | 3 | 1 | 0 | 0 |
| | 0 | 0 | 1/3 | 0 | 0 | 1 | ? | 0 | 5 | 0 |
| | 2 | 2 | 1 | 0 | 0 | 0 | 0 | 0 | 2 | 1 |
| Cladocera | 0 | 1 | 2 | 1 | 1 | 0 | 0 | - | 3 | - |
| | 0 | 1 | 1 | 0 | 0 | 1 | 0 | 0 | 1 | 0 |
| | 0 | 1 | 1 | ? | 0 | 3 | 4 | 3 | 1 | 2 |
| | 4 | 2 | 2 | 0 | 3 | 3 | 3 | 1 | 0/1 | 0/3 |
| | 0/1 | 0/1 | 0/1 | 1/2 | 4/5 | 0 | 0 | 1 | 0 | 0 |
| | 1 | 3 | 0 | 2 | 0 | - | - | - | 0 | 0 |
| | 0 | 2 | 0 | 0 | 4 | 7 | 3 | 1 | 0 | 0 |
| | 0 | 0 | 1 | 0 | 0 | 1 | ? | 1 | 5 | 0 |
| | 2 | 2 | 1 | 0 | 0 | 0 | 0 | 0 | 2 | 1 |
| Ostracoda | 0 | 0 | 0 | 1 | 0 | 0 | 0 | 0 | 3 | 0 |
| | 0/3 | 0 | 1 | 0 | 0 | 0 | 1 | 0 | 0 | 1 |
| | 0 | 0/1 | 1 | 0 | 0 | 2 | 3 | 0 | 1 | 2 |
| | 2/3 | 1/2 | 1 | 0 | 2 | 1/3 | 2 | 1 | 0 | 0 |
| | 0 | 0 | 0 | 2/3 | 3/6 | 0 | 0 | 1 | 0 | 0 |
| | 4 | 2 | 0 | 2 | 0 | - | - | - | 2 | 1 |
| | 0 | 1/2 | 0 | 0 | 3 | 4/7 | 3 | ? | 0 | 0 |
| | 0 | 0 | 3 | 0 | 0 | 0 | 1 | 1 | 5 | 0 |
| | 2 | 0 | 0 | 0 | 0 | 0 | 0/1 | 3 | 2 | 1 |
| Mystacocarida | 0 | 0 | 0 | 1 | 1 | 0 | 0 | 1 | 0 | 0 |
| | 3 | - | 0 | 0 | 0 | 0 | 0 | 0 | 0 | ? |
| | ? | ? | 1 | 1 | 1 | 1 | 1 | 0 | 1 | 1 |
| | 0 | 2 | 1 | 0 | 2 | 1 | 3 | 1 | 1 | 3 |
| | 0 | 1 | 1 | 2/3 | 4/6 | 0 | 0 | 0 | 0 | 0 |
| | 4 | 5 | 0 | 2 | 0 | - | - | - | 3 | 3 |
| | 0 | 0 | 0 | 0 | 3 | 4 | 0 | 0 | 0 | 0 |
| | 0 | 0 | 1 | 0 | 0 | 0 | 1 | ? | 0 | ? |
| | 1 | 0 | 0 | 0 | 0 | 0 | 1 | 3 | 2 | 2 |
| Branchiura | 0 | 1 | 2 | 0/1 | 2 | 1 | 0 | - | 3 | - |
| | 1 | 1 | 1 | 0 | 0 | 0 | 1 | 0 | 0 | 0 |
| | 0 | 1 | 1 | - | 0 | 3 | 4 | 6 | 1 | 3 |
| | 5 | 2 | 1 | 0 | 3 | 1 | 3 | 1 | 0 | 0 |
| | 0 | 0 | 0 | 2 | 6 | 0 | 0 | 1 | 0 | 0 |
| | 2 | 5 | 2 | 2 | 0 | - | - | - | 3 | 3 |

TABLE 6.3 *(continued)*

| | 1 | 2 | 3 | 4 | 5 | 6 | 7 | 8 | 9 | 10 |
|---|---|---|---|---|---|---|---|---|---|---|
| | 0 | 0 | 0 | 0 | 4 | 6 | 1 | 0 | 0 | 0 |
| | 0 | 0 | ? | 0 | 0 | 0 | ? | 1 | 0 | 0 |
| | 1 | 1 | 0 | 0 | 0 | 0 | 0/1 | 3 | 2 | 1 |
| Tantulocarida | 0 | 1 | 2 | 0 | 2 | 2 | 1 | - | 3 | ? |
| | 3 | - | ? | ? | ? | ? | - | ? | ? | ? |
| | ? | ? | 0 | ? | 2 | 3 | 4 | 6 | 3 | 3 |
| | 4 | 2 | 2 | 0 | 3 | 3 | 3 | ? | 0 | 3 |
| | 0 | 0 | 0 | 2 | 6 | 0 | 0 | 1 | 1 | 0 |
| | 2 | 5 | 0/1 | 2 | ? | - | - | - | 2 | 2 |
| | 0 | 0 | 0 | 0 | 3 | 6 | 0 | 0 | 0 | 0 |
| | 0 | 0 | ? | 0 | ? | ? | ? | ? | ? | 0 |
| | ? | ? | ? | ? | ? | ? | ? | ? | 2 | 1 |
| Copepoda | 0 | 0 | 0 | 1 | 1 | 0 | 0 | 0 | 0 | 0 |
| | 3 | - | 1 | 0 | 0 | 0 | 1 | 0 | 0 | 0/1 |
| | 0 | 0/1 | 1 | 0 | 0 | 1/2 | 1/2 | 0/3/4 | 1 | 2 |
| | 2/3 | 1 | 1 | 0 | 2 | 2 | 3 | 1 | 1 | 3 |
| | 0 | 0 | 1/3 | 3 | 2 | 1 | 0 | 1 | 0 | 0 |
| | 0 | 3 | 1 | 2 | 0 | - | - | - | 2 | 1 |
| | 0 | 0 | 0 | 0 | 3 | 4 | 0 | 0 | 0 | 0 |
| | 0 | 0 | 3 | 0 | 0 | 1 | 1 | 1 | 0/3 | 0 |
| | 2 | 0/1/2 | 0/1 | 0 | 0 | 0/1 | 1 | 0/3 | 2 | 1 |
| Rhizocephala | 0 | 0 | 2 | 0 | 2 | 2 | 1 | 0 | 3 | 0 |
| | 3 | - | 0 | 0 | 0 | 0 | - | - | ? | ? |
| | ? | ? | 0 | - | - | 3 | 4 | 6 | 3 | 3 |
| | 4 | 2 | 2 | 0 | 3 | 3 | 3 | 1 | 0 | 0 |
| | 0 | 0 | 0 | - | - | 0 | 0 | 1 | 0 | 0 |
| | - | - | 0 | - | - | - | - | - | - | - |
| | 0 | 0 | 0 | 0 | 3 | 7 | 1 | 0 | 0 | 0 |
| | - | 0 | 2 | 0 | 1 | - | 1 | - | 3 | 1 |
| | 1 | 0 | 0 | ? | 0 | 0 | 0 | 0 | 2 | - |
| *Cyclus* | 0 | 0 | 0 | 1 | 1 | 1 | 0 | ? | 3 | ? |
| | 1 | 0 | ? | ? | ? | ? | ? | ? | ? | ? |
| | ? | ? | 1 | ? | 0 | ? | ? | 5 | 1 | ? |
| | 1 | ? | 1 | 0 | 3 | 1 | 3 | ? | 1 | 0 |
| | 0 | 0 | 3 | 3 | 2 | ? | 0 | 0 | 0 | 0 |
| | 4 | 0 | 0 | 2 | ? | - | - | - | ? | ? |
| | 0 | 0 | 0 | 0 | 3 | 6 | 1 | 0 | 0 | 0 |
| | 0 | 0 | ? | ? | ? | ? | ? | ? | ? | ? |
| | ? | ? | ? | ? | ? | ? | ? | ? | 2 | 2 |
| *Skara* | 0 | 0 | 0 | 1 | 0 | 0 | 0 | 1 | 0 | ? |
| | 3 | - | ? | ? | ? | ? | ? | ? | ? | 0 |
| | ? | ? | 1 | ? | 0 | 0 | 0 | 0 | 1 | 0 |
| | 1 | 1 | 1 | 0 | 2 | 2 | 2 | 1 | 1 | 3 |
| | 0 | 0 | 0 | 2 | 4 | 0 | ? | ? | ? | ? |
| | ? | ? | ? | 2 | ? | - | - | - | ? | ? |

TABLE 6.3 *(continued)*

| | 1 | 2 | 3 | 4 | 5 | 6 | 7 | 8 | 9 | 10 |
|---|---|---|---|---|---|---|---|---|---|---|
| | 0 | 0 | 0 | 0 | 3 | 2 | 0 | 0 | 1 | 0 |
| | 0 | 0 | ? | 0 | 0 | 0 | ? | ? | ? | ? |
| | ? | ? | ? | ? | ? | ? | ? | ? | 2 | 1 |
| *Bredocaris* | 0 | 0 | 0 | 1 | 0 | 0 | 0 | 0 | 0 | 0 |
| | 0 | 0 | 1 | ? | ? | ? | ? | ? | ? | ? |
| | ? | ? | 1 | 1 | 0 | 0 | 0 | 0 | 1 | 2 |
| | 0 | 1 | 1 | 0 | 2 | 1 | 2 | 1 | 0 | 3 |
| | 0 | 0 | 0 | 2 | 0 | 1 | 0 | 1 | 1 | 0 |
| | 2 | 2/5 | 0 | 2 | ? | - | - | - | ? | ? |
| | 0 | 0 | 0 | 0 | 3 | 6 | 0 | 0 | 0 | 0 |
| | 0 | 0 | 3 | 0 | 0 | 1 | ? | ? | ? | 0 |
| | ? | ? | ? | ? | ? | ? | ? | ? | 2 | 1 |
| Ascothoracida | 0 | 0 | 2 | 0 | 2 | 2 | 1 | 1 | - | - |
| | 3 | - | 1 | ? | ? | ? | ? | 0 | ? | 0 |
| | 0 | 1 | 1 | - | 1 | 3 | 4 | 6 | 1 | ? |
| | 4 | 2 | 1 | 0 | ? | 3 | 3 | 1 | 0 | 0 |
| | 0 | 0 | 0 | 1 | 4/5 | 0 | 1 | 1 | 0 | 0 |
| | 1 | 3/4 | 1 | 2 | 0 | - | - | - | 2 | 2 |
| | 0 | 0 | 0 | 0 | 3 | 4 | 3 | 0 | 0 | 0 |
| | 0 | 0 | 3 | 0 | 1 | 0 | 1 | ? | 5 | 1 |
| | 0 | 0 | 0 | 0 | 0 | 0 | 0 | 0 | 2 | 1 |
| Acrothoracica | - | - | - | 0 | 2 | 2 | 1 | 0 | - | 0 |
| | 3 | - | 1 | 0 | 0 | 0 | 1 | 0 | 0 | 0 |
| | 0 | 1 | 1 | 0 | 0 | 2 | 4 | 5 | 1 | 2 |
| | 4 | 2 | 1 | 0 | 2 | 3 | 3 | 1 | 1 | 0 |
| | 0 | 0 | 1 | 0 | 0 | 1 | 0 | 1 | 0 | 0 |
| | 3 | 0 | 1 | ? | ? | ? | ? | ? | 2 | 2 |
| | 0 | 1 | 0 | 0 | 3 | 7 | 3 | ? | ? | 0 |
| | 0 | ? | 2 | 0 | 1 | 0 | 1 | ? | 5 | 1 |
| | 1 | 0 | 0 | 0 | 0 | 0 | 0 | 3 | 2 | 1 |
| Thoracica | - | - | - | 0 | 2 | 2 | 1 | 1 | - | 0 |
| | 3 | - | 1 | 0 | 0 | 0 | 1 | 0 | 0 | 0 |
| | 0 | 1 | 1 | 0 | 0 | 2 | 4 | 5 | 1 | 2 |
| | 4 | 2 | 1 | 0 | 2 | 3 | 3 | 1 | 0 | 0 |
| | 0 | 0 | 0 | 0 | 0 | ? | 0 | 1 | 0 | 0 |
| | 3 | 0 | 1 | ? | ? | ? | ? | ? | 2 | 2 |
| | 0 | 1 | 0 | 0 | 3 | 7 | 3 | ? | ? | 0 |
| | 0 | ? | 2 | 0 | 1 | 0 | 1 | 1 | 5 | 1 |
| | 1 | 0 | 0 | 0 | 0 | 0 | 0 | 3 | 2 | 1 |
| Facetotecta | ? | ? | ? | ? | ? | ? | ? | 1 | ? | 0 |
| | ? | ? | ? | ? | ? | ? | ? | ? | ? | ? |
| | ? | ? | ? | 0 | ? | ? | ? | ? | ? | ? |
| | ? | ? | ? | ? | ? | ? | ? | ? | ? | ? |
| | ? | ? | ? | ? | ? | ? | ? | ? | ? | ? |
| | ? | ? | ? | ? | ? | ? | ? | ? | ? | ? |

TABLE 6.3 *(continued)*

| | 1 | 2 | 3 | 4 | 5 | 6 | 7 | 8 | 9 | 10 |
|---|---|---|---|---|---|---|---|---|---|---|
| | ? | ? | ? | ? | ? | ? | ? | ? | ? | ? |
| | ? | ? | 3 | 0 | 1 | 0 | 1 | ? | ? | 1 |
| | ? | ? | ? | ? | ? | ? | ? | ? | 2 | ? |
| Phosphatocopina | 0 | 0 | 0 | 1 | 0 | 0 | 0 | ? | 0 | ? |
| | 3 | - | ? | ? | ? | ? | ? | ? | ? | ? |
| | ? | ? | 1 | ? | 0 | 0 | 0 | 0 | 0 | 2 |
| | 1 | 0 | 0 | 0 | 2 | 2 | 0 | ? | 0 | 0 |
| | 0 | 0 | 0 | 2 | ? | ? | 0 | 0 | 0 | 0 |
| | ? | ? | ? | ? | ? | ? | ? | ? | ? | ? |
| | ? | ? | 0 | 0 | 0 | 0 | 1 | 0 | 0 | 0 |
| | 0 | 0 | ? | ? | ? | ? | ? | ? | ? | ? |
| | ? | ? | ? | ? | ? | ? | ? | ? | 2 | 1 |
| *Cambrocaris* | 0 | 0 | 0 | 0 | ? | 0 | 0 | ? | 0 | ? |
| | ? | ? | ? | ? | ? | ? | ? | ? | ? | ? |
| | ? | ? | 0 | ? | ? | 0 | 1 | 0 | 0 | 3 |
| | 0 | 0 | 0 | 0 | 3 | 1 | 0 | 0 | 0 | 0 |
| | 0 | 0 | 0 | 0 | ? | 0 | 0 | 1 | 0 | 0 |
| | 0 | 2 | 0 | ? | ? | ? | ? | ? | ? | ? |
| | ? | ? | ? | ? | 0 | 0 | ? | ? | ? | 0 |
| | ? | ? | ? | ? | ? | ? | ? | ? | ? | ? |
| | ? | ? | ? | ? | ? | ? | ? | ? | 2 | 1 |
| *Martinssonia* | 0 | 0 | 0 | 0 | 0 | 0 | 0 | 0 | 0 | ? |
| | 3 | - | ? | ? | ? | ? | ? | ? | ? | ? |
| | ? | ? | 0 | ? | ? | 0 | 0 | 0 | 0 | 2 |
| | 0 | 0 | 0 | 0 | 2 | 0 | 0 | 0 | 0 | 3 |
| | 0 | 0 | 0 | 0 | 1 | 0 | 0 | 0 | 0 | 0 |
| | 4 | 5 | 0 | 2 | ? | - | - | - | ? | ? |
| | 1 | 1 | 0 | 0 | 0 | 0 | 0 | 0 | 1 | 0 |
| | ? | 0 | 0 | ? | ? | 0 | ? | ? | ? | ? |
| | ? | ? | ? | ? | ? | ? | ? | ? | 2 | 1 |
| *Goticaris* | 0 | 0 | ? | 0 | 0 | 0 | 0 | ? | 0 | ? |
| | 0 | 0 | ? | ? | ? | ? | ? | ? | ? | ? |
| | ? | ? | 0 | ? | ? | 0 | 3 | 0 | 0 | 2 |
| | 0 | 0 | 0 | 0 | 3 | 1 | 3 | 0 | 0 | 0 |
| | 0 | 0 | 0 | ? | ? | 0 | 0 | 0 | 0 | 0 |
| | 4 | 2 | 0 | ? | ? | ? | ? | ? | ? | ? |
| | 3 | 1 | ? | ? | 4 | 0 | 0 | ? | 0 | 0 |
| | ? | 1 | ? | ? | ? | ? | ? | ? | ? | ? |
| | ? | ? | ? | ? | ? | ? | ? | ? | 2 | 1 |
| *Cambropachycope* | 0 | 0 | ? | 0 | 0 | 0 | 0 | ? | 0 | ? |
| | 0 | 0 | ? | ? | ? | ? | ? | ? | ? | ? |
| | ? | ? | 0 | ? | ? | 0 | 3 | 0 | 0 | 2 |
| | 0 | 0 | 0 | 0 | 3 | 1 | 3 | 0 | 0 | 0 |
| | 0 | 0 | 0 | ? | ? | 0 | 0 | 0 | 0 | 0 |
| | 4 | 2 | 0 | ? | ? | ? | ? | ? | ? | ? |

TABLE 6.3 *(continued)*

| | 1 | 2 | 3 | 4 | 5 | 6 | 7 | 8 | 9 | 10 |
|---|---|---|---|---|---|---|---|---|---|---|
| | 3 | 1 | ? | ? | 5 | 0 | 0 | 0 | 0 | 0 |
| | ? | 1 | ? | ? | ? | ? | ? | ? | ? | ? |
| | ? | ? | ? | ? | ? | ? | ? | ? | 2 | 1 |
| *Henningsmoenicaris* | 0 | 1 | ? | 0 | 0 | 0 | 0 | ? | 0 | ? |
| | 1 | 0 | ? | ? | ? | ? | ? | ? | ? | ? |
| | ? | ? | 0 | ? | ? | 0 | 2 | 0 | 0 | 2 |
| | 1 | 0 | 0 | 0 | 2 | 2 | 2 | 1 | 0 | 0 |
| | 0 | ? | 0 | 2 | ? | 0 | 0 | ? | ? | ? |
| | ? | ? | ? | ? | ? | ? | ? | ? | ? | ? |
| | 1 | 1 | ? | ? | 5 | 0 | 0 | 0 | 0 | 0 |
| | ? | 0 | 0 | 0 | 0 | ? | ? | ? | ? | ? |
| | ? | ? | ? | ? | ? | ? | ? | ? | 2 | 1 |
| *Odaraia* | 0 | 1 | 0 | 1 | 1 | 1 | 1 | ? | 3 | ? |
| | 1 | 0 | ? | ? | ? | ? | ? | ? | ? | ? |
| | ? | ? | 1 | ? | 0 | 3 | 4 | 3 | 1 | 3 |
| | 0 | 2 | 1 | 0 | 0 | 0 | 3 | 0 | 4 | 0 |
| | 0 | 0 | 1 | 3 | 6 | 0 | 0 | 1 | 0 | 1 |
| | 3 | 0 | ? | ? | ? | ? | ? | ? | ? | ? |
| | 3 | 0 | 0 | 0 | 0 | 0 | 3 | 0 | 0 | 0 |
| | 0 | 0 | ? | ? | ? | ? | ? | ? | ? | ? |
| | ? | ? | ? | ? | ? | ? | ? | ? | 2 | 1 |
| *Waptia* | 0 | 0 | 0 | ? | ? | ? | ? | ? | ? | ? |
| | 1 | 0 | ? | ? | ? | ? | ? | ? | ? | ? |
| | ? | ? | ? | ? | ? | ? | ? | ? | ? | ? |
| | ? | ? | ? | ? | ? | ? | ? | ? | ? | ? |
| | ? | ? | ? | ? | ? | ? | 0 | 1 | ? | ? |
| | 3 | ? | ? | 2 | ? | - | - | - | ? | ? |
| | 0 | 0 | 0 | 0 | 3 | 3 | 1 | 0 | 1 | 0 |
| | 0 | 0 | ? | ? | ? | ? | ? | ? | ? | ? |
| | ? | ? | ? | ? | ? | ? | ? | ? | 2 | 1 |
| *Plenocaris* | 0 | 0 | 0 | ? | ? | ? | ? | ? | ? | ? |
| | ? | ? | ? | ? | ? | ? | ? | ? | ? | ? |
| | ? | ? | ? | ? | ? | ? | ? | ? | ? | ? |
| | ? | ? | ? | ? | ? | ? | ? | ? | ? | ? |
| | ? | ? | ? | ? | ? | ? | 0 | ? | ? | ? |
| | ? | 2 | ? | ? | ? | ? | ? | ? | ? | ? |
| | 0 | 0 | 0 | 0 | 4 | 1 | 1 | 0 | 2 | 0 |
| | 0 | 0 | ? | ? | ? | ? | ? | ? | ? | ? |
| | ? | ? | ? | ? | ? | ? | ? | ? | 2 | 1 |
| *Branchiocaris* | 0 | 0 | 0 | 1 | 0 | 1 | 0 | ? | 3 | ? |
| | 3 | - | ? | ? | ? | ? | ? | ? | ? | ? |
| | ? | ? | 0 | ? | ? | ? | ? | ? | 0 | 3 |
| | 0 | ? | 0 | 0 | 0 | 0 | 2 | 0 | 0 | 0 |
| | 0 | 0 | 0 | ? | 0 | 0 | 0 | 0 | 0 | 0 |
| | 2 | 0 | 0 | ? | ? | ? | ? | ? | ? | ? |

TABLE 6.3 *(continued)*

| | 1 | 2 | 3 | 4 | 5 | 6 | 7 | 8 | 9 | 10 |
|---|---|---|---|---|---|---|---|---|---|---|
| | 0 | 0 | 0 | 0 | 0 | 0 | 3 | 1 | 0 | 0 |
| | 0 | 0 | ? | ? | ? | ? | ? | ? | ? | ? |
| | ? | ? | ? | ? | ? | ? | ? | ? | 2 | 0 |
| *Marrella* | 0 | 0 | 0 | 1 | 1 | 1 | 0 | ? | 3 | ? |
| | 3 | ? | ? | ? | ? | ? | ? | ? | ? | ? |
| | ? | ? | 1 | ? | 0 | 0 | 0 | 0 | 0 | 3 |
| | 0 | ? | 0 | 0 | 0 | 1 | 0 | 0 | 0 | 0 |
| | 0 | 0 | 0 | 0 | 0 | 0 | 0 | 1 | 0 | 0 |
| | 3 | 1 | 0 | ? | ? | ? | ? | ? | ? | ? |
| | 0 | 1 | 0 | 0 | 0 | 0 | 0 | 0 | 0 | 0 |
| | 0 | 0 | ? | ? | ? | ? | ? | ? | ? | ? |
| | ? | ? | ? | ? | ? | ? | ? | ? | 2 | 1 |
| *Tesnusocaris* | 1 | 0 | 0 | 1 | 1 | 0 | 0 | ? | 1 | ? |
| | 0 | 0 | ? | ? | ? | ? | ? | ? | ? | ? |
| | ? | ? | 1 | ? | 0 | 3 | 4 | 3 | 2 | 2 |
| | 0 | 2 | 1 | 0 | 3 | 0 | 3 | 1 | 1 | 0 |
| | 0 | 0 | 3 | 0 | 0 | 1 | 0 | 0 | 0 | 0 |
| | 0 | 0 | 0 | ? | ? | ? | ? | ? | ? | ? |
| | 0 | 3 | 0 | 0 | 0 | 0 | 0 | 0 | 0 | ? |
| | ? | 0 | ? | ? | ? | ? | ? | ? | ? | ? |
| | ? | ? | ? | ? | ? | ? | ? | ? | 2 | 0 |

taxon Phyllopoda was suggested by the analysis of Schram (1986). Although the overall tree is different from that of Schram (1986), one can still discern the major groups of that earlier analysis: Malacostraca (although without the phyllocarids), Phyllopoda, Maxillopoda (albeit paraphyletic), and Remipedia near the base of the tree.

We next analyzed the matrix with all taxa included, fossil and Recent. We obtained 9,700 trees before an overflow occurred with trees of length 614 (CI = 0.316, HI = 0.684, RI = 0.653, RC = 0.206). We present the 50% majority rule in figure 6.8. This tree differs in regard to the indicated relationships of the modern groups seen in figure 6.6. The Remipedia still remain near the base of the crustacean crown group, but the next clade appearing on the tree is one that contains most of the phyllopods (namely, the Devonian fossil *Lepidocaris,* the cephalocarids, and the branchiopods). The Maxillopoda then emerge paraphyletically as two clades. The first clade contains essentially the Thecostraca (barnacles and allies) plus *Cyclus* and two Burgess Shale arthropods (*Waptia* and *Plenocaris*). The second clade has ostracodes and copepod-like taxa (mystacocarids and copepods) plus three Cambrian "Orsten" genera (*Skara, Bredocaris,* and *Rehbachiella*). One might have expected the latter two "Orsten" forms in this clade, but the occurrence of *Rehbachiella* here is at odds with what Walossek (1993) suggested. The Malacostraca *s.l.* then appear in a traditional arrangement with the Phyllocarida as a sister group to all other malacostracans. The syncarids emerge as clearly paraphyletic, and the stomatopods appear *within* the peracarids.

Quite surprising is the location of the majority of the Cambrian "Orsten" and Burgess Shale arthropods in a separate clade intermediate between the outgroup and the crown-group Crustacea. This clade contains the Euthycarcinoidea, which despite our command to treat this as an outgroup persistently appeared among these Cambrian forms. However, we must point out the low frequency value of only 52% for this clade. Figure 6.9 illustrates one possible variant on themes seen among the other 48% of the trees with a clear stem-group transition series of Cambrian arthropods as predicted by Walossek and Müller (1990).

The abrupt shifts of phyllocarids and stomatopods, the peculiar occurrences of some fossil groups, and the clumping of Cambrian forms in their own clade may in large part be due to what we call the *vraagteken* effect (from the Dutch for "question mark"), which we apply to the effect of missing data and inapplicable characters. To try and evaluate this, we analyzed the data excluding some of the soft anatomy (table 6.2), thus removing a large amount of uncertainty evident in the data set. The results (fig. 6.10) in part confirm our suspicion. The 50% majority rule now favors a transition series of stem forms with fairly high percentages of frequency of occurrence. Other interesting changes appear in this cladogram. The Phyllopoda shift again to a position above the paraphyletically arranged clades of Maxillopoda. Phyllocarida still reside within the Malacostraca but not as the basal-most clade within that group. The Hoplocarida now appear in a polytomy with some other fossils as a potential sister group to the Eumalacostraca, but two of the syncarid groups (Anaspidacea and Palaeocaridacea) are now among the peracarids. Clearly, while our data matrix can produce reasonable results, the nature of the fossils and the fact that the features themselves are not applicable across all taxa induce instability. Many clades appear with a high degree of frequency among the total array of clades in the trees produced by the matrix.

To obtain a more complete picture of the dynamics of our database, we also analyzed the fossils alone utilizing the complete character set (fig. 6.11). Interestingly, this cladogram mirrors the basic framework of the entire data set. The 50% majority rule consensus tree was derived from 10 000 equally parsimonious trees and contains many clades with a low percentage of occurrence. Even so, it reveals a pattern of possible relationships akin to that already seen for the entire character set. A stem-group transition series of Burgess Shale and "Orsten" arthropods leads to a crown-group Crustacea with the fossil remipede, *Tesnusocaris*, arising first, followed by a phyllopod/maxillopodan clade and then a clade containing an irresolution of Burgess arthropods that often appear among maxillopodans in other analyses and a malacostracan clade that includes phyllocarids and a clade of hoplocaridans and eumalacostracans. These trees have a length of 218, with the CI = 0.498, HI = 0.530,

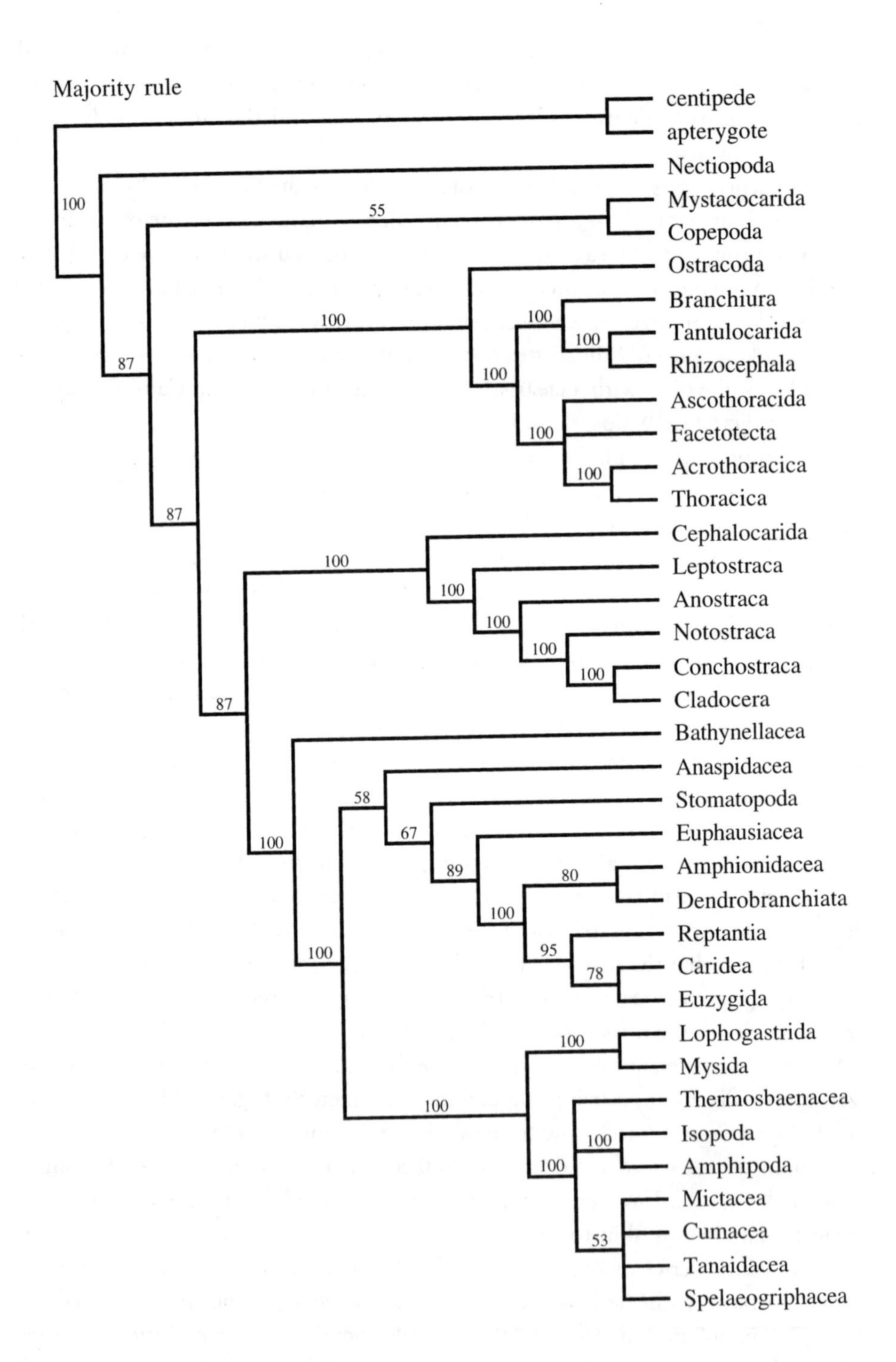


FIGURE 6.6

50% majority rule consensus of 165 trees that resulted from the analysis of our data set of 90 characters but using only the 35 taxa that are not exclusively fossil groups. Length = 553, CI = 0.541, HI = 0.673, RI = 0.657, and RC = 0.355.

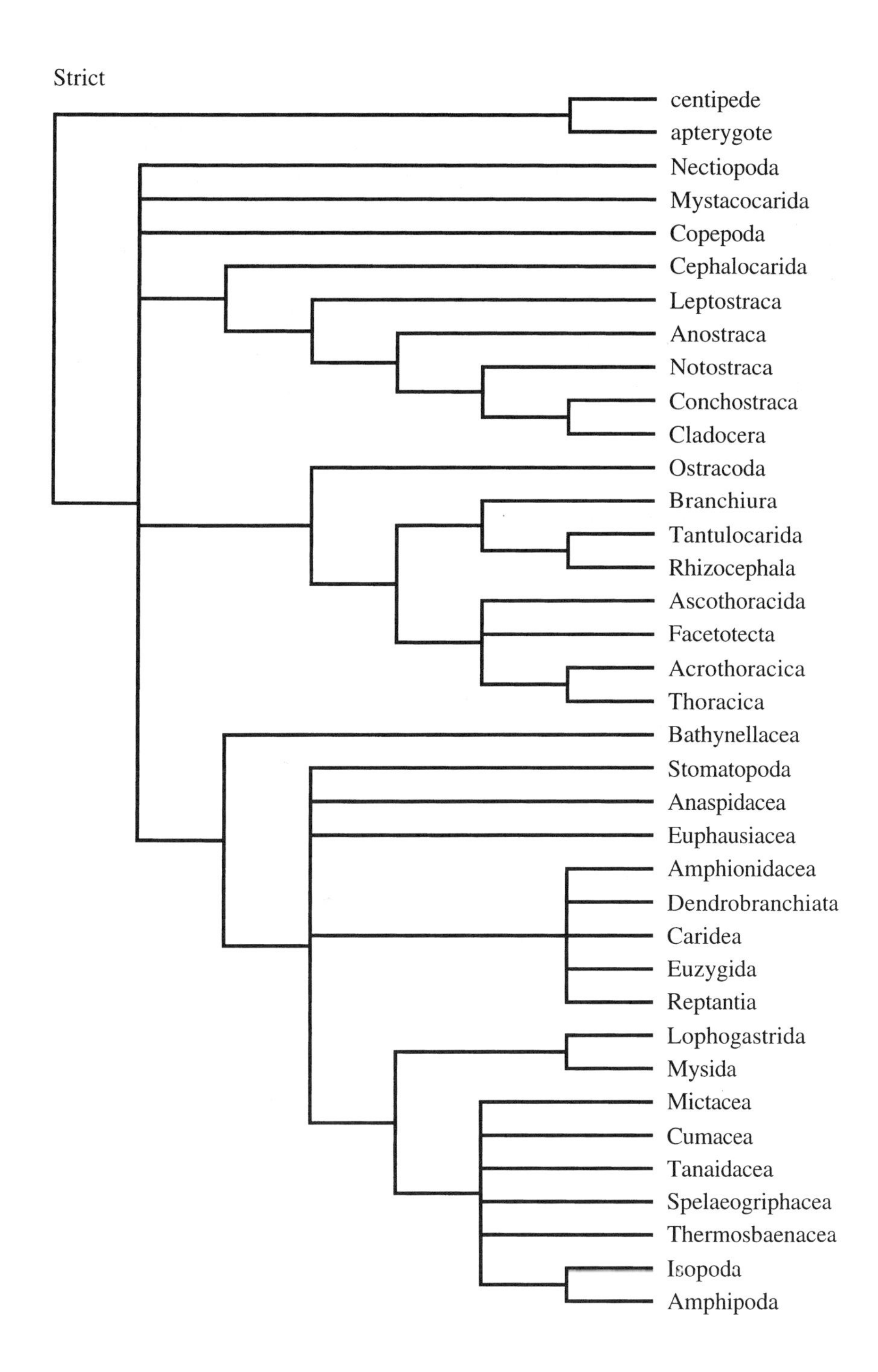


FIGURE 6.7.

Strict Consensus of 165 trees resulting from the analysis of 90 characters using only the 35 taxa that are not exclusively fossil groups (cf. fig. 6.6).

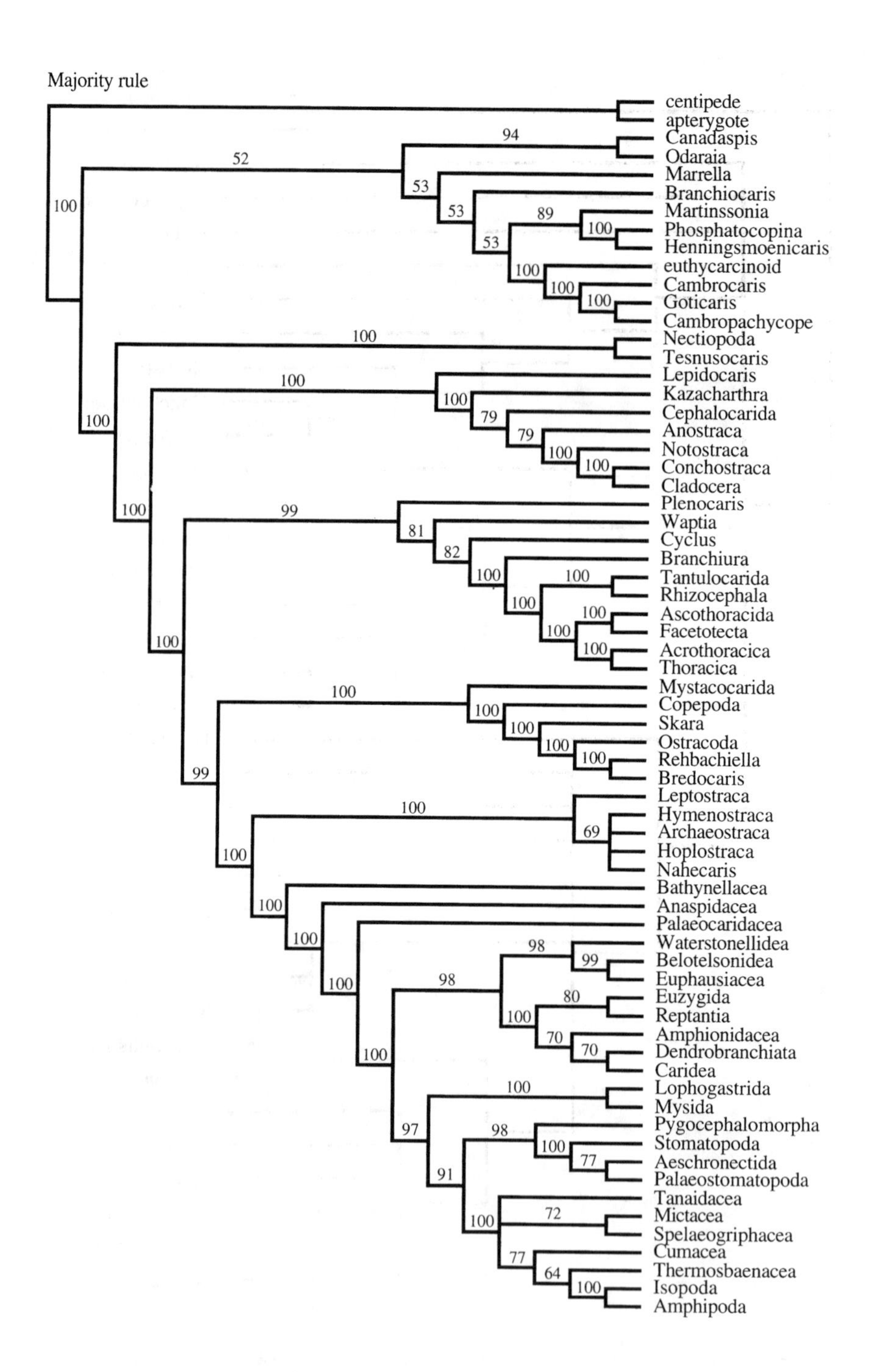


FIGURE 6.8

50% majority rule consensus derived from 9700 trees that resulted from analysis of our entire data set including 90 characters and 67 taxa. This majority rule consensus favored placing most of the Cambrian taxa in a single clade in 52% of the trees. Length = 614, CI = 0.316, HI = 0.684, RI = 0.653, RC = 0.206.

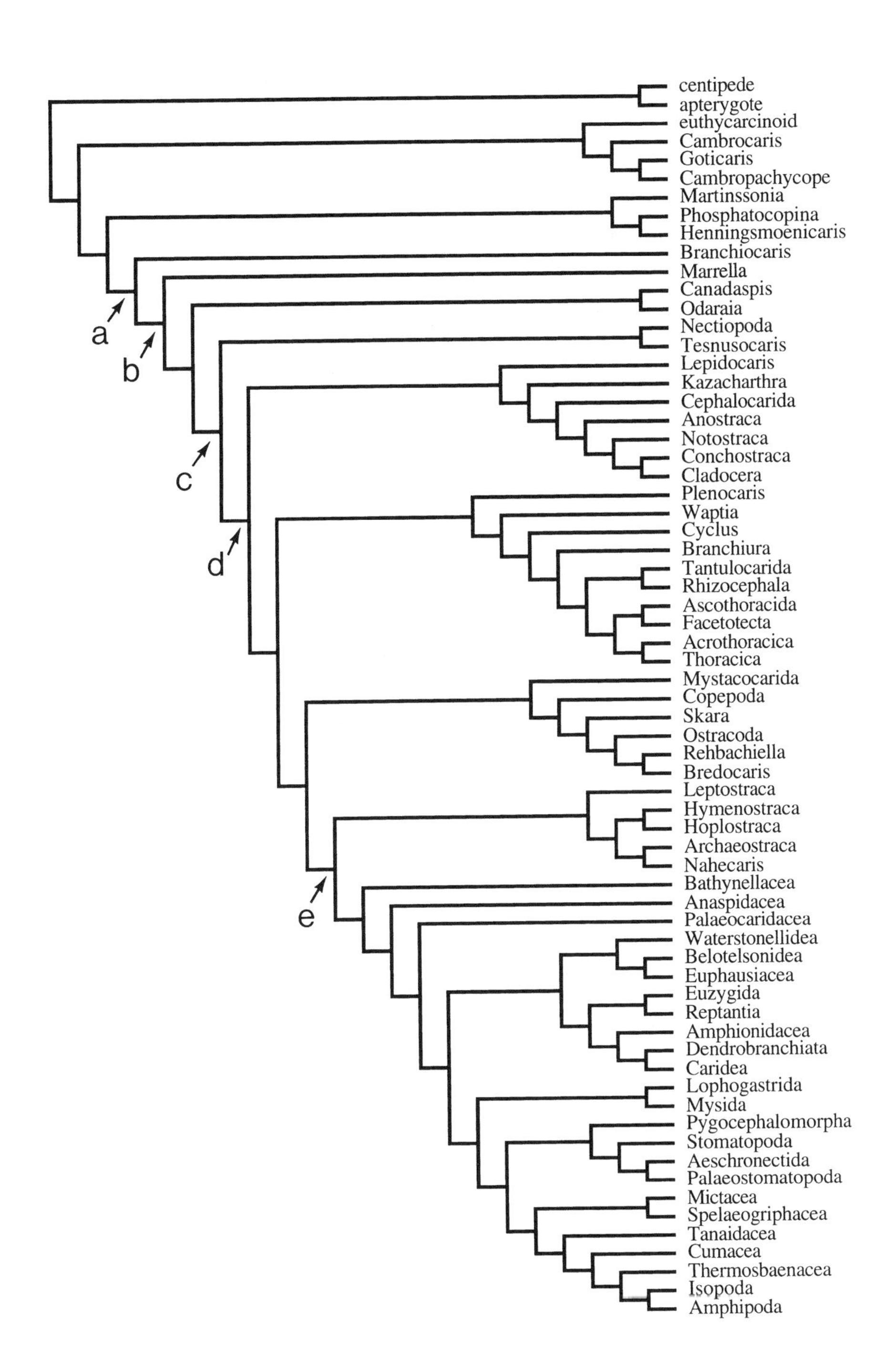


FIGURE 6.9

One of the 9700 cladograms from analysis of the entire data set (cf. fig. 6.8) that displays a variant with a Cambrian stem-group transition series. Points *a–e* mark stages in the evolution of the crown-group crustacean defining features; see text for discussion.

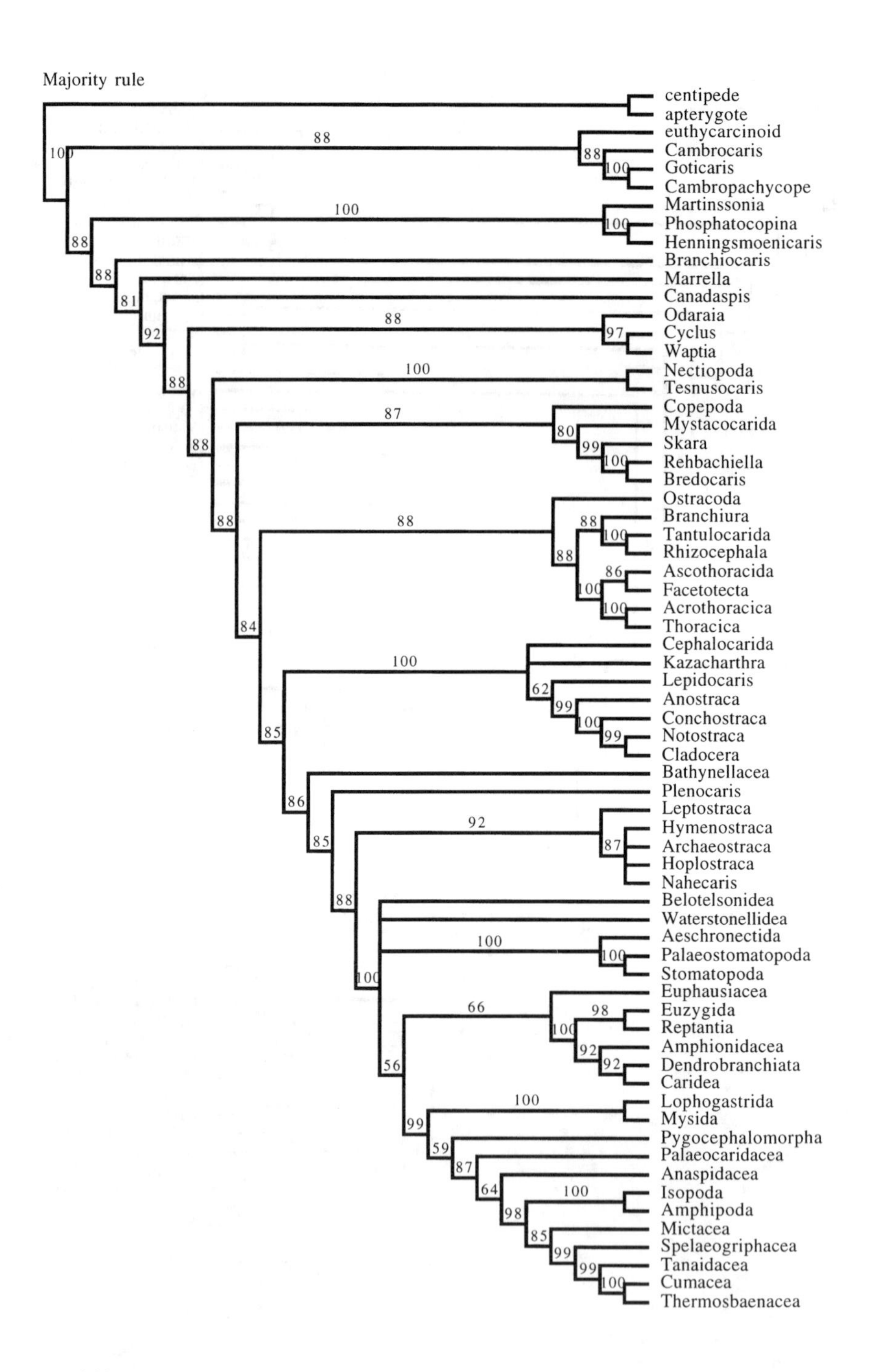


FIGURE 6.10

50% majority rule consensus derived from 9800 trees that resulted from our analysis when soft-anatomy characters are eliminated (see table 6.2). Length = 535, CI = 0.32, RI = 0.66, RC = 0.21. Note the higher percentage values for the Cambrian stem-group transition series than in figure 6.8 and the shifting positions of the Hoplocarida (Stomatopoda), Maxillopoda, and Phyllopoda clades. See text for discussion.

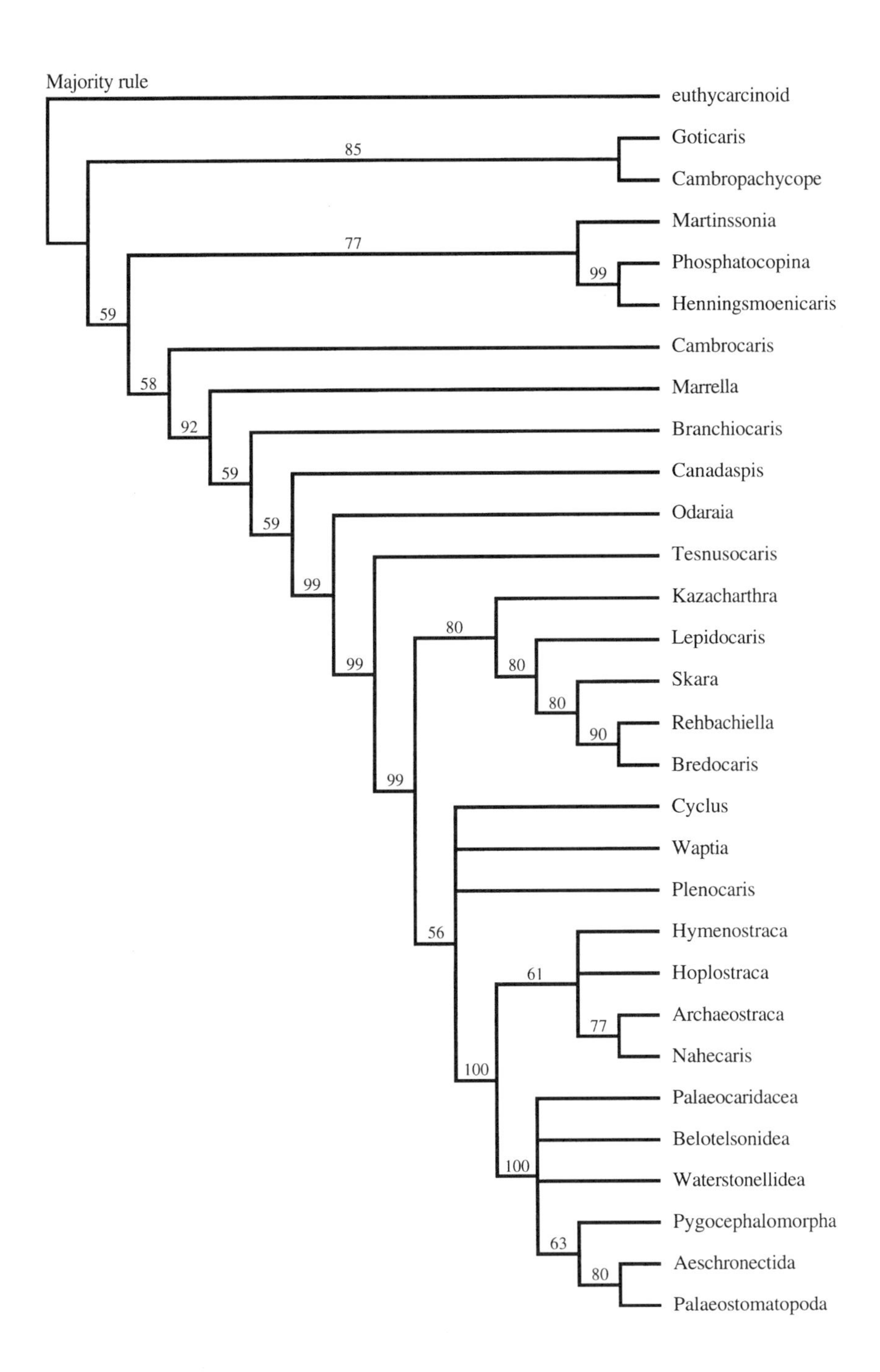


FIGURE 6.11

50% majority rule consensus derived from 10 000 trees when only the entirely extinct groups were analyzed. Length = 219, CI = 0.498, HI = 0.502, RI = 0.646, RC = 0.322. Note that despite the high degree of missing data in this analysis, the basic framework of the cladogram evident in other analyses in this study is confirmed. See text for discussion.

RI = 0.641, and RC = 0.319. While this analysis is not particularly robust and is subject to the great uncertainties engendered in the fossil part of the data matrix where there are a lot of gaps in the character scorings, nevertheless, the incipient pattern reinforces those seen in the somewhat more informative analyses above.

## Discussion

We can come to several conclusions. First, we believe that this database offers some substantial improvements over what has appeared in the literature to date. The inconsistencies of some of the way characters were handled in Schram (1986) have been removed. The limited focus of the character list of Wilson (1992) has been expanded. The result is a series of trees that appear "reasonable." Such a characterization is, of course, quite subjective, and begs the question, "What is reasonable?" However, there is some merit in being able to recognize many of the groups that occur in the tree. It provides some historical substance to the analysis. A difficulty with the single tree provided by Wilson (1992) was that some of its postulated relationships were totally unexpected; remipedes and cephalocarids among the maxillopodans, ostracodes allied with branchiopods and malacostracans, and *Canadaspis* as a sister group to ostracodes. Such a tree suggested that some serious mistakes had been made in the last one hundred years in the study of crustacean phylogeny. It is somewhat reassuring to realize that maybe this was not the case.

"Comfortableness," however, should not be a prime criterion for evaluating phylogenetic analyses. There is enough in the tree of figure 6.6 to arouse controversy. That Isopoda and Amphipoda continue to emerge as sister groups *must* give some pause to arguments against their alignment into the Edriophthalma. Wagner (1994) obtained a similar result. Clearly this issue of isopod/amphipod relationships now must be addressed more seriously.

The location of the hoplocaridans so far up into the eumalacostracan clade (figs. 6.8, 6.9) is very much at odds with what all previous authorities in one form or another have postulated (e.g., Calman 1909; Schram 1978, 1986; Dahl 1983c; Hessler 1983; Kunze 1983). This is undoubtedly due in part to the effects of missing data. When we removed features of soft anatomy so as to remove question marks from the data set, the Hoplocarida clade shifted to an unresolved position near the base of what we might characterize as the Malacostraca *s.str.* (without Bathynellacea in this case). Manipulation of the complete data set in MacClade revealed that to have a tree with hoplocaridans as a sister group to eucaridans and peracaridans requires six more steps, e.g., a tree that is about 1% longer than the most parsimonious — not that much really. So there is really no good reason to reject a possible sister group status of Hoplocarida and Eumalacostraca. However, clearly the issue bears further study.

Syncarida emerging as a paraphyletic group is probably due to a real effect of the data. The presence or absence of a carapace, when analyzed in the context of all crustaceaform arthropods reveals a character with a high degree of homoplasy. The lack of a carapace in syncarids is a plesiomorphic condition in this regard and not an apomorphic feature that could serve to unite the syncarid groups.

We also probably can conclude from this analysis that the ideas of Walossek and Müller (1990) concerning the stem-group transition to a crown-group Crustacea has considerable merit, although the issue is still somewhat clouded. Indeed, we can recognize in some instances some clear stages in the evolution through a stem-group transition series toward a crown-group Crustacea. The first important stage occurs at the point where *Branchiocaris* appears on the tree (point "a" in fig. 6.9) when apparently a distinct second antenna (character 4) arose. The next higher dichotomy at the level of *Marrella* (point "b" in fig. 6.9) marks the appearance of a labrum (character 23). The issue of just when the caudal rami flanking the terminal anus evolved (character 62) is not as clear as the discussion of Walossek and Müller (1990) implies. Although this feature appears at the base of the crown-group Crustacea (point "c" in fig. 6.9), it also occurs among some of the stem-group forms in the Burgess Shale arthropods *Branchiocaris* and *Odaraia*.

The crown-group Crustacea, however, is clearly marked by the appearance of several good characters. At this bifurcation (point "c" in fig. 6.9), the maxillule (character 29) and maxilla (character 33) occur as mouthparts (convergent to a similar situation lower in the tree in the outgroup centipede/apterygote clade), as does the fusion of the maxillary segment into the head (character 38). So despite the reservations and ambiguity noted above with regard to the five-segment crustacean head, it would seem that this complex of features is indeed diagnostic for crown Crustacea, co-occurring as it does with other crucial defining features of the group. In addition, this point "c" on the cladogram marks the stage of fixation of the female gonopores at distinctive midpoints on the body (character 60), although the status of this character is equivocal below this point because of lack of any knowledge as to gonopore locations in the Cambrian Burgess Shale and "Orsten" fossils.

The issue of the first appearance of the biramous antennules (character 1) has two possible interpretations. It could be viewed as a diagnostic feature appearing at point "c" with the other characters just noted. In that case, a character reversal occurs almost immediately to be re-reversed higher up in the clade at point "e" (fig. 6.9). Conversely, the feature could also be viewed as appearing independently in the lines leading to the remipedes and the Malacostraca and thus as not diagnostic for Crustacea *per se*.

Another critical point occurs within the crown group (point "d" in fig. 6.9). Several important features apparently became fixed at that stage. The most significant of these relates to the nauplius (character 73). At that point, only the egg nauplius appears. Since we do not know the state of this feature with regard to the remipedes, it seems that at this stage in our understanding it might be prudent to view the orthonauplius larva itself as a feature that appears independently in a number of groups, namely, in various maxillopodan clades and some eucarid eumalacostracans. At this same point in the clade, however, we can note the appearance of a regionalized trunk with distinct thorax (character 65) and abdomen (character 66). Finally, this point sees a fixation of the male gonopores (character 59) in relation to that of the female, but the location of this pore is a dynamic one that shifts repeatedly in special cases higher up in the crustacean clade. Thus it seems clear that the evolution of "defining" crustacean characters, both in the sense of Walossek and those we ourselves recognized above, is a gradual one that extends into the crown Crustacea clade.

We believe, finally, that the fossils definitely are necessary for the elucidation and understanding of crustacean relationships. We might draw conclusions without the fossils that would be quite different from those we derive with them; or conversely, if we focus too much on fossils and do not consider the whole array of crustaceans we can also come to different conclusions. This can be seen when considering alternative hypotheses for individual group or species affinities (table 6.4). We can manipulate the cladograms of figures 6.8 and 6.9 in MacClade and test the effects of alternative relationships on the overall length of the tree. We have already alluded to the fact that if we moved Hoplocarida to a sister group position with Eumalacostraca, six more steps would be added. Likewise, creation of a clade Phyllopoda (in the sense of Schram 1986) would involve seven extra steps; placing Pygocephalomorpha in closer proximity to the Lophogastrida and Mysida entails three extra steps. Forming a single clade of Maxillopoda would add only two more steps; making Cephalocarida a separate clade just above the Remipedia takes three additional steps. Finally, *Rehbachiella* is placed by Walossek (1993) among the branchiopods, a position not found in this study. To move it there instead of where it occurs on our trees among maxillopodans would entail twelve extra steps.

Even so, we still have some serious problems. The *vraagteken* effect — missing data — is one we must deal with in the future in some way. It is not clear at this point whether the total morphospace occupied by the entire array of Recent and fossil forms really is substantially bigger than that occupied by the crown-group alone. While the alignment of many of these forms here would seem to confirm in part the suggestions concerning stem-group crustaceans put forth by Walossek and Müller (1990) and Walossek (1993), other interpretations are equally likely.

There are many things we could and indeed shall do but have not had the time to pursue here. Selectively weighting and ordering characters might prove instructive. Some preliminary experiments by us along these lines have indicated intriguing results, but clearly much more needs to be carefully undertaken in this regard. Another avenue that needs to be explored is the degree of homoplasy evident in the data. Consistency indices hovering around 0.5 indicate that caution needs to be used in interpreting the trees. Testing the results after deleting characters that are too homoplastic would be a first step in this direction. Even so, we should examine such characters closely to determine whether they are in fact homoplastic, or whether they might currently combine in fact distinct features. Heijerman (1992, 1993) found that if homoplasy is not too high, then parsimony methods such as PAUP can produce reliable results. However, if the consistency of the data is too low, he found that more numerical taxonomic methods of analysis such as UPGMA (unweighted pair-group methods using arithmetic averages) appear more reliable in producing trees that may approach what the "real" tree may have been like.

We also must be alert to the discovery of new and potentially interesting features that could elucidate crustacean relationships and add them into future versions of

TABLE 6.4
Overall parsimony of the cladograms when exploring the trees of figure 6.8 with McClade

| FORCED CLADE | NEW LENGTH | ADDED STEPS |
|---|---|---|
| Hoplocarida/Eumalacostraca | 620 | 6 |
| Pygocephalomorpha/Mysida/Lophogastrida | 617 | 3 |
| Phyllopoda (*sensu* Schram 1986) | 621 | 7 |
| Cephalocarida (as a distinct clade) | 617 | 3 |
| A single clade Maxillopoda | 616 | 2 |
| *Rehbachiella* into Branchiopoda (Walossek 1993) | 626 | 12 |

*Note:* Certain traditional clades do not appear in our trees. Forcing them adds extra steps to the basic trees here of length 614.

the matrix. We have acted here to merge data based on naupliar eye and frontal organs (Elofsson 1963, 1965, 1966), sperm (Jamieson 1991 and other authors), strong emphasis on mouthparts (Wilson 1992), and larvae and body tagma (Schram 1986). It is *not* time to propose any new classification of the Crustacea. To do so might result in some interesting but not particularly acceptable taxonomic scheme (e.g., see Starobogatov 1988). We believe that we do not yet have the definitive database. Further investigations of the development, larvae, and internal soft anatomy still have much to tell us. Molecules will have much to contribute in this regard as well. The results of Wheeler et al. (1993) remain an object lesson as to what happens when one merges molecular and traditional morphological data into a single database.

In short, our effort here serves only to organize the information we now have at hand. The science of crustacean studies has moved rapidly in the last several years and does not show any signs of slowing down. Contrary to what some have written, cladograms do not obscure but in fact utilize the fundamental biological attributes of animals. Problems arise only when we focus on what we may believe are the "important" features in crustacean evolution to the exclusion of the entire array of data available to us. In this regard, the fossils have served, and will continue to serve, as important sources of information.

## ACKNOWLEDGMENTS

We would like to thank the following individuals for reading earlier drafts of this manuscript: Profs. Barry Jamieson and Dieter Walossek, Drs. Mark Grygier, Jens Høeg, and George Wilson. We welcomed their suggestions for improvements, but this does not imply, of course, that they necessarily agreed with our conclusions. Dr. Willem Ellis checked our cladistic analyses.

## REFERENCES

Abele, L. G., ed. 1982. *The Biology of Crustacea.* Vol. 1, *Systematics, the Fossil Record, and Biogeography.* New York: Academic Press.

Abele, L. G. 1991. Comparison of morphological and molecular phylogeny on the Decapoda. *Memoirs of the Queensland Museum* 31:101–108.

Abele, L. G., M. Applegate, and T. Spears. 1990. Molecular phylogeny of the Crustacea based on 18S rRNA and PCR-amplified data rDNA nucleotide sequences. *American Zoologist* 30 (4): 6A.

Abele, L. G., W. Kim, and B. E. Felgenhauer. 1989. Molecular evidence for inclusion of the phylum Pentastomida in the Crustacea. *Molecular Biology and Evolution* 6:685–691.

Abele, L. G., T. Spears, W. Kim, and M. Applegate. 1992. Phylogeny of selected maxillopodan and other crustacean taxa based on 18S ribosomal nucleotide sequences: A preliminary analysis. *Acta Zoologica* 73:373–382.

Anderson, D. T. 1973. *Embryology and Phylogeny in Annelids and Arthropods.* Oxford: Pergamon.

Ax, P. 1985. Stem species and the stem lineage concept. *Cladistics* 1:279–287.

Bachmann, K. and E. L. Rheinsmith. 1973. Nuclear DNA amounts in Pacific Crustacea. *Chromosoma* 43:225–236.

Barnes, R. D. and F. W. Harrison. 1992. Introduction. In F. W. Harrison and A. G. Humes, eds., *Microscopic Anatomy of the Invertebrates,* 9:1–8. New York: Wiley-Liss.

Bergström, J. 1992. The oldest arthropods and the origin of the Crustacea. *Acta Zoologica* 73:287–291.

Borradaile, L. A. 1917. On the structure and function of the mouthparts of the palaemonid prawns. *Proceedings of the Zoological Society of London* 1917:69–72.

Borradaile, L. A. 1926. Notes upon crustacean limbs. *Annals and Magazine of Natural History* (9) 17:193–213.

Boxshall, G. A. and R. Huys. 1989. A new tantulocarid, *Stygotantulus stocki,* parasitic on harpacticoid copepods, with an analysis of the phylogenetic position within the Maxillopoda. *Journal of Crustacean Biology* 9:126–140.

Boxshall, G. A., J.-O. Strömberg, and E. Dahl, eds. 1992. The Crustacea: Origin and Evolution. Proceedings of a workshop held at Kristeneberg Marine Biological Station, The Royal Swedish Academy of Sciences, 22–28 September 1990. *Acta Zoologica* 73:271–392.

Briggs, D. E. G. 1976. The arthropod *Branchiocaris* n. gen., Middle Cambrian Burgess Shale, British Columbia. *Geological Survey of Canada Bulletin* 264:1–29.

Briggs, D. E. G. 1978. The morphology, mode of life, and affinities of *Canadaspis perfecta* (Crustacea: Phyllocarida), Middle Cambrian, Burgess Shale, British Columbia. *Philosophical Transactions of the Royal Society of London,* series B, 281:439–487.

Briggs, D. E. G. 1981. The arthropod *Odaraia alata* Walcott, Middle Cambrian, Burgess Shale, British Columbia. *Philosophical Transactions of the Royal Society of London,* series B, 291:541–585.

Briggs, D. E. G. 1983. Affinities and early evolution of the Crustacea: The evidence of the Cambrian fossils. In F. R. Schram, ed., *Crustacean Issues.* Vol. 1, *Crustacean Phylogeny,* pp. 1–22. Rotterdam: Balkema.

Briggs, D. E. G. 1990. Early arthropods: Dampening the Cambrian radiation. In D. Mikulic, ed., *Short Courses in Paleontology.* Vol. 3, *Arthropod Paleobiology,* pp. 1–17. Knoxville: Paleontological Society.

Briggs, D. E. G. 1992. Phylogenetic significance of the Burgess Shale crustacean *Canadaspis.* *Acta Zoologica* 73:293–300.

Briggs, D. E. G. and R. A. Fortey. 1989. The early radiation and relationships of the major arthropod groups. *Science* 246:241–243.

Briggs, D. E. G., R. A. Fortey, and M. A. Wills. 1992. Morphological disparity in the Cambrian. *Science* 256:1670–1673.

Briggs, D. E. G., R. A. Fortey, and M. A. Wills. 1993. How big the Cambrian explosion? A taxonomic and morphologic comparison of Cambrian and Recent arthropods. In D. R. Lees and D. Edwards, eds., *Evolutionary Patterns and Processes,* pp. 33–44. London: Linnean Society Symposium Series, Academic Press.

Briggs, D. E. G. and R. A. Robison. 1984. Exceptionally preserved nontrilobite arthropods and *Anomalocaris* from the Middle Cambrian of Utah. *University of Kansas Paleontological Contributions* 111:1–23.

Briggs, D. E. G. and H. B. Whittington. 1981. Relationships of arthropods from the Burgess Shale and other Cambrian sequences. In M. E. Taylor, ed., *Short Papers for the Second International Symposium on the Cambrian System: United States Department of the Interior, Geological Survey Open-File Report* 81–743:38–41.

Brooks, H. K. 1962. The Paleozoic Eumalacostraca of North America. *Bulletins of American Paleontology* 44:160–338.

Brusca, R. C. and G. D. F. Wilson. 1991. A phylogenetic analysis of the Isopoda with some classificatory recommendations. *Memoirs of the Queensland Museum* 31:143–204.

Burkenroad, M. D. 1981. The higher taxonomy of the Decapoda. *Transactions of the San Diego Society of Natural History* 19:251–268.

Burnett, B. R. 1987. The cirripede circulatory system and its evolution. In A. J. Southward, ed., *Crustacean Issues.* Vol. 5, *Barnacle Biology,* pp. 175–190. Rotterdam: Balkema.

Calman, W. T. 1909. Crustacea. In R. Lankester, ed., *A Treatise on Zoology, Part 7: Appendiculata,* 3:1–346. London: Adam & Charles Black.

Cannon, H. G. and S. M. Manton. 1927. On the feeding mechanism of a mysid crustacean, *Hemimysis lamornae. Transactions of the Royal Society of Edinburgh* 55:355–369.

Casanova, B. 1993. L'origine protocéphalique de la carapace chez les Thermosbaenacés, Tanaïdacés, Cumacés, et Stomatopodes. *Crustaceana* 65:144–150.

Cisne, J. L. 1981. *Triarthrus eatoni* (Trilobita): Comparative anatomy of its exoskeletal, skeletomuscular, and digestive systems. *Palaeontographica Americana* 9:95–142.

Cisne, J. L. 1982. Origin of Crustacea. In L. G. Abele, ed., *The Biology of Crustacea.* Vol. 1, *Systematics, the Fossil Record, and Biogeography,* pp. 65–92. New York: Academic Press.

Collins, D. H. and D. M. Rudkin. 1981. *Priscansermarinus barnetti,* a probable lepadomorph barnacle from the Middle Cambrian Burgess Shale of British Columbia. *Journal of Paleontology* 55:1006–1015.

Dahl, E. 1956. Some crustacean relationships. In K. G. Wingstrand, ed., *Bertil Hanström: Zoological Papers in Honour of his Sixty-fifth Birthday,* pp. 138–147. Lund: Zoological Institute.

Dahl, E. 1963. Main evolutionary lines among recent Crustacea. In H. B. Whittington and W. D. I. Rolfe, eds., *Phylogeny and Evolution of Crustacea,* pp. 1–15. *Special Publication, Museum of Comparative Zoology.* Cambridge, Mass.

Dahl, E. 1983a. Phylogenetic systematics and the Crustacea Malacostraca: A problem of prerequisites. *Abhandlungen des naturwissenschaftlichen Vereins in Hamburg (NF)* 26:355–371.

Dahl, E. 1983b. Malacostracan phylogeny and evolution. In F. R. Schram, ed., *Crustacean Issues*. Vol. 1, *Crustacean Phylogeny*, pp. 189–212. Rotterdam: Balkema.

Dahl, E. 1983c. Alternatives in malacostracan evolution. *Memoirs of the Australian Museum* 18:1–5.

Dahl, E. 1984. The subclass Phyllocarida (Crustacea) and the status of some early fossils: A neontologist's view. *Videnskabelige Meddelelser fra dansk naturhistorisk Forening* 145:61–76.

Dahl, E. 1991. Crustacea Phyllopoda and Malacostraca: A reappraisal of cephalic and thoracic shield and fold systems and their evolutionary significance. *Philosophical Transactions of the Royal Society of London*, series B, 334:1–26.

Dahl, E. 1992. Aspects of malacostracan evolution. *Acta Zoologica* 73:339–346.

Elofsson, R. 1963. The nauplius eye and frontal organs in Decapoda (Crustacea). *Sarsia* 12:1–68.

Elofsson, R. 1965. The nauplius eye and frontal organs in Malacostraca (Crustacea). *Sarsia* 19:1–54.

Elofsson, R. 1966. The nauplius eye and frontal organs of the non-Malacostraca (Crustacea). *Sarsia* 25:1–128.

Emerson, M. J. and F. R. Schram. 1990. The origin of crustacean biramous appendages and the evolution of Arthropoda. *Science* 250:667–669.

Emerson, M. J. and F. R. Schram. 1991. Remipedia, part 2: Paleontology. *Proceedings of the San Diego Society of Natural History* 7:1–52.

Emerson, M. J. and F. R. Schram. 1997. Theories, patterns, and reality: game plan for arthropod phylogeny. In R. A. Fortey and R. H. Thomas, eds., *Arthropod Relationships*, pp. 67–86. London: Chapman and Hall.

Felgenhauer, B. E. and L. G. Abele. 1991. Morphological diversity of decapod spermatozoa. In R. T. Bauer and J. W. Martin, eds., *Crustacean Sexual Biology*, pp. 322–341. New York: Columbia University Press.

Felgenhauer, B. E., L. G. Abele, and D. L. Felder. 1992. Remipedia. In F. W. Harrison and A. G. Humes, eds., *Microscopic Anatomy of the Invertebrates*, 9:225–247. New York: Wiley-Liss.

Fryer, G. 1988. A new classification of the branchiopod Crustacea. *Zoological Journal of the Linnean Society* 91:357–383.

Fryer, G. 1992. The origin of Crustacea. *Acta Zoologica* 73:273–286.

Gould, S. J. 1991. The disparity of the Burgess Shale arthropod fauna and the limits of cladistic analysis: Why we must strive to quantify morphospace. *Paleobiology* 17:411–423.

Grygier, M. J. 1987. New records, external and internal anatomy, and the systematic position of Hansen's y-larvae (Crustacea:Maxillopoda:Facetotecta). *Sarsia* 72:261–278.

Hansen, H. J. 1925. *Studies on Arthropoda*. Vol. 2. Copenhagen: Gyldendalske Boghandel.

Heijerman, T. 1992. Adequacy of numerical taxonomic methods: A comparative study based on simulation experiments. *Zeitschrift für zoologische Systematik und Evolutionsforschung* 30:1–20.

Heijerman, T. 1993. Adequacy of numerical taxonomic methods: Further experiments using simulation data. *Zeitschrift für zoologische Systematik und Evolutionsforschung* 31:81–97.

Hessler, R. R. 1983. A defence of the caridoid facies: Wherein the early evolution of the Eumalacostraca is discussed. In F. R. Schram, ed., *Crustacean Issues*. Vol. 1, *Crustacean Phylogeny*, pp. 145–164. Rotterdam: Balkema.

Hessler, R. R. 1992. Reflections on the phylogenetic position of the Cephalocarida. *Acta Zoologica* 73:315–316.

Hessler, R. R., B. M. Marcotte, W. A. Newman, and R. F. Maddocks. 1982. Evolution within the Crustacea. In L. G. Abele, ed., *The Biology of Crustacea.* Vol. 1, *Systematics, the Fossil Record, and Biogeography,* pp. 149–239. New York: Academic Press.

Hessler, R. R. and W. A. Newman. 1975. A trilobitomorph origin for the Crustacea. *Fossils and Strata* 4:437–459.

Høeg, J. T. 1992. The phylogenetic position of the Rhizocephala: Are they truly barnacles? *Acta Zoologica* 73:323–326.

Holmquist, R. 1983. Transitions and transversions in evolutionary descent: An approach to understanding. *Journal of Molecular Evolution* 19:277–290.

Hou X.-G. and J. Bergström. 1991. The arthropods of the Lower Cambrian Chengjiang fauna, with relationships and evolutionary significance. In A. M. Simonetta and S. Conway Morris, eds., *The Early Evolution of Metazoa and the Significance of Problematic Taxa,* pp. 179–187. Cambridge: Cambridge University Press.

Hou X.-G., D. J. Siveter, M. Williams, D. Walossek, and J. Bergström. 1996. Appendages of the arthropod *Kunmingella* from the Early Cambrian of China: Its bearing on the systematic position of the Bradoriida and the fossil record of the Ostracoda. *Philosophical Transactions of the Royal Society of London,* series B, 351:1131–1145.

Huo, S.-C. and D.-G. Shu. 1983. On the phylogeny and ontogeny of Bradoriida with discussion of the origin of Crustacea. *Journal of Northwest University* 1983:82–88.

Itô, T. 1989. Origin of the limb basis in copepod limbs, with reference to remipedian and cephalocarid limbs. *Journal of Crustacean Biology* 9:85–103.

Itô, T. and F. R. Schram. 1988. Gonopores and the reproductive system of nectiopodan Remipedia. *Journal of Crustacean Biology* 8:250–253.

Jamieson, B. G. M. 1987. *The Ultrastructure and Phylogeny of Insect Spermatozoa.* Cambridge: Cambridge University Press.

Jamieson, B. G. M. 1991. Ultrastructure and phylogeny of crustacean spermatozoa. *Memoirs of the Queensland Museum* 31:109–142.

Jensen, P. G. 1993. Ultrastructure and phylogenetic significance of lattice organs in thecostracan larvae. *American Zoologist* 33 (5): 6A.

Jensen, P. G., J. Moyse, J. T. Høeg, and H. Al-Yahya. 1993. Comparative SEM studies of lattice organs: Putative sensory structures on the carapace of larvae from Ascothoracida and Cirripedia. *Acta Zoologica* 75:125–142.

Jones, P. J. and K. G. McKenzie. 1980. Queensland Middle Cambrian Bradoriida (Crustacea): New taxa, palaeobiogeography and biological affinities. *Alcheringa* 4:203–225.

Kim, W. and L. G. Abele. 1990. Molecular phylogeny of selected decapod crustaceans based on 18S rRNA nucleotide sequences. *Journal of Crustacean Biology* 10:1–13.

Kunze, J. 1983. Stomatopoda and the evolution of the Hoplocarida. In F. R. Schram, ed., *Crustacean Issues.* Vol. 1, *Crustacean Phylogeny,* pp. 165–188. Rotterdam: Balkema.

Manton, S. M. 1977. *The Arthropoda: Habits, Functional Morphology, and Evolution.* Oxford: Clarendon Press.

Martin, J. W. and M. S. Laverack. 1992. The distribution of the crustacean dorsal organ. *Acta Zoologica* 73:357–368.

McKenzie, K. G. 1983. On the origin of the Crustacea. *Memoirs of the Australian Museum* 18:21–43.

McKenzie, K. G. 1991. Crustacean evolutionary events: Sequences and consequences. *Memoirs of the Queensland Museum* 31:19–38.

McKenzie, K. G., K. J. Müller, and M. N. Gramm. 1983. Phylogeny of Ostracoda. In F. R. Schram, ed., *Crustacean Issues.* Vol. 1, *Crustacean Phylogeny,* pp. 29–46. Rotterdam: Balkema.

Müller, K. J. 1979. Phosphatocopine ostracodes with preserved appendages from the Upper Cambrian of Sweden. *Lethaia* 12:1–27.

Müller, K. J. 1982. *Hesslandona unisulcata* sp. nov. (Ostracoda) with phosphatized appendages from the Upper Cambrian "Orsten" of Sweden. In R. H. Bate, E. Robinson, and L. Shepard, eds., *A Research Manual of Fossil and Recent Ostracodes,* pp. 276–307. Chichester: Ellis Horwood.

Müller, K. J. 1983. Crustacea with preserved soft parts from the Upper Cambrian of Sweden. *Lethaia* 16:93–109.

Müller, K. J. and D. Walossek. 1985. Skaracarida: A new order of Crustacea from the Upper Cambrian of Västergötland, Sweden. *Fossils and Strata* 17:1–65.

Müller, K. J. and D. Walossek. 1986a. *Martinssonia elongata* gen. et sp. n.: A crustacean-like euarthropod from the Upper Cambrian "Orsten" of Sweden. *Zoologica Scripta* 15:73–92.

Müller, K. J. and D. Walossek. 1986b. Arthropod larvae from the Upper Cambrian of Sweden. *Transactions of the Royal Society of Edinburgh, Earth Sciences* 77:157–179.

Müller, K. J. and D. Walossek. 1988. External morphology and larval development of the Upper Cambrian maxillopod *Bredocaris admirabilis. Fossils and Strata* 23:1–70.

Newman, W. A. and M. D. Knight. 1984. The carapace and crustacean evolution: A rebuttal. *Journal of Crustacean Biology* 4:682–687.

Paulus, H. F. 1979. Eye structure and the monophyly of Arthropoda. In A. P. Gupta, ed., *Arthropod Phylogeny,* pp. 299–383. New York: Van Nostrand Reinhold.

Pinna, G., P. Arduini, C. Pesarini, and G. Teruzzi. 1985. Some controversial aspects of the morphology and anatomy of *Ostenocaris cypriformis* (Crustacea: Thylacocephala). *Transactions of the Royal Society of Edinburgh, Earth Sciences* 76:373–379.

Robison, R. A. 1984. New occurrences of the unusual trilobite *Naraoia* from the Cambrian of Idaho and Utah. *University of Kansas Paleontological Contributions* 112:1–8.

Roderigo, A. G., R. R. Bergquist, P. L. Bergquist, and R. A. Reeves. 1994. Are sponges animals? An investigation into the vagaries of phylogenetic inference. In R. W. M. van Soest, T. M. G. van Kempen, and J. C. Braekman, eds., *Sponges in Time and Space: Proceedings of the 4th International Porifera Congress,* pp. 47–54. Rotterdam: Balkema.

Rolfe, W. D. I. 1992. Not yet proven Crustacea: The Thylacocephala. *Acta Zoologica* 73:301–304.

Sanders, H. L. 1957. The Cephalocarida and crustacean phylogeny. *Systematic Zoology* 6:112–129.

Sanders, H. L. 1963a. Significance of the Cephalocarida. In H. B. Whittington and W. D. I. Rolfe, eds., *Phylogeny and Evolution of Crustacea,* pp. 163–176. *Special Publication, Museum of Comparative Zoology.* Cambridge, Mass.

Sanders, H. L. 1963b. The Cephalocarida: Functional morphology, larval development, comparative external anatomy. *Memoirs of the Connecticut Academy of Arts and Sciences* 15:1–80.

Schram, F. R. 1978. Arthropods: A convergent phenomenon. *Fieldiana, Geology* 39:61–108.

Schram, F. R. 1982. The fossil record and evolution of Crustacea. In L. G. Abele, ed., *The Biology of Crustacea*. Vol. 1, *Systematics, the Fossil Record, and Biogeography*, pp. 93–147. New York: Academic Press.

Schram, F. R., ed. 1983a. *Crustacean Issues*. Vol. 1, *Crustacean Phylogeny*. Rotterdam: Balkema.

Schram, F. R. 1983b. Remipedia and crustacean phylogeny. In F. R. Schram, ed., *Crustacean Issues*. Vol. 1, *Crustacean Phylogeny*, pp. 23–28. Rotterdam, Balkema.

Schram, F. R. 1984. Relationships within eumalacostracan Crustacea. *Transactions of the San Diego Society of Natural History* 20:301–312.

Schram, F. R. 1986. *Crustacea*. New York: Oxford University Press.

Schram, F. R. 1990. On Mazon Creek Thylacocephala. *Proceedings of the San Diego Society of Natural History* 3:1–16.

Schram, F. R. and M. J. Emerson. 1991. Arthropod pattern theory: A new approach to arthropod phylogeny. *Memoirs of the Queensland Museum* 31:1–18.

Schram, F. R. and C. A. Lewis. 1989. Functional morphology of feeding in the Nectiopoda. In B. E. Felgenhauer, L. Watling, and A. B. Thistle, eds., *Crustacean Issues*. Vol. 6, *Functional Morphology of Feeding and Grooming in Crustacea*, pp. 15–26. Rotterdam: Balkema.

Schram, F. R., R. Vonk, and C. H. J. Hof. 1997. Mazon Creek Cycloidea. *Journal of Paleontology* 71:261–284.

Schram, F. R., J. Yager, and M. J. Emerson. 1986. Remipedia, part 1: Systematics. *Memoir of the San Diego Society of Natural History* 15:1–60.

Scourfield, D. J. 1926. On a new type of crustacean from the Old Red Sandstone, *Lepidocaris rhyniensis*. *Philosophical Transactions of the Royal Society of London*, series B, 214:153–187.

Scourfield, D. J. 1940. Two new nearly complete specimens of young stages of the Devonian fossil crustacean *Lepidocaris rhyniensis*. *Proceedings of the Linnean Society, London* 152:290–298.

Secretan, S. 1964. La carapace des crustacés: Différents modes d'adaptation aux segments du corps. *Annales de Paléontologie* 50:191–208.

Secretan, S. 1985. Conchyliocarida, a class of fossil crustaceans: Relationships to Malacostraca and postulated behavior. *Transactions of the Royal Society of Edinburgh, Earth Sciences* 76:381–389.

Sieg, J. 1983. Evolution of Tanaidacea. In F. R. Schram, ed., *Crustacean Issues*. Vol. 1, *Crustacean Phylogeny*, pp. 229–256. Rotterdam: Balkema.

Spears, T., L. G. Abele, and W. Kim. 1992. The unity of the Brachyura: A phylogenetic study based on rRNA and rDNA sequences. *Systematic Biology* 41:446–461.

Starobogatov, Y. I. 1988. Systematics of Crustacea. *Journal of Crustacean Biology* 8:300–311.

Wagner, H. P. 1994. A monographic review of the Thermosbaenacea (Crustacea: Peracarida). *Zoologische Verhandelingen* 291:1–338.

Walossek, D. 1993. The Upper Cambrian *Rehbachiella* and the phylogeny of the Branchiopoda and Crustacea. *Fossils and Strata* 32:1–202.

Walossek, D., I. Hinz-Schallreuter, J. H. Shergold, and K. J. Müller. 1993. Three-dimensional preservation of arthropod integument from the Middle Cambrian of Australia. *Lethaia* 26:7–15.

Walossek, D. and K. J. Müller. 1990. Upper Cambrian stem-lineage crustaceans and their bearing on the monophyly of Crustacea and the position of *Agnostus. Lethaia* 23:409–427.

Walossek, D. and K. J. Müller. 1992. The "alum-shale window": Contribution of "Orsten" arthropods to the phylogeny of Crustacea. *Acta Zoologica* 73:305–312.

Walossek, D. and H. Szaniawski. 1991. *Cambrocaris baltica* n. gen., n. sp.: A possible stem-lineage crustacean from the Upper Cambrian of Poland. *Lethaia* 24:363–378.

Watling, L. 1983. Peracaridan disunity and its bearing on Eumalacostracan phylogeny with a redefinition of eumalacostracan superorders. In F. R. Schram, ed., *Crustacean Issues*. Vol. 1, *Crustacean Phylogeny*, pp. 213–228. Rotterdam: Balkema.

Wheeler, W. C., P. Cartwright, and C. Y. Hayashi. 1993. Arthropod phylogeny: A combined approach. *Cladistics* 9:1–39.

Whittington, H. B. 1959. Ontogeny of Trilobita. In R. C. Moore, ed., *Treatise on Invertebrate Paleontology, Part O, Arthropoda 1*, pp. O127–O145. Boulder, Lawrence: Geological Society of America, University of Kansas Press.

Whittington, H. B. 1974. *Yohoia* Walcott and *Plenocaris* n. gen.: Arthropods from the Burgess Shale, Middle Cambrian, British Columbia. *Geological Survey of Canada Bulletin* 231:1–21.

Wilson, G. D. F. 1992. Computerized analysis of crustacean relationships. *Acta Zoologica* 73:383–389.

Wingstrand, K. G. 1972. Comparative spermatology of a pentastomid, *Raillietiella hemidactyli,* and a branchiuran crustacean, *Argulus foliaceous:* With a discussion of pentastomid relationships. *Kongelige Danske Videnskaberner Selskab, Biologiske Skrifter* 19:1–72.

Wingstrand, K. G. 1978. Comparative spermatology of the Crustacean Entomostraca, part 1: Subclass Branchiopoda. *Kongelige Danske Videnskaberner Selskab, Biologiske Skrifter* 22:1–66.

Wingstrand, K. G. 1988. Comparative spermatology of the Crustacean Entomostraca, part 1: Subclass Ostracoda. *Kongelige Danske Videnskaberner Selskab, Biologiske Skrifter* 32:1–79.

Yager, J. 1981. Remipedia: A new class of Crustacea from a marine cave in the Bahamas. *Journal of Crustacean Biology* 1:328–333.

Yager, J. 1989. The male reproductive system, sperm, and spermatophores of the primitive, hermaphroditic, remipede crustacean *Speleonectes benjamini. International Journal of Reproduction and Development* 15:75–81.

# CHAPTER 7

## Fossil Taxa and Relationships of Chelicerates

*Paul A. Selden and Jason A. Dunlop*

## Abstract

Chelicerata belong in Arachnomorpha and comprise Xiphosura, Eurypterida, and Arachnida. Pycnogonida, Aglaspidida, Chasmataspida, and a number of merostomoid taxa, e.g., *Sanctacaris*, are regarded as possible chelicerate relatives but are not included in Chelicerata. Autapomorphies are presented for xiphosurans, eurypterids, and the recognized arachnid orders, and the fossil record, mode of life, and probable phylogenetic affinities of each group are discussed. The fossil record of the chelicerate taxa is compared to published phylogenies. Current controversies in arachnid phylogeny include whether scorpions are closer to eurypterids than to other arachnids, though in either case scorpions and non-scorpion arachnids terrestrialized independently. Some phylogenies predict aquatic Silurian opilionids. The problems of recognizing early occurrences of crown groups, because their stem groups acquired their autapomorphies through geological time, are discussed. A strong consensus among published arachnid phylogenies recognizes a taxon, Tetrapulmonata, comprising Trigonotarbida (extinct), Araneae, Amblypygi, Uropygi, and Schizomida.

Chelicerata is a major arthropod clade defined by an anterior prosoma bearing six pairs of appendages including pre-oral chelicerae, an opisthosoma of twelve segments, and a post-anal telson. Chelicerates have been traditionally divided into two classes: the aquatic Merostomata (Xiphosura + Eurypterida) and the terrestrial Arachnida, though many authors consider Merostomata to be a paraphyletic grade of aquatic chelicerates (Kraus 1976). Chelicerates underwent an initial radiation in the Cambrian, with the arachnids radiating later, in the Silurian (Lindquist 1984); most arachnid orders were established by Carboniferous times. Dunlop and Selden

(1997) discussed the relationships of chelicerates with other arthropods; in this chapter we provide brief sketches of each chelicerate group (fig. 7.1) and some chelicerate relatives, and then we discuss how the fossil record accords with recent phylogenies of the Chelicerata.

## Chelicerate Relatives

Heymons (1901) introduced the name Chelicerata for arachnids, eurypterids, and xiphosurans. From a modern perspective, chelicerates appear to be a distinctive, clearly definable taxon, with the exception of the questionable inclusion of the Pycnogonida (see below). When fossil taxa are considered, however, the limits of Chelicerata are more vague. A number of Paleozoic forms, such as Aglaspidida Raasch, 1939, are superficially similar to xiphosurans but lack diagnostic features such as chelicerae; they may bear antennae, or their appendages are unknown. The case for inclusion or exclusion of these groups in Chelicerata is discussed below.

### Pycnogonida Latreille, 1810

Pycnogonids (sea spiders) are exclusively marine arthropods that resemble Chelicerata in having a chelate first appendage (chelifore). The prosoma consists of a series of tubular segments with lateral extensions each bearing a walking leg; there is a minute opisthosoma. An anterior proboscis precedes the chelifore; the second appendage is palplike. Appendage 3 (oviger) is modified in the male to carry ova during incubation. There are usually four pairs of walking legs, but some species bear one or two additional walking legs and corresponding trunk segments. The first prosomal segment bears, in addition to the first three pairs of appendages, a dorsal tubercle with eyes. Some deep-sea forms are blind.

King (1973) summarized the three main hypotheses about pycnogonid affinities. Pycnogonids may be related to: (1) crustaceans, since their protonymph larva resembles the nauplius larva, and they share certain embryological features; (2) chelicerates, because of their pre-oral chelifores and gut diverticula that extend into the leg coxae; or (3) neither of these groups.

Hedgpeth (1955a) referred Recent pycnogonids to Pantopoda and the monotypic *Palaeopantopus* Broili, 1928, from the Lower Devonian of Germany to Palaeopantopoda. Another monotypic arthropod, *Palaeoisopus*, also from the Lower Devonian of Germany, resembles pycnogonids, but its affinities are obscure (Hedgpeth 1955b). The fossil record does not help to determine the exact systematic position of the pycnogonids. Bergström et al. (1980) redescribed the fossil pycnogonids and suggested that they were derived from primitive merostomes.

### Aglaspidida

Aglaspidida is a distinct group of Lower Paleozoic arthropods that show similarities to both xiphosurans and trilobites. Aglaspidids have a phosphatic exoskeleton and

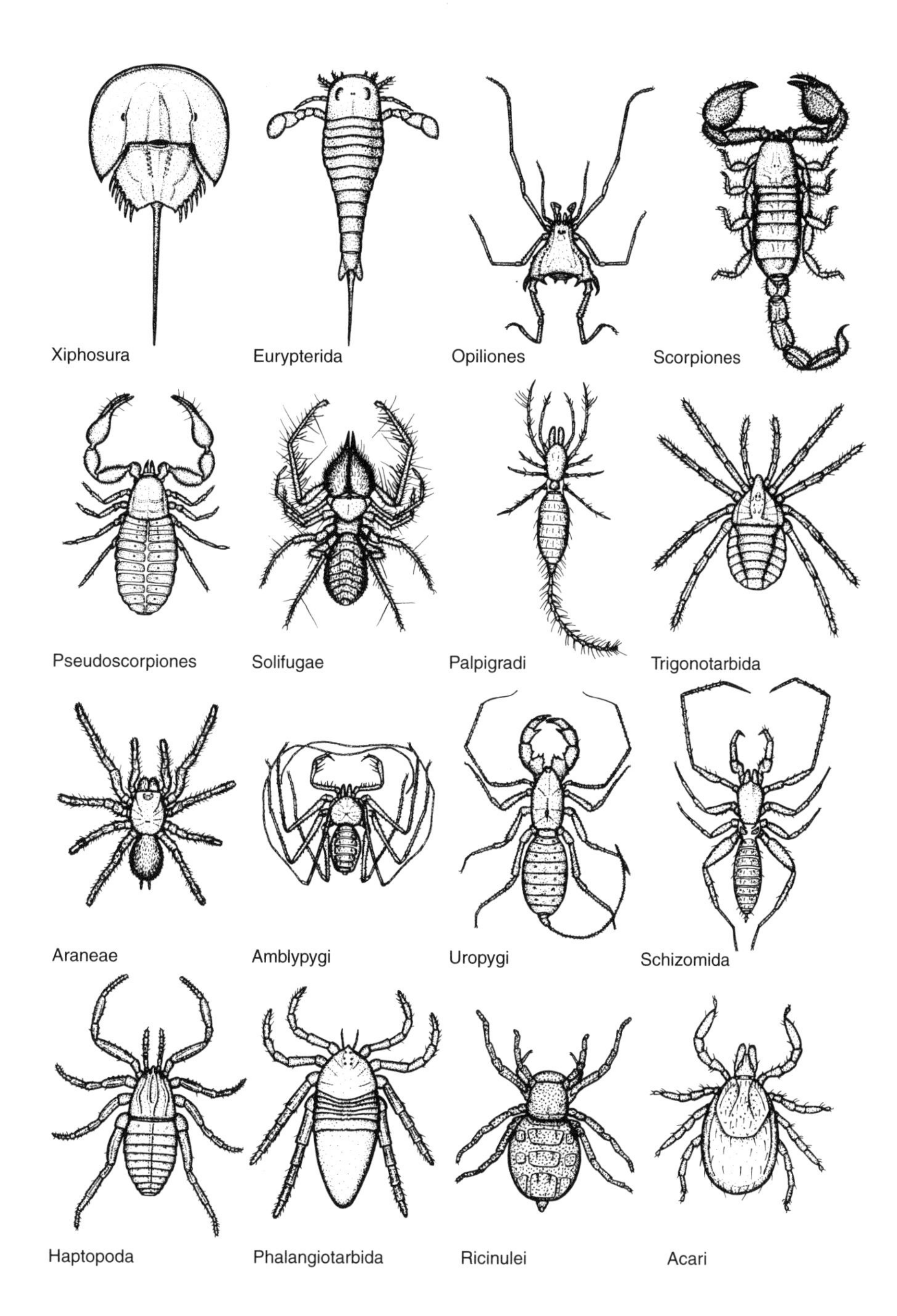


FIGURE 7.1

Sketch diagrams of a number of taxa described in the text. Xiphosura: *Limulus polyphemus,* Eurypterida: *Baltoeurypterus tetragonophthalmus,* Opiliones: *Gonyleptus* sp., Scorpiones: *Scorpio martus,* Pseudoscorpiones: *Lasiochernes pilosus,* Solifugae: *Galeodes arabs,* Palpigradi: *Koenenia mirabilis,* Trigonotarbida: *Lissomatus schucherti,* Araneae: *Poecilotheria regalis,* Amblypygi: *Charinus milloti,* Uropygi: *Mastigoproctus giganteus,* Schizomida: *Schizontus peradyenensis,* Haptopoda: *Plesiosiro madeleyi,* Phalangiotarbida: *Goniotarbus tuberculatus,* Ricinulei: *Ricinoides feae,* Acari: *Ixodes ricinus.*

leglike appendages on both the prosoma and opisthosoma. The carapace has genal spines and the lateral margins of the tergites curve strongly backward. The group was classified as merostomes by Raasch (1939) and regarded as a separate order of the Xiphosura by Størmer (1955). Bristowe (1971), Manton (1977), and Weygoldt and Paulus (1979) placed aglaspidids within the Chelicerata. Weygoldt and Paulus (1979) regarded Aglaspidida as the sister taxon to all other chelicerates (Euchelicerata). Aglaspidids were removed from Chelicerata by Briggs et al. (1979) on the grounds that there were only four or five pairs of cephalic appendages, the first of which could not be demonstrated to be chelate. They did not refer them to another higher taxon. Their removal from Chelicerata was criticized by Bergström (1980) who argued that the number of head segments (and hence head appendages) was variable in early chelicerates and other arthropods. The extinct aglaspidids are probably not chelicerates since they bear neither chelicerae nor other features that would ally them more closely with the Chelicerata than with any other arthropod group (Briggs et al. 1979).

A number of Lower Paleozoic taxa representing potential candidates for basal chelicerates, e.g., *Paleomerus,* were referred to the Aglaspidida (Raasch 1939). This referral was disputed by Bergström (1971) since these genera do not have phosphatic exoskeletons, tergites drawn laterally into pleural spines, or genal spines like typical aglaspidids. The appendages of *Paleomerus, Strabops,* and *Neostrabops* are unknown.

## Chasmataspida Caster and Brooks, 1956

Chasmataspids are a monotypic group from the Ordovician and were originally referred to the Xiphosura (Caster and Brooks 1956). Chasmataspids have a semicircular prosoma, a short pre-abdomen (buckler) of three segments and an elongate post-abdomen of nine segments. The appendages are poorly known. The Devonian fossil *Diploaspis* was referred to Chasmataspida by Størmer (1972). *Diploaspis* has a pair of paddles, but also appears to have a 3 + 9 opisthosomal segmentation, which argues against it being a eurypterid. Bergström (1980) removed *Diploaspis* from Chasmataspida, suggesting that it might represent a group from which arachnids evolved, leaving Chasmataspida as a monotypic order consisting only of the genus *Chasmataspis.* The prosoma of *Chasmataspis* resembles that of aglaspidids, and initial observations suggest these taxa may be sister groups, though restudy of this material, together with *Diploaspis* and the merostomoids, is required.

## Merostomoidea Størmer, 1944

A range of Lower Paleozoic aquatic arthropods that superficially resemble xiphosurans in having a body comprised of two tagmata was referred to the class Merostomoidea by Størmer (1944). These include *Sidneyia, Helmetia, Emeraldella, Ha-*

*belia, Molaria,* and *Cheloniellon.* They show variable numbers of segments in each tagma and a range of appendages including antennae in some forms and leglike opisthosomal limbs in others (Bergström 1980). Merostomoids almost certainly do not form a monophyletic group; recent studies have indicated they are scattered within the group Arachnomorpha, which includes Chelicerata and Trilobita (Wills et al. 1995).

*Sanctacaris* Briggs and Collins, 1988, from the Middle Cambrian Burgess Shale of British Columbia, was suggested as plesiomorphic sister taxon (plesion) to all other Chelicerata on the grounds of having two tagmata with six pairs of appendages on the first tagma, a cardiac lobe, and the anus at the rear of the last trunk segment. Briggs and Collins (1988) proposed a new diagnosis of Chelicerata, more appropriate to fossil material. Arguing against a close association of *Sanctacaris* with Chelicerata are: lack of chelicerae, an opisthosoma of eleven segments (twelve is the number regarded as plesiomorphic for Chelicerata [Shultz 1990]) and a broad, rather than styliform, telson, which is generally considered to be the plesiomorphic state. In a more recent cladistic analysis, *Sanctacaris* was grouped with other Burgess Shale merostomoids (Wills et al. 1995).

## Chelicerata Heymons, 1901

### Xiphosura Latreille, 1802

Xiphosura (horseshoe crabs) are the only extant, primarily aquatic chelicerates. They are widely regarded as the most primitive chelicerates and commonly cited as outstanding examples of "living fossils" because of their apparent conservative morphological change since the Paleozoic. One living species (*Limulus polyphemus*), which can grow up to 0.6 m long, occurs on the Atlantic coast of North America, and three species of *Tachypleus* and *Carcinoscorpius* occur in southeast Asia and the Philippines. Xiphosurans are characterized by a horseshoe-shaped carapace, two to five chelate appendages, modification of appendage 6 into a "pusher" and modification of the first opisthosomal appendages into chilaria.

Though previously allied with Eurypterida in the Merostomata Dana, 1852, most authors now place Xiphosura with either the Scorpiones (Bergström 1979, 1980; Bergström et al. 1980; van der Hammen 1985, 1986) or as the sister group to all other chelicerates (except *Sanctacaris*) (Grasshoff 1978; Boudreaux 1979; Paulus 1979; Weygoldt and Paulus 1979; Weygoldt 1980), thereby rendering Merostomata an unnatural group (Kraus 1976).

About thirty fossil genera are known. A carapace dating from the Lower Cambrian from Öland, Sweden, *Eolimulus alatus* (Moberg 1892), was referred to Xiphosura, though it might equally well be an aglaspidid or some other arthropod. Flower (1968) placed *Lemoneites,* from the Ordovician of New Mexico, in the

Aglaspidida but remarked on the number of similarities with Synziphosurina Packard, 1886. Synziphosurines (Ordovician — Upper Devonian) are primitive xiphosurans with a horseshoe-shaped carapace and a segmented opisthosoma of freely articulating segments divided into pre- and post-abdominal regions. A recent cladistic analysis (Anderson and Selden 1997) has shown that Synziphosurina is paraphyletic.

## Eurypterida Burmeister, 1843

Eurypterids flourished from Ordovician to Permian times in aquatic environments worldwide. Most eurypterids were marine, but some Silurian and later forms lived in fresh water and some may have been amphibious. Approximately 300 species have been described, ranging from 0.1 m to 2 m in length; the latter were the largest arthropods that ever lived. Most were predators, and their prey included the early vertebrates. Eurypterids are characterized by an elongate body bearing a telson, a carapace with median ocelli and a pair of compound eyes, three-segmented chelicerae and five further pairs of prosomal appendages with gnathobasic coxae, and five pairs of opisthosomal appendages (*Blattfüsse*) modified as opercula for respiratory organs. Autapomorphies of the Eurypterida are: a median opisthosomal (genital) appendage, a platelike metastoma (probably homologous with the first pair of opisthosomal appendages, i.e., the xiphosuran chilaria) and so-called gill tracts (*Kiemenplatten*) on opisthosomal sternites 1–5, within branchial chambers.

Eurypterids were placed by Woodward (1865) with the Xiphosura in the class Merostomata Dana, 1852, but in most recent studies Xiphosura have been considered to be the sister group to all other Chelicerata (except *Sanctacaris*) and not to Eurypterida alone (e.g., Kraus 1976). Eurypterida is a monophyletic group, though some authors (e.g., Tollerton 1989) have regarded the hibbertopteroids (= Cyrtoctenida) as a sister group of equal rank to all other eurypterids. Most authors (e.g., Clarke and Ruedemann 1912; Caster and Kjellesvig-Waering 1964) divided the main eurypterid line into two taxa: Pterygotina and Eurypterina. Pterygotines have enormous chelicerae for food capture and simple walking legs; eurypterines have small chelicerae and commonly spinose anterior limbs for food gathering. An alternative division into those forms in which the sixth appendage is leglike (Stylonuracea) and those in which it is formed into a paddle for swimming (Eurypteracea) was proposed by Størmer (1955). A cladistic analysis was attempted by Plotnick (1983) at the generic level, and Tollerton (1989) has produced the most recent systematic account of the group.

## Opiliones Sundevall, 1833

Opilionids are among the most diverse extant chelicerates with approximately 4000–5000 living species worldwide. They range in size from less than 1 mm to about 22 mm. They are omnivorous animals and are the only arachnids known to ingest solid food (Petrunkevitch 1955). Their morphology ranges from the round-bodied, long-legged phalangioids to the shorter-legged, spiny laniatores and the mitelike cyphopthalmids. Opilionids are characterized by a prosoma that is broadly joined to the opisthosoma, the division between the two often being indistinct. The chelicerae are three-segmented and chelate, and in some taxa the pedipalps are raptorial. Opilionids respire with tracheae opening through a single pair of spiracles on the second opisthosomal segment. Autapomorphies proposed for the Opiliones are: an elongate and tactile leg pair 2, a trochanterofemoral joint with a vertical bicondylar articulation, tracheal stigmata on the genital segment, a male penis and female ovipositor, loss of the lateral eyes, and prosomal defensive glands.

Though a distinctive group, the position of Opiliones within Chelicerata is poorly defined. Some authors (e.g., Yoshikura 1975; Weygoldt and Paulus 1979) placed them in a derived position as the sister group to Acari. Van der Hammen (1989) argued that opilionids were related to Xiphosura and Scorpiones, and more recently Shultz's (1990) analysis placed them as the plesiomorphic sister taxon to Scorpiones, Pseudoscorpiones, and Solifugae (see below). Of particular phylogenetic interest are the cyphopthalmid opilionids and the opilioacarid mites. Some authors have cited these similar groups as providing strong evidence for Acari and Opiliones representing sister groups, though Acari are normally placed as the sister group of Ricinulei.

Opiliones occur in the Lower Carboniferous rocks of East Kirkton, near Edinburgh, Scotland (Wood et al. 1985), in Upper Carboniferous strata of France (Thevenin 1901; undescribed fossils from Montceau-les-Mines) and Illinois (Petrunkevitch 1913), and in the Lower Cretaceous of Koonwarra, Victoria, Australia (Jell and Duncan 1986). They are much better known from Tertiary ambers (Koch and Berendt 1854). The order Kustarachnida Petrunkevitch, 1913, should be included with Opiliones, following reassessments by Beall (1986) and Dunlop (1996).

## Scorpiones Hemprich and Ehrenburg, 1810

Approximately 1500 living species of scorpions inhabit tropical and temperate parts of the world, and about 110 fossil species are known. Scorpions are the oldest known arachnids, ranging from Silurian (Llandovery) strata onward, and have a good fossil record, especially in the Paleozoic. Living scorpions range from 9 mm to 210 mm, and some fossil scorpions may have reached 1 m. All living scorpions are terrestrial predators, but some Silurian to Carboniferous forms were aquatic. Scorpion biology has been summarized by Polis (1990).

Scorpions are characterized by three-segmented, chelate chelicerae (the observation that some fossil forms had four segments is erroneous), chelate pedipalps, and an opisthosoma divided into a broad mesosoma and a narrow metasoma, the latter with ankylosed segments. Scorpions respire through four pairs of book lungs opening on opisthosomal segments 2–5. Autapomorphies proposed for the Scorpiones are pectines, spermatozoa with free flagella throughout their development, opisthosomal venom glands, a pretarsal levator muscle originating in the tibia, and large endites on the coxae of walking legs 1 and 2.

A monograph of fossil scorpions was prepared by Kjellesvig-Waering (1986), but this posthumous publication was marred by compilation defects, and the classification scheme proposed therein was controversial. For example, the linchpin of Kjellesvig-Waering's classification was the supposed Devonian gilled scorpion described as *Tiphoscorpio hueberi*. Restudy of this material (Selden and Shear 1992; Shear and Selden 1995) revealed that it is not a scorpion but an arthropleurid myriapod. Stockwell (1989), in an unpublished thesis, produced a more acceptable classification scheme of Scorpiones, based on an exhaustive cladistic analysis that included fossils. This was adopted by Selden (1993a,b) and Jeram (1994).

Scorpions are the oldest arachnid group; the oldest known scorpion is *Dolichophonus loudonensis* (Laurie, 1889) from the Llandovery of the Pentland Hills, near Edinburgh, Scotland. The Early Silurian record of scorpions could be interpreted as representing the earliest terrestrial animals because all modern scorpions are terrestrial. However, all Silurian fossil scorpions occur in marine or marginal marine sediments, and morphological features suggest an aquatic mode of life. Petrunkevitch (e.g., 1953) considered that all fossil scorpions were terrestrial, but other workers (e.g., Pocock 1902; Wills 1947; Størmer 1970; Rolfe and Beckett 1984; Kjellesvig-Waering 1986) argued for an aquatic habitat for Silurian scorpions at least. Evidence for aquatism among fossil scorpions includes: gills and digitigrade tarsi as well as the absence of terrestrial modifications such as coxal apophyses, stigmata, book lungs, trichobothria, highly developed pectines, and plantigrade tarsi. There is considerable overlap in the ranges of aquatic and terrestrial scorpions, but the first terrestrial forms had appeared by the Devonian (Selden and Jeram 1989). It is not always easy to decide whether a given fossil had an aquatic or terrestrial mode of life; the original environment of the enclosing sediment is commonly the best clue. Well-preserved book lungs have been found in a Carboniferous (Viséan) scorpion from East Kirkton, near Edinburgh, Scotland (Jeram 1990). Few new records of fossil scorpions have turned up in recent years although in the otherwise sparsely recorded Mesozoic scorpions from the Triassic of France (Gall 1971) and the Cretaceous of Brazil (Campos 1986) are currently under study.

## Pseudoscorpiones de Geer, 1778

Pseudoscorpions (= Chelonethi) are small (1 mm to 7 mm), predatory arachnids with large, chelate pedipalps. Approximately 2500 living species have been described from around the world. Pseudoscorpions can be found typically in leaf litter, moss, and under stones. Their method of dispersal, phoresy, involves hitching a ride on a flying insect using the pedipalps. Pseudoscorpions have cheliceral silk glands with which they construct elaborate brood chambers. Pseudoscorpion biology has been summarized by Weygoldt (1969).

Some workers have placed scorpions and pseudoscorpions together on account of their overall similarity (e.g., Yoshikura 1975). However most recent studies have all placed pseudoscorpions and solifuges as sister groups united by the synapomorphies of two-segmented chelate chelicerae (Weygoldt and Paulus 1979), lack of a patella, presence of a rostrum in the mouthparts (van der Hammen 1989), and a chelicero-carapacal articulation (Shultz 1990).

Pseudoscorpions are characterized by their small, often rounded bodies and large chelate pedipalps. These animals lack median eyes and Malpighian tubules. They possess a brood sac and respire with tracheae opening through spiracles on the first two opisthosomal segments. The female has a ram's-horn organ for use in taking up the spermatophore. Autapomorphies proposed for Pseudoscorpionida are: absence of an anterior patellotibial muscle in the walking legs, cheliceral silk glands, and complex brood care.

Many pseudoscorpions are known from the Tertiary (mainly in ambers, e.g., listed in Schawaller 1982, table 1), and some are known from Cretaceous ambers of the Lebanon (Whalley 1980) and Manitoba (Schawaller 1991). An important find of fossil pseudoscorpions was the discovery of exceptionally well preserved specimens in the Upper Devonian mudstones of Gilboa, New York, described as *Dracochela deprehendor* (Shear et al. 1989; Schawaller et al. 1991). Only protonymphs and tritonymphs are known, which were referred to the superfamily Chthonioidea (Harvey 1992).

## Solifugae Sundevall, 1833

Solifugae (= Solpugida) are medium-sized (7 mm to 70 mm), swift-running, predatory arachnids. Approximately 900 living species are known from all arid and semiarid regions of the world except Australia. They have a tracheal system rivaling that of insects, allowing them to be among the most active of the arachnids. With their large, chelate chelicerae, they have probably the largest jaws relative to body size of any animal, and large forms can overpower vertebrate prey.

The placement of Solifugae and Pseudoscorpiones as sister taxa was discussed above, though Grasshoff (1978) placed Solifugae as the sister group to all other arachnids + eurypterids + xiphosurans. This interpretation was based partly on the interpretation of their divided carapace as a plesiomorphic character, though this could also be seen as a secondary adaptation associated with small size and/or prosomal mobility.

Solifuges are characterized by a divided carapace, huge chelicerae, loss of lateral eyes, a constriction between the prosoma and opisthosoma, and an extremely hairy body and legs. They respire using an extensive tracheal system opening through spiracles on the midline of the abdomen and also in the prosoma. Solifuge pedipalps are stouter than the walking legs and have a tactile function. Autapomorphies proposed for the Solpugida include their complex respiratory organs and prosomal stigmata, the presence of malleoli (racquet organs) on the underside of the posterior legs, loss of the endosternite, and a monocondylar femur-patella joint.

The Carboniferous solifuge *Protosolpuga carbonaria* Petrunkevitch, 1913, is in a poor state of preservation, but its morphology places it in the Solifugae rather than any other arachnid order (Selden and Shear 1996). Only two other fossil solifuges are known, both described recently: *Happlodontus proterus* Poinar and Santiago-Blay, 1989, from Oligocene Dominican amber, and *Cratosolpuga wunderlichi* Selden, 1996, from the Cretaceous of Brazil.

## Palpigradi Latreille, 1810

Palpigradi are tiny (up to 3 mm body length) arachnids found mostly in soil and interstitial environments (Monniot 1966). Approximately 125 living species are known. Palpigrades have a thin, colorless cuticle. They are blind, have no respiratory organs or Malpighian tubules, and have a divided carapace, a narrow prosoma-opisthosoma junction, and an opisthosoma ending in a long, jointed flagellum. Autapomorphies proposed for Palpigradi are: simple coxosternal region with a terminal mouth, paired anteromedial sensory organ and trochanterofemoral joint formed by a dorsal hinge in the walking legs.

Another name for Palpigradi is Microthelyphonida, which suggests a relationship with Thelyphonida (Uropygi and Schizomida), which they superficially resemble. However, palpigrades have three-segmented chelicerae, and their pedipalps are not chelate. Some authors have regarded palpigrades as primitive arachnids that approximate the hypothetical "archaearachnid" (Savory 1971), but many of their characters (e.g., divided carapace and separate sternites) may be a consequence of miniaturization rather than of the retention of plesiomorphic characters and should be treated with caution in phylogenetic analysis. Even so, most studies (Weygoldt and Paulus 1979; Shultz 1990) have classified palpigrades as relatively primitive arachnids.

The preservation potential of palpigrades is low; their small size, thin cuticle, and interstitial habitats make them difficult objects of study. *Sternarthron zitteli* Haase, 1890, from the Jurassic lithographic limestone of Solnhofen, Germany, was synonymized with the heteropteran insect *Propygolampis* by Carpenter (1992), following Handlirsch (1906). Thus, the only good fossil palpigrade is *Palaeokoenenia mordax* Rowland and Sissom, 1980, from the "Onyx Marble" quarries (Pliocene) of Arizona.

## Trigonotarbida Petrunkevitch, 1949

Trigonotarbids are medium-sized (1 mm to 50 mm), extinct, spiderlike arachnids characterized by opisthosomal tergites divided into median and lateral plates. They respired with two pairs of book lungs opening on opisthosomal segments 2 and 3. Autapomorphies proposed for the Trigonotarbida are: loss of sternite 1, divided tergites, and a locking mechanism between the prosoma and opisthosoma (a similar mechanism is found in Ricinulei, but may be convergent). Studies of well-preserved specimens of their chelicerae suggest they were active predators and probably ran down or ambushed prey. Some later forms had spinous legs for prey capture. Early trigonotarbids tended to be small, with stout legs. Later forms were often larger, spinous, and heavily tuberculated on their dorsal surface. Trigonotarbids represent probably the most diverse Paleozoic arachnids, second only to scorpions, and show a range of body forms reflected in their division into some ten families.

The first trigonotarbids to be described were referred to Pseudoscorpiones; later the group was allied to Opiliones. The order Anthracomarti was erected by Karsch (1882), and the order Trigonotarbi was carved from it by Petrunkevitch (1949). In the latter paper, Petrunkevitch recognized a fundamental division between Haptopoda (see below) and Anthracomarti (= subclass Stethostomata Petrunkevitch, 1949), defined by the fixed state of their characters, and Trigonotarbi (= subclass Soluta Petrunkevitch, 1949), defined on a combination of characters in a labile state. It has been pointed out by Shear et al. (1987) and Selden (1993b) that many of the differences between Soluta and Stethostomata suggested by Petrunkevitch (1949) are due to poor preservation and/or interpretation of the fossils, and the others compare to familial, or at most subordinal, differences in other arachnid groups. On this basis, Dunlop (1996) reunited Anthracomartida with the better defined Trigonotarbida under the latter name. Thus trigonotarbids show a division of opisthosomal tergites into three or more plates; anthracomartids are trigonotarbids with opisthosomal tergites divided into five plates. Following the discovery of two pairs of book lungs in some exceptionally preserved trigonotarbid material, their placement within the Tetrapulmonata Shultz, 1990 (see below), became apparent. Shear et al. (1987) classified Trigonotarbida as the sister group to all other tetrapulmonates.

Approximately fifty trigonotarbid species are known, from Silurian (Přídolí) to Permian (Asselian) strata of Euramerica (one fossil is known from Gondwana). Trigonotarbids are the commonest Paleozoic arachnids and one of the best known groups. They are among the first known land animals (Jeram et al. 1990). First described from Upper Carboniferous rocks (e.g., Buckland 1837; Fritsch 1901; Pocock 1902, 1911), Hirst (1923) described the first Devonian specimens, from the Rhynie Chert of Scotland, and Størmer (1970) described forms from the Middle Devonian of Alken an der Mosel, Germany. Trigonotarbids are one of the few arachnid groups found relatively frequently in Paleozoic terrestrial rocks; forms have been described recently from Argentina (Pinto and Hünicken 1980), Spain (Selden and Romano 1983), the Czech Republic (Oplustil 1985), and Germany (Jux 1982).

## Araneae Clerck, 1757

Araneae (spiders) are the most familiar of all arachnids and have generated the most research. Spiders are a very diverse group, with approximately 35 000 living species described. They range from 1 mm to 90 mm in length (250 mm leg span) and have been found in many different habitats, from high latitudes and altitudes to tropical forests and deserts; some species live much of their lives submerged in marine or fresh water. The use of silk for prey capture in many taxa has doubtless contributed to the spiders' success. Foelix (1982) provided an excellent review of the biology of spiders. With the exception of the Mesothelae, Recent spiders are characterized by a lack of opisthosomal segmentation. All spiders show a narrow pedicel between the prosoma and opisthosoma, and their appendages are less modified than in other arachnid taxa (with the exception of the pedipalp of the adult male, which is greatly modified for sperm transfer). The plesiomorphic respiratory organs in spiders are a pair of book lungs on opisthosomal segments 2 and 3 in mesotheles, mygalomorphs, and one araneomorph family. Most araneomorphs have modified the second pair of lungs into tracheal systems, and some have replaced both lungs with tracheae. Autapomorphies proposed for the Araneae are opisthosomal silk-producing spinnerets, a naked cheliceral fang, cheliceral venom glands, a copulatory device on the male pedipalp, and the absence of the trochanterofemoral depressor muscle in the walking legs. Recent spider systematics were reviewed by Coddington and Levi (1991). Araneae is divided into two suborders: Mesothelae and Opisthothelae, the latter further subdivided into infraorders Mygalomorphae and Araneomorphae. Mesotheles show the greatest number of plesiomorphic character states, araneomorphs the most derived.

Approximately 600 fossil spider species have been described. The oldest spider is *Attercopus fimbriunguis* Shear, Selden, and Rolfe, 1987, from the Devonian of Gilboa, New York; supposed Devonian spiders from Rhynie (Hirst 1923) and Alken an der Mosel (Størmer 1976) have been disproved (Selden et al. 1991). *Attercopus* is sister to all other spiders; the patella-tibia joint is a rocking joint but in a more plesiomorphic state than in other spiders, lacking the "compression zone Y" of Manton (1977). Autapomorphies of *Attercopus* are: fimbriate paired claws, spinules on the palpal femur, and lack of trichobothria; the latter feature is puzzling. Many Carboniferous spiders were originally referred to the Mesothelae on account of their segmented opisthosomas, but since this is a plesiomorphic character, and autapomorphies of mesotheles are absent in the fossils, this conclusion is unwarranted. However, a true fossil mesothele was described recently (Selden 1996), the first and oldest of this suborder. The oldest recorded mygalomorph is Triassic in age (Selden and Gall 1992). Reported Paleozoic araneomorphs (e.g., Pocock 1911; Størmer 1976) are actually misidentified other arachnids or other arthropods (Selden et al. 1991). The oldest described araneomorph is Jurassic in age (Eskov 1984). Recent finds of Cretaceous araneomorphs have emphasized the diversity of a spider fauna of modern aspect during this period. Some show too little morphological detail to be of value (Jell and Duncan 1986), but Selden (1990b) described some specimens from the Lower Cretaceous of northeast Spain, beautifully preserved in lithographic limestone. The specimens included a deinopoid and a tetragnathid, demonstrating that both cribellate and ecribellate orb weavers were in existence at this time. In broad terms, by the Tertiary, the spider fauna was almost identical to that of today, and only three families are known to have become extinct since the Paleogene (Eskov 1990).

## Amblypygi Thorell, 1883

Amblypygi are medium-sized (15 mm to 45 mm) arachnids with flattened bodies and elongate legs that move in the horizontal plane and thus enable them to live in narrow crevices under bark, stones, and leaf litter. Approximately eighty extant species of Amblypygi live in the tropical regions of the world. They detect prey with the antenniform leg 1 and grab it with huge, spiny pedipalps. Few autapomorphies exist for Amblypygi (Shultz 1990); these are: a pretarsal depressor muscle without a patella head, vestigial labrum, large anterior coxal apodemes on all legs, divided tibiae, and an immovable patello-tibia joint. Amblypygi respire with book lung pairs on the second and third opisthosomal segments.

Amblypygids have long been recognized as close relatives of Uropygi, with which they were originally placed in the order Pedipalpi. Subsequently, they have been interpreted as the sister group of spiders (e.g., Weygoldt and Paulus 1979; van der Hammen 1989) on the synapomorphies of a pedicel and a sucking stomach. Shultz (1990) and Selden et al. (1991) placed Amblypygi as sister group to Uropygi + Schizomida principally on the synapomorphies of an antenniform leg pair 1 and subchelate pedipalps (see these authors for further synapomorphies).

Four fossil amblypygid species have been described from the Upper Carboniferous (Westphalian) of Europe and North America, with a gross morphology similar to modern forms (Dunlop 1994). Fossils are also known from the Cretaceous of Brazil (unpublished) and Tertiary ambers (e.g., Schawaller 1979). Amblypygi may be present in the Devonian of Gilboa; a possible pedipalp tarsus was figured by Shear et al. (1984) and *Ecchosis pulchribothrium* Selden and Shear, 1991, may belong in this group (Selden et al. 1991).

## Uropygi Thorell, 1882

Eighty-five living species of Uropygi (whip scorpions, vinegaroons) occur in tropical and subtropical zones, with the exception of Europe and Australia. They are closely related to schizomids. Uropygids are 15 mm to 80 mm in body length and are characterized by a pentagonal carapace, stout, chelate pedipalps, and slender legs with segmented tarsi. Leg 1 is antenniform. They respire using book lung pairs on opisthosomal segments 2 and 3. Autapomorphies proposed for the Uropygi by Weygoldt and Paulus (1979) are: a camerostome (fused palpal coxae), development involving a prenymph and four nymphal instars, and the male grabbing the female's opisthosoma during mating. However, these authors proposed these characters for what is essentially Uropygi + Schizomida. Shultz (1990) proposed pygidial ommatoids, two pairs of small accessory lateral eyes, and internal musculature for the movable finger of the pedipalp as autapomorphies of Uropygi alone. Uropygids typically live in burrows and hunt using their sub-chelate pedipalps. The name whip scorpion comes partly from their habit of raising the opisthosoma in a defensive display called aggressive posturing by which they mimic a scorpion. The name vinegaroon comes from their other defensive mechanism, spraying acetic acid from a pair of pygidial glands. Uropygi, with Amblypygi and Schizomida, have been united in Pedipalpi by some authors. The similarities and differences between Uropygi and Schizomida are discussed below.

Six well-preserved fossil species of Uropygi are known from Carboniferous (Namurian–Westphalian) rocks of Europe (e.g., Brauckmann and Koch 1983) and North America. All are placed in the modern family Thelyphonidae. Fossil uropygids are very similar in morphology to Recent forms.

## Schizomida Petrunkevitch, 1945a

Schizomids are small (2 mm to 15 mm) arachnids resembling the larger whip scorpions, but they have only a short flagellum, a single pair of book lungs, and a divided carapace. Schizomids lack eyes and have an antenniform leg 1 and semi-raptorial pedipalps. Approximately eighty species live in the temperate and tropical regions of Asia, Africa, and the Americas. They are predators in leaf litter or soil environments. Autapomorphies proposed for the Schizomida are a specialized pygidial flagellum in the male and an enlarged leg 4 femur.

Schizomida are included in Uropygi by some authors; for example, Shultz (1990) divided Uropygi into Thelyphonida and Schizomida. There is little doubt that schizomids represent miniaturized Uropygi.

Three species of schizomids have been described from the Pliocene "Onyx Marble" quarries of Arizona (Petrunkevitch 1945b) and one from the Oligocene of China (Lin et al. 1988). Undescribed specimens are known from the Oligocene Dominican amber.

## Haptopoda Pocock, 1911

This monotypic order was established on the basis of the subdivided tarsus of the first leg. Petrunkevitch (1949) cleaned and reexamined the specimens and redefined the order based on a new interpretation of the abdominal segmentation. Only nine specimens of the 12 mm long single species, *Plesiosiro madeleyi* Pocock, 1911, are known, all from a single British Upper Carboniferous (Westphalian B) locality. Haptopoda have an opisthosoma with a segment pattern similar to that of the Uropygi and a carapace with what may be a pair of median eyes and two lateral eye tubercles. The anterior legs are slightly spinous, and all legs have subdivided tarsi. The pedipalps are pediform, and the animal has a broad prosoma-opisthosoma junction.

Cheliceral morphology and respiratory organs are not clearly preserved, making referral to any higher taxon difficult. No distinct autapomorphies have been proposed for the Haptopoda. Some features are reminiscent of Uropygi, and an alternative possibility is that this is an unusual form of opilionid (Shear and Kukalová-Peck 1990). Since it shows the autapomorphies of neither group, its status as a distinct order appears deserved.

## Phalangiotarbida Haase, 1890

Phalangiotarbida (= Architarbida) is an extinct group known only from a few Upper Carboniferous (Westphalian) Euramerican localities (Petrunkevitch 1953, 1955), but where they do occur they can be locally abundant. Phalangiotarbids have 10 mm to 20 mm long, flattened, oval bodies with stout walking legs, small pediform palps, and tiny chelicerae, possibly hidden in a pre-oral chamber above the coxae of leg 1. The sternum is divided, and the opisthosomal segmentation pattern with abbreviated anterior tergites and longitudinally divided sternites is unique among arachnids. Specific autapomorphies have not been proposed for the Phalangiotarbida, but they could include: abbreviated anterior tergites, a median eye tubercle bearing six eye lenses, and extreme reduction of the pedipalps and chelicerae. Twenty-six species are recorded. Their mode of life is unclear, but they have been interpreted as ambush predators on account of the orientation of their stout legs. It has been suggested that their morphology mimicked lycopod leaves (Beall 1984). The Phalangiotarbida were renamed Architarbi by Petrunkevitch (1945a), but this change was not justified (Selden 1993b).

Phalangiotarbids have at times been likened to Opiliones (Pocock 1911), Haptopoda (Petrunkevitch 1913), Ricinulei, and anactinotrichid mites (van der Hammen 1979). Shultz (1990) tentatively suggested that they belonged in his taxon Micrura, comprising arachnids with a pygidium. An unpublished manuscript by Kjellesvig-Waering figured a supposed clasp-knife chelicera on one specimen. If this observation is correct, then these animals could belong in Tetrapulmonata. The lack of convincing synapomorphies with other arachnids makes placing phalangiotarbids difficult, and restudy of this material to try to identify new characters would be helpful.

## Ricinulei Thorell, 1876

Ricinuleids are blind, slow-moving, predatory arachnids with a body length of 10 mm to 20 mm. Twenty-five living species in three genera are known from leaf litter and caves in tropical regions of West Africa and the Americas. The mouthparts are two-segmented, sub-chelate chelicerae, which are hidden beneath a movable hood (cucullus) that is hinged onto the anterior margin of the carapace. Ricinuleids have a thick cuticle, anteriorly located gonopore, pedipalps with a small terminal chela, and a double trochanter on legs 3 and 4. Respiration is through tracheae, the spiracles opening at the rear of the prosoma. Autapomorphies proposed for the Ricinulei are the cucullus, a male copulatory organ on leg 3, an endosternite with one segmental component, divided tergites, and a coupling device between the prosoma and opisthosoma. The latter two characters have analogues in the Trigonotarbida.

Selden (1992) split the Ricinulei into the extant Neoricinulei (vestigial eyes; coxa 2 larger than coxae 3 or 4) and fossil Palaeoricinulei (eyed; coxa 2 smaller than coxae 3 or 4). Palaeoricinulei are further subdivided into Poliocheridae, with opisthosomal tergites similar to Neoricinulei, and Curculioididae, without opisthosomal tergites, but instead a single median dorsal dividing line or sulcus. Most recent authors have placed ricinuleids as the sister group to the Acari on account of similar mouthparts, comprising a "gnathosoma," and a hexapodous larva.

The first fossil ricinuleid was illustrated by Buckland in 1837, one year before the first extant species was described (Guerin-Méneville 1838). Fourteen fossil species of fossil Ricinulei are now known from the Upper Carboniferous (Namurian-Westphalian) of Europe and North America. A revision of the fossils (Selden 1992) revealed a greater diversity in the Carboniferous than today, but based on an essentially similar bodyplan. None is yet known from any other geological period. It appears that the group has remained in warm, humid habitats (equatorial forest litter and caves) throughout its geological history.

## Acari Sundevall, 1833

The Acari (mites, ticks) may be the most diverse of living chelicerate orders, although more spiders have been named. Thirty thousand species have been described worldwide, but this may represent only a small proportion of their total diversity. Acari are small (80 µm to 30 mm), primarily terrestrial arachnids that occupy a wide range of niches and include predators, parasites (of plants and animals), and detritivores. Acari are characterized by poorly defined body segmentation and a broad prosoma-opisthosoma junction. They respire with tracheae that can open on any body segment. Acari have an ovipositor and a hexapodous larva (see Lindquist [1984] for additional characters). The only autapomorphy proposed for Acari is the presence of a gnathosoma, a modified region of the prosoma bearing the chelicerae and pedipalps (Weygoldt and Paulus 1979). This character was considered synapomorphic for Acari + Ricinulei by Shultz (1990). The mouthparts of Acari are variable within the group, which reflects their diverse feeding habits.

There has been debate about whether or not the Acari are monophyletic (Lindquist 1984) or polyphyletic (van der Hammen 1989). Most authors agree that Anactinotrichida (Opilioacarida + Parasitiformes) and Actinotrichida form two fundamental branches of the Acari, but in a series of papers on the subject van der Hammen developed the idea of a separate origin of these groups (summarized in van der Hammen 1972). He argued from comparative morphology that Anactinotrichida were more closely allied to Ricinulei than to other mites (van der Hammen 1979) and that Actinotrichida were closer to Palpigradi (van der Hammen 1982). The cladistic analyses of Lindquist (1984) and Shultz (1990) concluded that Acari is monophyletic and that Ricinulei is the sister group.

The oldest mites are Actinedida (Prostigmata) from the Lower Devonian (Pragian) Rhynie Chert of Scotland (Hirst 1923). Other Devonian Actinotrichida are known from Gilboa, New York (Norton et al. 1988; Kethley et al. 1989). A few Jurassic and Cretaceous Actinotrichida are known (e.g., Bulanova-Zachvatina 1974; Krivolutsky and Ryabinin 1976; Sivhed and Wallwork 1978), but the majority of fossil mites are oribatids from Baltic amber (e.g., Koch and Berendt 1854; Sellnick 1918, 1931). Anactinotrichida are very poorly represented in the fossil record; there are no fossil Opilioacarida or Holothyrida and only a few records of Ixodida (Scudder 1890, somewhat suspect) and Gamasida (Hirschmann 1971; Blaszak et al. 1995).

## Chelicerate Fossils and Phylogenies

There have been many attempts to elucidate the evolutionary relationships within the Chelicerata. Some authors included fossil and Recent chelicerates in the same scheme, for example, Størmer (1944), Petrunkevitch (1949), Bristowe (1971), Savory (1971), Manton (1977), Grasshoff (1978), Beall and Labandeira (1990), and Starobogatov (1990). Others, for example, Firstman (1973), Yoshikura (1975), and van der Hammen (1989), excluded fossils from their analyses. The first attempt at a cladistic analysis of Chelicerata was made by Weygoldt and Paulus (1979). Recent studies of exceptionally well preserved Paleozoic chelicerates (Selden 1981; Shear et al. 1987; Jeram et al. 1990; Selden et al. 1991) have provided much greater detail of the morphology of these forms, such as Eurypterida, Scorpiones, and Trigonotarbida, and consequently they can be included in cladistic analyses with greater confidence (Shultz 1989, 1990; Selden 1990a, 1993b).

Figure 7.2 shows cladograms derived from the analyses of Shultz (1989, 1990), Weygoldt and Paulus (1979), and van der Hammen (1989). Details of the character states can be found in these papers. One point of disagreement between the cladograms lies in the placement of Opiliones. Weygoldt and Paulus's (1979) analysis (fig. 7.2c) placed Opiliones in a derived position as sister group to Ricinulei + Acari, while in the analyses of van der Hammen (1989) and Shultz (1990) Opiliones occupies a basal position and is sister group to Scorpiones + other groups. Whereas earlier authors (e.g., Savory 1971) considered Palpigradi to be primitive arachnids, it is interesting that all of the studies depicted in figure 7.2 show Palpigradi not to be primitive, though there is no agreement concerning their relationship to any other arachnid group. There is agreement concerning the close relationship of the following: Ricinulei + all or some Acari; Tetrapulmonata (Araneae, Amblypygi, Uropygi,

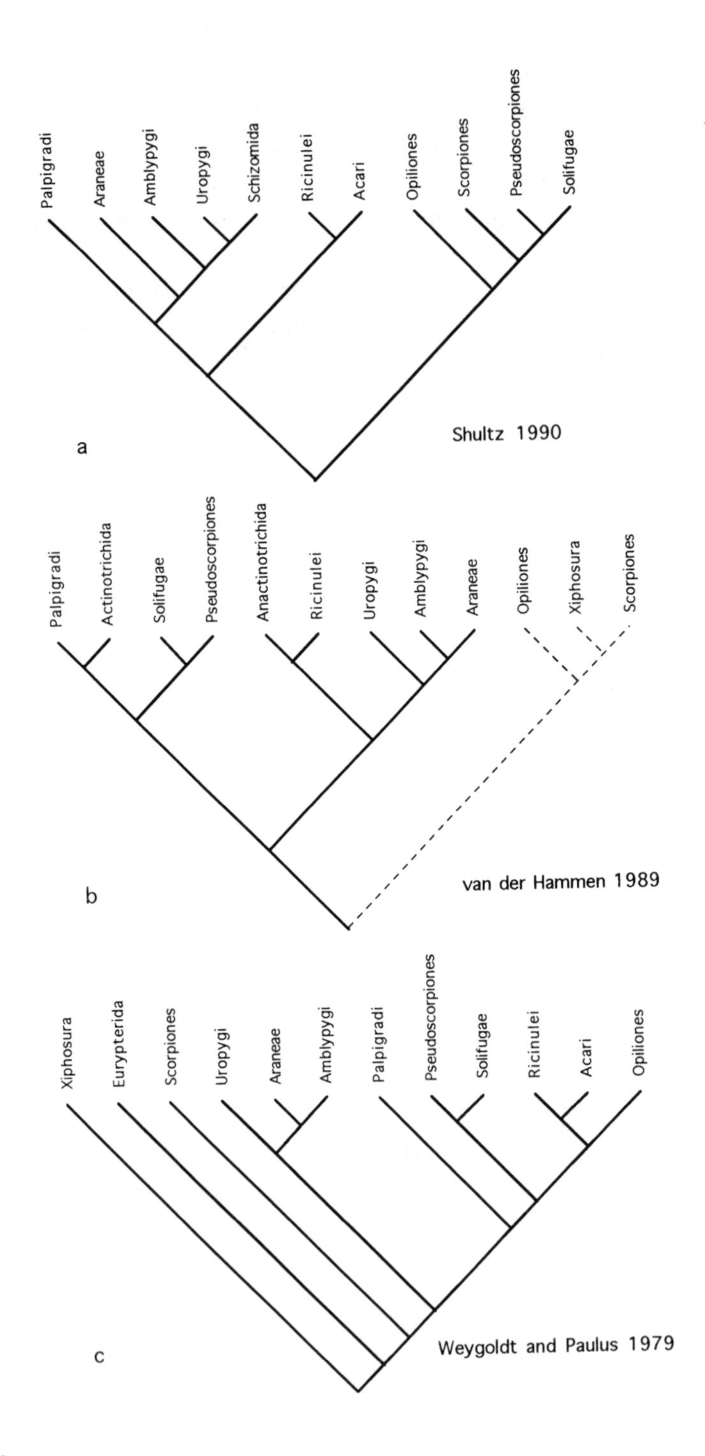


FIGURE 7.2

Cladograms of relationships among the arachnid groups as viewed by *(a)* Shultz (1990); *(b)* van der Hammen (1989); and *(c)* Weygoldt and Paulus (1979). Interrupted lines indicate uncertainty.

Schizomida); Solifugae + Pseudoscorpiones. How do these analyses correspond to the fossil record of Chelicerata?

Figure 7.3 shows an evolutionary tree produced by combining the fossil record of Chelicerata with a cladogram. The fossil record is the most up-to-date possible and uses both published and unpublished data. The cladogram combines the analyses of Wills et al. (1995) for major chelicerate groups and Shultz (1990) for arachnids. Fine vertical lines show range extensions of named taxa and ghost lineages (see Smith 1994 for explanation) that are predicted by the cladogram. The longest range extensions are predicted for Palpigradi, Anactinotrichida, and Schizomida (this is predicted by every analysis shown in fig. 7.2). This is not surprising since all of these animals are tiny, lightly sclerotized, and generally occur in habitats that are poorly represented in the fossil record. Thus, they not only have low fossilization potential but are also easily overlooked. Other predictions of this evolutionary tree are considered in detail below.

## Scorpions, Pseudoscorpions, and Solifuges

There are four opinions concerning the position of scorpions within the Chelicerata: Arachnida is monophyletic and scorpions are true arachnids (e.g., Shultz 1990); Scorpiones is the sister group to all other arachnids (e.g., Weygoldt and Paulus 1979); Scorpiones is the sister group to Eurypterida (e.g., Grasshoff 1978; Kjellesvig-Waering 1986); Scorpiones is the sister group to Xiphosura (van der Hammen 1989). Van der Hammen's scheme appears the most unusual; his analysis was not cladistic in its approach, and these relationships were considered tentative by its author. In its favor, however, is the finding of Anderson (1973) that scorpion embryology differs from that of other arachnids and is closer to that of Xiphosura and that while other arachnids could be derived from a xiphosuran-like embryology, they could not be derived from the scorpion pattern. In Shultz's (1989, 1990) scheme (fig. 7.2a), scorpions are sister to pseudoscorpions + solifuges, and Opiliones is the sister group to (Scorpiones (Pseudoscorpionida + Solifugae)). Shultz (1990) proposed the names Novogenuata for (Scorpiones (Pseudoscorpionida + Solifugae)), and Dromopoda for (Opiliones (Scorpiones (Pseudoscorpionida + Solifugae))). Dromopoda has the following synapomorphies: extensor muscles and specialized joints at femoropatellar and patellotibial joints, distinct transverse furrows on carapace (representing primitive tergite boundaries), reduced intercoxal sternal region, prosomal endosternite with two components, undivided leg 3 and 4 femora, pretarsal depressor muscle with a patellar head, and stomotheca (pre-oral chamber formed by pedipalp and anterior walking leg endites). A number of these characters are further modified in Solifugae. Weygoldt and Paulus (1979) proposed eye rhabdome morphology, sperm morphology, the presence of lyriform organs, and tactile leg 1 as synapomorphies of the non-scorpion arachnids. Kjellesvig-Waering (1986) argued for a common scorpion-eurypterid ancestor on the basis of similar morphology, e.g., abdominal plates, multifaceted lateral eyes (in early scorpions) and three-segmented

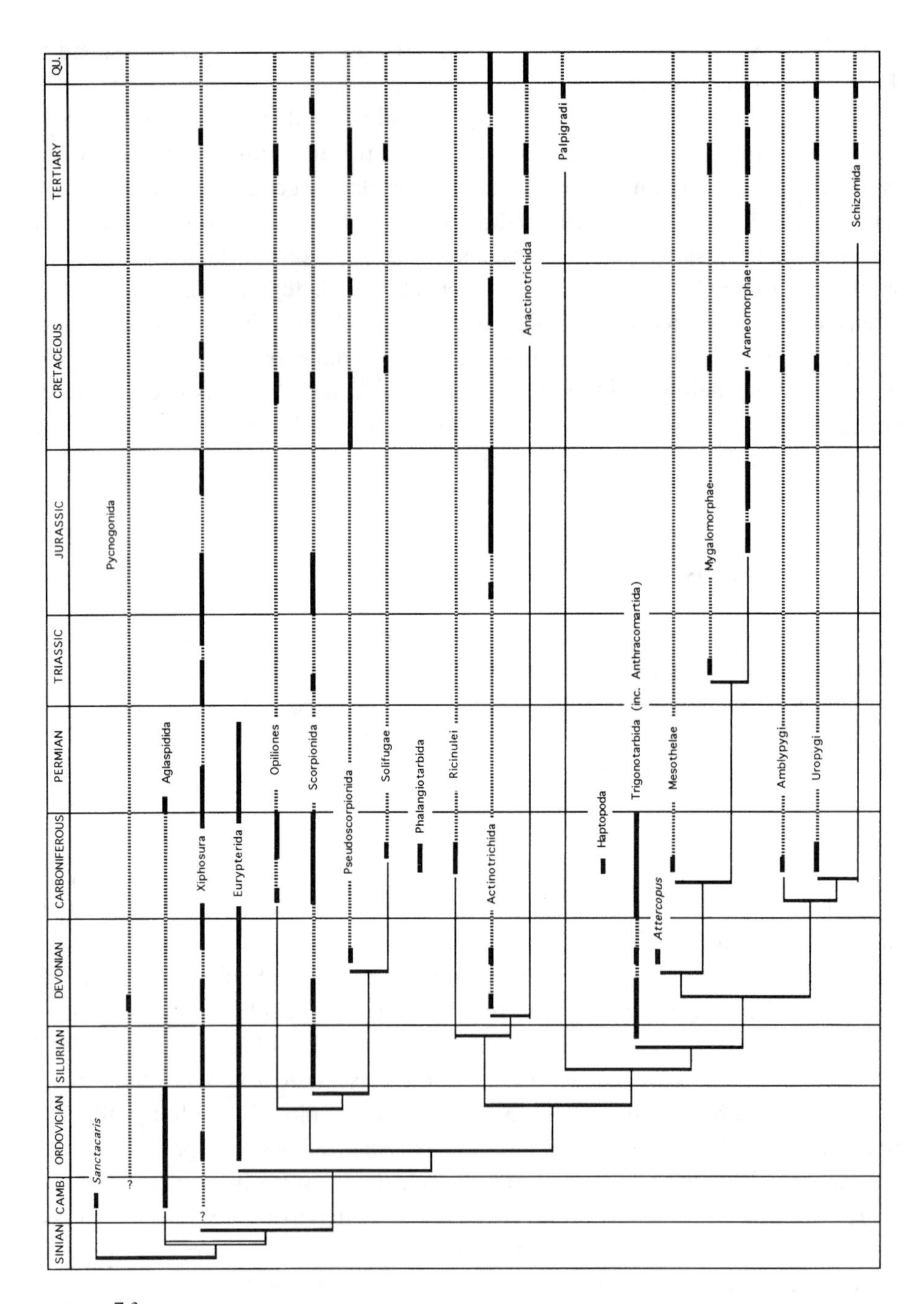


FIGURE 7.3

Evolutionary tree combining results of the most recent cladistic analyses of Chelicerata (Shultz 1990; Wills et al. 1995) with current knowledge of the fossil record of Chelicerata and relatives. Thick vertical lines denote actual occurrence in the stratigraphic stage(s) concerned; interrupted lines indicate presumed occurrence in intervening stages. *?* denotes doubtful record. Open horizontal lines show relationships derived from cladistic analyses; thin vertical lines denote range extensions and ghost lineages predicted from the analyses. Note that taxon ranks are not equivalent; stratigraphic axis not to scale.

chelate chelicerae, which are plesiomorphic character states. Grasshoff (1978) placed scorpions and eurypterids together on account of their overall body shape, which he interpreted as an adaptation for swimming.

The light the fossil record sheds on this problem is that scorpions are the oldest arachnid group; the earliest specimens known are from Lower Silurian strata. The early scorpions were aquatic (see above), but since they clearly show scorpion apomorphies, they must have emerged onto land separately from other arachnid groups (see Selden and Jeram [1989] for a discussion). Thus, terrestrial arachnid characters, such as book lungs and trichobothria, are convergent adaptations to life on land.

## Opiliones

In Shultz's (1989, 1990) phylogenetic scheme, Opiliones are the sister group to his Novogenuata (see above), so their fossils should occur in strata at least as old as Lower Silurian. Since scorpions were aquatic at that time, presumably opilionids would have been, too. In van der Hammen's (1989) phylogeny, Opiliones came out as the sister group to Xiphosura + Scorpiones; since the oldest xiphosuran is Lower Ordovician or older, Opiliones would have been around in aquatic environments in the Lower Ordovician in this scheme. In Weygoldt and Paulus's (1979) scheme, Opiliones occupy a rather derived position.

The prediction of Silurian aquatic Opiliones was discussed by Selden (1990a). Shultz (1994), in a critique of the use of stratigraphic evidence to assess phylogenetic hypotheses, pointed out that the earliest members of the Opiliones clade, after it had diverged from Novogenuata, may not have had all of the apomorphies of the clade as now recognized. This is a common problem with fossils, since apomorphies are collected during evolution of the stem group, and the full complement might occur only in the crown group. Whether the name Opiliones is used for the total group (stem group + crown group) or the crown group alone is debatable. However, at the divergence of two clades, there must be at least one defining apomorphy for the two clades to be recognized; Shultz's (1994) contention that "the earliest members of two sister lineages may be indistinguishable from one another and from their immediate ancestors in morphology, behavior, genetics, etc." is unsustainable.

## Tetrapulmonata

This clade, comprising the orders Trigonotarbida, Araneae, Amblypygi, Uropygi, and Schizomida, is probably the best supported and least contentious area of arachnid phylogeny. As its name suggests, it is diagnosed by two pairs of book lungs that are situated on opisthosomal segments 2 and 3 (subsequently modified in some groups, e.g., many spiders and all schizomids have only one pair of lungs and some spiders have tracheae only). Tetrapulmonata is strongly supported by a range of other synapomorphies, including two-segmented, clasp-knife-like chelicerae, plagula ventralis, a sucking stomach, constriction of segment 7 (pedicel), 9 + 2 axoneme arrangement in the spermatozoa, and prosomal endosternite with four segmental components.

Almost all phylogenetic studies (Firstman 1973; Grasshoff 1978; Weygoldt and Paulus 1979; van der Hammen 1989; Shultz 1990) have placed spiders, Amblypygi, Uropygi, and Schizomida together, and a number of names have been proposed for this taxon (see Shear et al. 1987), of which Tetrapulmonata Shultz, 1990, is the most appropriate and best defined. These studies differed only in whether Amblypygi was placed closer to Araneae (Firstman 1973; Grasshoff 1978; Weygoldt and Paulus 1979; van der Hammen 1989) or to Uropygi + Schizomida (Shultz 1990). Two pairs of book lungs were identified in the Rhynie Chert trigonotarbid arachnids (Claridge and Lyon 1961). Shear et al. (1987) and Selden et al. (1991) included Trigonotarbida in Tetrapulmonata. These authors' analyses supported placing Amblypygi as the sister group of Uropygi + Schizomida and placed Trigonotarbida as the sister taxon to all other tetrapulmonates.

Trigonotarbids are among the oldest terrestrial animals (Jeram et al. 1990), and the evolutionary tree (fig. 7.3) predicts that other tetrapulmonates should be found alongside trigonotarbids in early terrestrial ecosystems. The oldest spider is *Attercopus* from the Devonian of New York (Shear et al. 1987; Selden et al. 1991), and it is sister to other spiders that arose in the Devonian or earlier. The first mesothele occurs in the Late Carboniferous (Selden 1996), which predicts that opisthotheles (Mygalomorphae + Araneomorphae) should have originated by then. The long gap in the fossil record between the Late Carboniferous and the Triassic (first mygalomorph) and Jurassic (first described araneomorph) occurrences of fossil spiders is due largely to the poor preservation generally of fossils in Permian strata in parts of the world where specimens are regularly collected.

## ACKNOWLEDGMENTS

We thank Lyall Anderson for information on Xiphosura, Simon Braddy for information on Eurypterida, and both for useful discussions. We thank Bill Shear for reading and commenting on the manuscript. We acknowledge support from the Natural Environment Research Council.

## REFERENCES

Anderson, D. T. 1973. *Embryology and Phylogeny in Annelids and Arthropods.* Oxford: Pergamon Press.

Anderson, L. I. and P. A. Selden. 1997. Opisthosomal fusion and phylogeny of Palaeozoic Xiphosura. *Lethaia* 30:19–31.

Beall, B. S. 1984. Functional and ecological analyses of phalangiotarbid arachnids. *Geological Society of America, Abstracts with Programs* 15:440.

Beall, B. S. 1986. Reinterpretation of the Kustarachnida. *American Arachnology* 34:4.

Beall, B. S. and C. C. Labandeira. 1990. Macroevolutionary patterns of the Chelicerata and Tracheata. In D. G. Mikulic, ed., *Short Courses in Paleontology.* Vol. 3, *Arthropod Palaeobiology,* pp. 257–284. Knoxville: Paleontological Society.

Bergström, J. 1971. *Palaeomerus:* Merostome or merostomoid. *Lethaia* 4:393–401.

Bergström, J. 1979. Morphology of fossil arthropods as a guide to phylogenetic relationships. In A. P. Gupta, ed., *Arthropod Phylogeny,* pp. 3–56. New York: Van Nostrand Reinhold.

Bergström, J. 1980. Morphology and systematics of early arthropods. *Abhandlungen des naturwissenschaftlichen Vereins in Hamburg (NF)* 23:7–42.

Bergström, J., W. Stürmer, and G. Winter. 1980. *Palaeoisopus, Palaeopantopus,* and *Palaeothea:* Pycnogonid arthropods from the Lower Devonian Hunsrück Slate, West Germany. *Paläontologische Zeitschrift* 54:7–54.

Blaszak, C., J. C. Cokendolpher, and V. J. Polyak. 1995. *Paleozercon cavernicolus,* n. gen., n. sp.: Fossil mite from a cave in the southwestern U.S.A. (Acari, Gamasida: Zerconidae), with a key to Nearctic genera of Zerconidae. *International Journal of Acarology* 21:253–259.

Boudreaux, H. B. 1979. Significance of intersegmental tendon system in arthropod phylogeny and a monophyletic classification of Arthropoda. In A. P. Gupta, ed., *Arthropod Phylogeny,* pp. 551–586. New York: Van Nostrand Reinhold.

Brauckmann, C. and L. Koch. 1983. *Prothelyphonus naufragus* n. sp.: Ein neuer Geisselscorpion (Arachnida: Thelyphonida: Thelyphonidae) aus dem Namurium (unteres Ober-Karbon) von West-Deutschland. *Entomologica Generalis* 9:63–73.

Briggs, D. E. G., D. L. Bruton, and H. B. Whittington. 1979. Appendages of the arthropod *Aglaspis spinifer* (Upper Cambrian, Wisconsin) and their significance. *Palaeontology* 22:167–180.

Briggs, D. E. G. and D. Collins. 1988. A Middle Cambrian chelicerate from Mount Stephen, British Columbia. *Palaeontology* 31:779–798.

Bristowe, W. S. 1971. *The World of Spiders.* 2d Edition. London: Collins.

Broili, F. 1928. Crustaceenfunde aus dem rheinischen Unterdevon. *Sitzungsberichte Bayerischen Akademie der Wissenschaften* 1928:197–201.

Buckland, W. 1837. *The Bridgewater Treatises on the Power, Wisdom, and Goodness of God as Manifested in the Creation: Treatise IV., Geology and Mineralogy Considered with Reference to Natural Theology.* 2d Edition. London: William Pickering.

Bulanova-Zachvatina, E. M. 1974. New genera of oribatid mites from the Upper Cretaceous of Taimyr. *Paleontologicheskii Zhurnal* 2:141–144.

Burmeister, H. 1843. *Die Organisation der Trilobiten, aus ihren lebenden Verwandten entwickelt; nebst systematischer Uebersicht aller zeither beschribenen Arten.* Berlin: G. Riemer.

Campos, D. R. B. 1986. Primeiro registro fóssil de Scorpionoidea na Chapado do Araripe (Cretaceo Inferior), Brasil. *Anais da Academia Brasileira de Ciencias* 58:135–137.

Carpenter, F. M. 1992. Hexapoda. *Treatise on Invertebrate Paleontology, Part R, Arthropoda 4.* Vol. 3. Boulder, Lawrence: Geological Society of America, University of Kansas Press.

Caster, K. E. and H. K. Brooks. 1956. New fossils from the Canadian-Chazyan (Ordovician) hiatus in Tennessee. *Bulletins of American Paleontology* 36:157–199.

Caster K. E. and E. N. K. Kjellesvig-Waering. 1964. Upper Ordovician eurypterids of Ohio. *Palaeontographica Americana* 4:301–358.

Claridge, M. F. and A. G. Lyon. 1961. Lung-books in the Devonian Palaeocharinidae (Arachnida). *Nature* 191:1190–1191.

Clarke J. M. and R. Ruedemann. 1912. The Eurypterida of New York. *New York State Museum Natural History Memoirs* 14:1–628.

Clerck, C. 1757. *Svenska spindlar (Aranea Suecici) L. Salvii* (in Swedish and Latin). Stockholm.

Coddington, J. A. and H. W. Levi. 1991. Systematics and evolution of spiders (Araneae). *Annual Reviews of Ecology and Systematics* 22:565–592.

Dana, J. D. 1852. Crustacea. *United States Exploring Expedition,* 1838–1842 13:21.

de Geer, C, 1778. *Mémoires pour servir a l'histoire des Insectes.* Vol. 7. Stockholm.

Dunlop, J. A. 1994. An Upper Carboniferous amblypygid from the Writhlington Geological Nature Reserve. *Proceedings of the Geologists' Association* 105:245–250.

Dunlop, J. A. 1996. Systematics of the fossil arachnids. *Revue Suisse de Zoologie* vol. hors série: 173–184.

Dunlop, J. A. and P. A. Selden. 1997. The early history and phylogeny of the chelicerates. In R. A. Fortey and R. H. Thomas, eds., *Arthropod Relationships,* pp. 221–235. London: Chapman and Hall.

Eskov, K. 1984. A new fossil spider family from the Jurassic of Transbaikalia (Araneae: Chelicerata). *Neues Jahrbuch für Geologie und Paläontologie, Monatshefte* 1984:645–653.

Eskov, K. 1990. Spider Paleontology: Present trends and future expectations. *Acta Zoologica Fennica* 190:123–127.

Firstman, B. 1973. The relationship of the chelicerate arterial system to the evolution of the endosternite. *Journal of Arachnology* 1:1–54.

Flower, R. 1968. Merostomes from a Cotter horizon of the El Paso Group. *New Mexico State Bureau of Mines and Mineral Resources Memoir* 22:35–44.

Foelix, R. F. 1982. *Biology of Spiders.* Cambridge, Mass.: Harvard University Press.

Fritsch, A. 1901. *Fauna der Gaskohle und der Kalksteine der Permformation Böhmens,* 4:64. Prague.

Gall, J.-C. 1971. Faunes et paysages du Grès a Voltzia du nord des Vosges: Essai paléoécologique sur le Buntsandstein supérieur. *Mémoires du Service de la Carte Géologique d'Alsace et de Lorraine* 34:1–318.

Grasshoff, M. 1978. A model of the evolution of the main chelicerate groups. *Symposia of the Zoological Society of London* 42:273–284.

Guerin-Méneville, M. 1838. Note sur l'Acanthodon et sur le Cryptostemme, nouveaux génères d'Arachnides. *Revue Zoologique Société Cuvier* 1:10–12.

Haase, E. 1890. Beiträge zur Kenntnis der fossilen Arachniden. *Zeitschrift der deutschen geologischen Gesellschaft* 42:629–657.

Hammen, L. van der. 1972. A revised classification of the mites (Arachnidea, Acarida) with diagnoses, a key, and notes on phylogeny. *Zoologische Mededeelingen, Leiden* 47: 273–292.

Hammen, L. van der. 1979. Comparative studies in Chelicerata, part 1: The Cryptognomae (Ricinulei, Architarbi, and Anactinotrichida). *Zoologische Verhandelingen, Leiden* 174:1–62.

Hammen L. van der. 1982. Comparative studies in Chelicerata, part 2: Epimerata (Palpigradi and Actinotrichida). *Zoologische Verhandelingen, Leiden* 196:1–70.

Hammen, L. van der. 1985. Functional morphology and affinities of extant Chelicerata in evolutionary perspective. *Transactions of the Royal Society of Edinburgh, Earth Sciences* 76:137–146.

Hammen, L. van der. 1986. Comparative studies in Chelicerata, part 4: Apatellata, Arachnida, Scorpionida, Xiphosura. *Zoologische Verhandelingen, Leiden* 226:1–52.

Hammen, L. van der. 1989. *An Introduction to Comparative Arachnology.* The Hague: SPB Academic Publishing.

Handlirsch, A. 1906. *Die fossilen Inseckten und die Phylogenie der rezenten Formen.* Leipzig: Wilhelm Engelman.

Harvey, M. S. 1992. The phylogeny and classification of the Pseudoscorpionida (Chelicerata: Arachnida). *Invertebrate Taxonomy* 6:1373–1435.

Hedgpeth, J. W. 1955a. Pycnogonida. In R. C. Moore, ed., *Treatise on Invertebrate Paleontology, Part P, Arthropoda 2,* pp. P163–P170. Boulder, Lawrence: Geological Society of America, University of Kansas Press.

Hedgpeth, J. W. 1955b. *Palaeoisopus.* In R. C. Moore, ed., *Treatise on Invertebrate Paleontology, Part P, Arthropoda 2,* pp. P171–P173. Boulder, Lawrence: Geological Society of America, University of Kansas Press.

Heymons, R. 1901. Die Entwicklungsgeschichte der Scolopender. *Zoologica* 33:1–244.

Hirschmann, W. 1971. A fossil mite of the genus *Dendrolaelaps* (Acarina, Mesostigmata, Digamasellidae) found in amber from Chiapas, Mexico. In A. Petrunkevitch, ed., *Studies of Fossiliferous Amber Arthropods of Chiapas, Mexico.* University of California Publications, Entomology 63:69–70.

Hirst, S. 1923. On some arachnid remains from the Old Red Sandstone (Rhynie Chert bed, Aberdeenshire). *Annals and Magazine of Natural History* 9:455–474.

Jell, P. A. and P. M. Duncan. 1986. Invertebrates, mainly insects, from the freshwater, Lower Cretaceous, Koonwarra Fossil Bed (Korumburra Group), South Gippsland, Victoria. *Memoirs of the Association of Australasian Palaeontologists* 3:111–205.

Jeram, A. J. 1990. Book-lungs in a Lower Carboniferous scorpion. *Nature* 342:360–361.

Jeram, A. J. 1994. Carboniferous Orthosterni and their relationship to living scorpions. *Palaeontology* 37:513–550.

Jeram, A. J., P. A. Selden, and D. Edwards. 1990. Land animals in the Silurian: Arachnids and myriapods from Shropshire, England. *Science* 250:658–661.

Jux, U. 1982. *Somaspidion hammapheron* n. gen. n. sp.: Ein Arachnide aus dem Oberkarbon der subvaristischen Saumsenke NW Deutschlands. *Paläontologische Zeitschrift* 56:77–86.

Karsch, F. 1882. Über einige neues Spinnenthier aus der Schlesischen Steinkohle und die Arachnoiden der Steinkohlenformation überhaupt. *Zeitschrift der deutschen geologischen Gesellschaft* 34:556–561.

Kethley, J. B., R. A. Norton, P. M. Bonamo, and W. A. Shear. 1989. A terrestrial alicorhagiid mite (Acari: Acariformes) from the Devonian of New York. *Micropaleontology* 35:367–373.

King, P. E. 1973. *Pycnogonids.* London: Hutchinson.

Kjellesvig-Waering, E. N. 1986. A restudy of the fossil Scorpionida of the world. *Palaeontographica Americana* 55:1–299.

Koch, C. L. and G. C. Berendt. 1854. *Die im Bernstein befindlichen Crustaceen, Myriapoden, Arachniden, und Apteren der Vorwelt.* Berlin.

Kraus, O. 1976. Zur phylogenetischen Stellung und Evolution der Chelicerata. *Entomologica Germanica* 3:1–12.

Krivolutsky, D. A. and N. A. Ryabinin. 1976. Oribatid mites in Siberian and Far East amber. *Reports of the Academy of Science, USSR* 230:945–948.

Latreille, P. A. 1802. *Histoire naturelle, générale, et particulière: Crustacés et des Insectes.* Vol. 3. Paris: Dufart.

Latreille, P. A. 1810. *Considerations générales sur l'ordre naturel des animaux composant la classe des Crustacés, des Arachnides, et des Insectes.* Paris: Dufart.

Laurie, M. 1889. On a Silurian scorpion and some additional eurypterid remains from the Pentland Hills. *Transactions of the Royal Society of Edinburgh* 39:515–590.

Lin, Q.-B., Y.-M. Yao, W.-D. Xiang, and Y.-R. Xia. 1988. An Oligocene micropalaeoentomofauna from Gubei District of Shandong and its ecological environment. *Acta Micropalaeontologica Sinica* 5:331–345.

Lindquist, E. E. 1984. Current theories on the evolution of the major groups of Acari and on their relationships with other groups of Arachnida, with consequent implications for their classification. In D. A. Griffiths and C. E. Brown, eds., *Acarology,* 1:28–62. New York: John Wiley.

Manton, S. M. 1977. *The Arthropoda: Habits, Functional Morphology, and Evolution.* Oxford: Clarendon.

Moberg, J. C. 1892. Om en nyupptäckt fauna i block af kambrisk sandsten, insamlade af dr N. O. Holst. *Geologiska Föreningens i Stockholm Förhandlingar* 14:103–120.

Monniot, F. 1966. Un palpigrade interstitiel: *Leptokoenenia scurra,* n. sp. *Revue d'Écologie et de Biologie du Sol* 3:41–64.

Norton, R. A., P. M. Bonamo, J. D. Grierson, and W. A. Shear. 1988. Oribatid mite fossils from a terrestrial Devonian deposit near Gilboa, New York. *Journal of Paleontology* 62:259–269.

Oplustil, S. 1985. New findings of Arachnida from the Bohemian Upper Carboniferous. *Véstník ústředního ústavu geologického* 60:35–42.

Packard, A. S. 1886. On the Carboniferous xiphosurous fauna of North America. *Memoirs of the National Academy of Sciences* 3:143–157.

Paulus, H. F. 1979. Eye structure and the monophyly of Arthropoda. In A. P. Gupta, ed., *Arthropod Phylogeny,* pp. 299–383. New York: Van Nostrand Reinhold.

Petrunkevitch, A. 1913. A monograph of the terrestrial Palaeozoic Arachnida of North America. *Transactions of the Connecticut Academy of Arts and Sciences* 18:1–137.

Petrunkevitch, A. 1945a. Palaeozoic Arachnida of Illinois: An inquiry into their evolutionary trends. *Illinois State Museum, Scientific Papers* 3:1–72.

Petrunkevitch, A. 1945b. *Calcitro fisheri:* A new fossil arachnid. *American Journal of Science* 243:320–329.

Petrunkevitch, A. 1949. A study of Palaeozoic Arachnida. *Transactions of the Connecticut Academy of Arts and Sciences* 37:69–315.

Petrunkevitch, A. 1953. Palaeozoic and Mesozoic Arachnida of Europe. *Memoirs of the Geological Society of America* 53:1–128.

Petrunkevitch, A. 1955. Arachnida. In R. C. Moore, ed., *Treatise on Invertebrate Paleontology, Part P, Arthropoda 2,* pp. P42–P162. Boulder, Lawrence: Geological Society of America, University of Kansas Press.

Pinto, I. D. and M. A. Hünicken. 1980. *Gondwanarachne:* A new genus of the Order Trigonotarbida (Arachnida) from Argentina. *Boletin de la Academia Nacional de Ciencias, Córdoba, Argentina* 53:307–315.

Plotnick, R. E. 1983. Patterns in the evolution of the eurypterids. Ph.D. diss., University of Chicago.

Pocock, R. I. 1902. *Eophrynus* and allied Carboniferous Arachnida. Part 1. *Geological Magazine* 9:439–448.

Pocock, R. I. 1911. A monograph of the terrestrial Carboniferous Arachnida of Great Britain. *Monographs of the Palaeontographical Society* 64:1–84.

Poinar, G. O. and J. A. Santiago-Blay. 1989. A fossil solpugid, *Happlodontus proterus,* new genus, new species (Arachnida: Solpugida) from Dominican amber. *Journal of the New York Entomological Society* 97:125–132.

Polis, G. A., ed. 1990. *Biology of Scorpions.* Stanford: Stanford University Press.

Raasch, G. O. 1939. Cambrian Merostomata. *Special Papers of the Geological Society of America* 19:1–146.

Rolfe, W. D. I. and C. M. Beckett. 1984. Autecology of Silurian Xiphosurida, Scorpionida, and Phyllocarida. In M. G. Bassett and J. D. Lawson, eds., *Autecology of Silurian Organisms: Special Papers in Palaeontology* 32:27–37.

Rowland J. M. and W. D. Sissom. 1980. Report on a fossil palpigrade from the Tertiary of Arizona, and a review of the morphology and systematics of the order (Arachnida: Palpigradida). *Journal of Arachnology* 8:69–86.

Savory, T. H. 1971. *Evolution in the Arachnida.* Watford: Merrow.

Schawaller, W. 1979. Erstnachweis der Ordnung Geisselspinnen in Dominicanischem Bernstein (Stuttgarter Bernsteinsammlung: Arachnida, Amblypygi). *Stuttgarter Beiträge zur Naturkunde aus dem Staatlichen Museum für Naturkunde in Stuttgart,* series B, 50:1–12.

Schawaller, W. 1982. Spinnen der Familien Tetragnathidae, Uloboridae, und Dipluridae in Dominicanischem Bernstein und allgemeine Gesichtspunkte (Arachnida, Araneae). *Stuttgarter Beiträge zur Naturkunde aus dem Staatlichen Museum für Naturkunde in Stuttgart,* series B, 89:1–19.

Schawaller, W. 1991. The first Mesozoic pseudoscorpion, from Cretaceous Canadian amber. *Palaeontology* 34:971–976.

Schawaller, W., W. A. Shear, and P. M. Bonamo. 1991. The first Palaeozoic pseudoscorpions (Arachnida, Pseudoscorpionida). *American Museum Novitates* 3009:1–17.

Scudder, S. H. 1890. Illustrations of the Carboniferous Arachnida of North America. *Memoirs of the Boston Society of Natural History* 4:443–456.

Selden, P. A. 1981. Functional morphology of the prosoma of *Baltoeurypterus tetragonophthalmus* (Fischer) (Chelicerata: Eurypterida). *Transactions of the Royal Society of Edinburgh, Earth Sciences* 72:9–48.

Selden, P. A. 1990a. Fossil history of the arachnids. *Newsletter of the British Arachnological Society* 58:4–6.

Selden, P. A. 1990b. Lower Cretaceous spiders from the Sierra de Montsech, northeast Spain. *Palaeontology* 33:257–285.

Selden, P. A. 1992. Revision of the fossil ricinuleids. *Transactions of the Royal Society of Edinburgh, Earth Sciences* 83:595–634.

Selden, P. A. 1993a. Arthropoda (Aglaspidida, Pycnogonida, and Chelicerata). In M. J. Benton, ed., *The Fossil Record 2,* pp. 297–320. London: Chapman & Hall.

Selden, P. A. 1993b. Fossil arachnids: Recent advances and future prospects. *Memoirs of the Queensland Museum* 33:389–400.

Selden, P. A. 1996. Fossil mesothele spiders. *Nature* 379:498–499.

Selden, P. A. and J.-C. Gall. 1992. A Triassic mygalomorph spider from the northern Vosges, France. *Palaeontology* 35:211–235.

Selden, P. A. and A. J. Jeram. 1989. Palaeophysiology of terrestrialisation in the Chelicerata. *Transactions of the Royal Society of Edinburgh, Earth Sciences* 80:303–310.

Selden, P. A. and M. Romano. 1983. First Palaeozoic arachnid from Iberia: *Aphantomartus areolatus* Pocock, (Stephanian; Prov. León, N.W. Spain), with remarks on aphantomartid taxonomy. *Boletin del Instituto de Geología e Mineralogía de Espana* 94:106–112.

Selden, P. A. and W. A. Shear. 1992. A myriapod identity for the Devonian "scorpion" *Tiphoscorpio hueberi. Berichte des naturwissenschaftlich-medizinischen Vereins in Innsbruck, Supplement* 10:35–36.

Selden, P. A. and W. A. Shear. 1996. First Mesozoic solpugid (Arachnida), from the Cretaceous of Brazil, and a redescription of the Palaeozoic solpugid. *Palaeontology* 39:583–604.

Selden, P. A., W. A. Shear, and P. M. Bonamo. 1991. A spider and other arachnids from the Devonian of New York, and reinterpretations of Devonian Araneae. *Palaeontology* 34:241–281.

Sellnick, M. 1918. Die Oribatiden der Bernsteinsammlung der Universität Königsberg. *Schriften der physikalisch-ökonomischen Gesellschaft zu Königsberg* 59:21–42.

Sellnick, M. 1931. Milben im Bernstein. *Bernsteinforschungen* 2:148–180.

Shear, W. A., P. M. Bonamo, J. D. Grierson, W. D. I. Rolfe, E. L. Smith, and R. A. Norton. 1984. Early land animals in North America. *Science* 224:492–494.

Shear, W. A. and J. Kukalová-Peck. 1990. The ecology of Palaeozoic terrestrial arthropods: The fossil evidence. *Canadian Journal of Zoology* 68:1807–1834.

Shear, W. A., W. Schawaller, and P. M. Bonamo. 1989. Record of Palaeozoic pseudoscorpions. *Nature* 341:527–529.

Shear, W. A. and P. A. Selden. 1995. *Eoarthropleura* (Arthropoda, Arthropleurida) from the Silurian of Britain and the Devonian of North America. *Neues Jahrbuch für Geologie und Paläontologie, Abhandlungen* 196:347–375.

Shear, W. A., P. A. Selden, W. D. I. Rolfe, P. M. Bonamo, and J. D. Grierson. 1987. New terrestrial arachnids from the Devonian of Gilboa, New York (Arachnida, Trigonotarbida). *American Museum Novitates* 2901:1–74.

Shultz, J. W. 1989. Morphology and locomotor appendages in Arachnida: Evolutionary trends and phylogenetic implications. *Zoological Journal of the Linnean Society* 108:335–365.

Shultz, J. W. 1990. Evolutionary morphology and phylogeny of Arachnida. *Cladistics* 6:1–38.

Shultz, J. W. 1994. The limits of stratigraphic evidence in assessing phylogenetic hypotheses of Recent arachnids. *Journal of Arachnology* 22:169–172.

Sivhed, U. and J. A. Wallwork. 1978. An Early Jurassic oribatid mite from southern Sweden. *Geologiska Föreningens i Stockholm Förhandlingar* 100:65–70.

Smith, A. B. 1994. *Systematics and the Fossil Record.* Oxford: Blackwell Scientific Publications.

Starobogatov, Y. I. 1990. The systematics and phylogeny of the lower chelicerates (a morphological analysis of the Paleozoic groups). *Paleontologicheskii Zhurnal* 1:4–17.

Stockwell, S. A. 1989. Revision of the phylogeny and higher classification of scorpions (Chelicerata). Ph.D. diss., University of California, Berkeley.

Størmer, L. 1944. On the relationships and phylogeny of fossil and recent Arachnomorpha: A comparative study on Arachnida, Xiphosura, Eurypterida, Trilobita, and other fossil Arthropoda. *Skrifter Utgitt av Det Norske Videnskaps-Akademi i Oslo, I. Matematisk-Naturvidenskapelig Klasse* 5:1–158.

Størmer, L. 1955. Merostomata. In R. C. Moore, ed., *Treatise on Invertebrate Paleontology, Part P, Arthropoda 2,* pp. P4–P41. Boulder, Lawrence: Geological Society of America, University of Kansas Press.

Størmer, L. 1970. Arthropods from the Lower Devonian (Lower Emsian) of Alken an der Mosel, Germany, part 1: Arachnida. *Senckenbergiana Lethaea* 51:335–69.

Størmer, L. 1972. Arthropods from the Lower Devonian (Lower Emsian) of Alken an der Mosel, Germany, part 2: Xiphosura. *Senckenbergiana Lethaea* 53:1–29.

Størmer, L. 1976. Arthropods from the Lower Devonian (Lower Emsian) of Alken an der Mosel, Germany, part 5: Myriapoda and additional forms, with general remarks on fauna and problems regarding invasion of land by arthropods. *Senckenbergiana Lethaea* 57:87–183.

Sundevall, J. C. 1833. *Onceptus Arachnidum.* Londini, Gothorum.

Thevenin, A. 1901. Sur le découverte d'arachnides dans le terrain houiller de Commentry. *Bulletin de la Société Géologique de France* 4:605–611.

Thorell, T. H. 1876. Sopra alcuni Opilioni (Phalangidea) d'Europa e dell'Asia occidentale, con un quadro dei generi europei di quest'ordine. *Anali del Museo Civico di Storia Naturale di Genova,* series 1, 8:452–408.

Tollerton, V. P. 1989. Morphology, taxonomy, and classification of the Order Eurypterida Burmeister, 1843. *Journal of Paleontology* 63:642–57.

Weygoldt, P. 1969. *The Biology of Pseudoscorpions.* Cambridge, Mass.: Harvard University Press.

Weygoldt, P. 1980. Towards a cladistic classification of the Chelicerata. *8th International Congress of Arachnology, Wien,* 1980:331–334.

Weygoldt, P. and H. F. Paulus. 1979. Untersuchungen zur Morphologie, Taxonomie, und Phylogenie der Chelicerata. *Zeitschrift für zoologische Systematik und Evolutionsforschung* 17:85–116.

Whalley, P. E. S. 1980. Neuroptera (Insecta) in amber from the Lower Cretaceous of Lebanon. *Bulletin of the British Museum (Natural History), Geology* 33:157–164.

Wills, L. J. 1947. A monograph of British Triassic scorpions. *Monograph of the Palaeontographical Society* 101:1–137.

Wills, M. A., D. E. G. Briggs, R. A. Fortey, and M. Wilkinson. 1995. The significance of fossils in understanding arthropod evolution. *Verhandlungen der deutschen zoologischen Gesellschaft* 88:203–215.

Wood, S. P., A. L. Panchen, and T. R. Smithson. 1985. A terrestrial fauna from the Scottish Lower Carboniferous. *Nature* 314:355–356.

Woodward. H. 1865. A monograph of the British Fossil Crustacea. *Monograph of the Palaeontographical Society* 19:1–43.

Yoshikura, M. 1975. Comparative embryology and phylogeny of Arachnida. *Kumamoto Journal of Science, Biology* 12:71–142.

# Contributors

Jan Bergström
Department of Palaeozoology
Swedish Museum of Natural History
Box 50007, S-104 05, Stockholm, Sweden

Derek E. G. Briggs
Department of Geology
University of Bristol
Queen's Road, Bristol BS8 1RJ, UK

Chen Junyuan
Nanjing Institute of Geology and Palaeontology
Academia Sinica, Chi-Ming-Ssu
Nanjing 210008, People's Republic of China

Jason A. Dunlop
Institut für Systematische Zoologie
Museum für Naturkunde
Invalidenstrasse 43, D-10115, Berlin, Germany

Gregory D. Edgecombe
Centre for Evolutionary Research
Australian Museum, 6 College Street,
Sydney South, NSW 2000, Australia

Richard A. Fortey
Department of Palaeontology
The Natural History Museum
Cromwell Road, London SW7 5BD, UK

Cees H. J. Hof
Institute for Systematics and Population Biology
University of Amsterdam, P.O. Box 94766
1090 GT Amsterdam, The Netherlands

Hou Xianguang
Nanjing Institute of Geology and Paleontology
Academia Sinica, Chi-Ming-Ssu
Nanjing 210008, People's Republic of China

Klaus J. Müller
Institut für Paläontologie
Nussallee 8, D-53115 Bonn, Germany

Lars Ramsköld
Museum of Palaeontology
University of Uppsala
Norbyvägen 22, Uppsala, Sweden

Frederick R. Schram
Institute for Systematics and Population Biology
University of Amsterdam, P.O. Box 94766
1090 GT Amsterdam, The Netherlands

Paul A. Selden
Department of Earth Sciences
University of Manchester
Manchester M13 9PL, UK

Peter H. A. Sneath
Microbiology and Immunology Department
Leicester University
Leicester LE1 7RH, UK

Dieter Walossek
Section for Biosystematic Documentation
University of Ulm
Listsrasse 3, D-89079 Ulm, Germany

Ward Wheeler
Department of Invertebrates
American Museum of Natural History
New York, NY, USA 10024

Mark Wilkinson
School of Biological Sciences
University of Bristol
Bristol BS8 1UG, UK

Matthew A. Wills
Department of Geology
University of Bristol
Queen's Road, Bristol BS8 1RJ, UK

# Index